Ion Visa · Anca Duta · Macedon Moldovan ·
Bogdan Burduhos · Mircea Neagoe

Solar Energy Conversion Systems in the Built Environment

Springer

Ion Visa
R&D Centre: Renewable Energy Systems and Recycling
Transilvania University of Braşov
Braşov, Romania

Anca Duta
R&D Centre: Renewable Energy Systems and Recycling
Transilvania University of Braşov
Braşov, Romania

Macedon Moldovan
R&D Centre: Renewable Energy Systems and Recycling
Transilvania University of Braşov
Braşov, Romania

Bogdan Burduhos
R&D Centre: Renewable Energy Systems and Recycling
Transilvania University of Braşov
Braşov, Romania

Mircea Neagoe
R&D Centre: Renewable Energy Systems and Recycling
Transilvania University of Braşov
Braşov, Romania

ISSN 1865-3529 ISSN 1865-3537 (electronic)
Green Energy and Technology
ISBN 978-3-030-34828-1 ISBN 978-3-030-34829-8 (eBook)
https://doi.org/10.1007/978-3-030-34829-8

This Springer imprint is published by the registered company Springer Nature Switzerland AG
The registered company address is: Gewerbestrasse 11, 6330 Cham, Switzerland

Green Energy and Technology

Climate change, environmental impact and the limited natural resources urge scientific research and novel technical solutions. The monograph series Green Energy and Technology serves as a publishing platform for scientific and technological approaches to “green”—i.e. environmentally friendly and sustainable—technologies. While a focus lies on energy and power supply, it also covers “green” solutions in industrial engineering and engineering design. Green Energy and Technology addresses researchers, advanced students, technical consultants as well as decision makers in industries and politics. Hence, the level of presentation spans from instructional to highly technical. **Indexed in Scopus**.

More information about this series at http://www.springer.com/series/8059

Preface

The concept of sustainable development was never contested since 1987 when it was firstly introduced. However, the measures required to implement this concept were not quickly identified and were even more slowly implemented in the development strategies, all over the world. The main reasons are relying in the complexity and specificity of these measures that are mainly focusing on the energy required to feed the production processes, the built environment, the agriculture and in any other activities in any society. This specificity depends on objective factors, as the availability of certain natural resources that can be used to produce energy, mainly renewable energy sources, but also on the features of the implementation location and of the implementing society. This is why only recently the sustainable development implementation was practically approached, as e.g. by planning and developing sustainable energy buildings or sustainable communities.

In these applications, the role of solar energy is highly important, as this is a cost-free energy source that is readily available in any location on the Earth. However, the daily and seasonal variability of the solar energy, that does not necessarily follow the rhythm of the humans' life and the associated energy demand, raises significant barriers in implementing and operating a sustainable built environment based only on solar energy conversion systems, in any location in the world. Thus, solutions have to be identified to solve specific issues that can support the development of a built environment with a high solar energy share and these solutions have to be feasible, affordable and well suited to communities with different development levels and different levels of financial resources.

Considering the real need for optimized solutions based on solar energy conversion systems implemented in the built environment, the authors of this book developed a complex material, focusing on the mostly used systems (photovoltaic and solar-thermal systems) and on the specific problems that are faced when implementing these in various locations. Further on, case studies are analysed as these can be further improved and replicated to get a positive result in the quest for sustainability.

Thus, this book addresses the professionals working in the design and development of solar energy conversion systems to be implemented in the built environment and aims at providing structured information on the current knowledge and main issues that do not have well-accepted solutions yet.

The book content was developed by a group of authors that are working on these type of systems since many years and who are familiar with the trends and the expectations for the future in the field of solar energy conversion systems. The book includes results recorded on the indoor and outdoor testing rigs in the R&D Centre Renewable Energy Systems and Recycling (RESREC), in the Transilvania University of Braşov, Romania, where all the authors are working. Relevant in-field results are included, coming from other research groups all over the world, with many of which the authors have direct collaboration activities. Each three years, teams working on sustainable energy present their results in the frame of the *International Conference on Sustainable Energy*, CSE.

We invite you to read this book and contact us for any new idea or any cooperation you might consider, as this represents the viable path towards the extended implementation of the sustainable development concept, at different community levels.

Braşov, Romania
2019

Ion Visa
Anca Duta
Macedon Moldovan
Bogdan Burduhos
Mircea Neagoe

Contents

Chapter 1
The Built Environment

1.1 Building, Built Environment, Community

According to the *Merriam-Webster* dictionary [90], a building represents a structure usually consisting of roof, walls, floor(s) and openings (windows and doors) where activities can be permanently run.

The International Code Council [57] uses two indicators for describing the buildings: (a) use and occupancy and (b) type of construction. Considering the use and occupancy, the categories of buildings synthetically presented in Table 1.1 were identified.

There are also buildings with a mixed occupancy, e.g. an office building with a parking place at the ground floor that shall follow the requirements of both B and S groups. Mixed occupancy can also be found in residential buildings (R) that include, e.g. a level of shops (M) or one or more levels with business occupancy (B).

The type of construction divides the buildings considering the materials used and their fire resistance. According to the International Code Council, there are five types of buildings, as outlined in Table 1.2.

There is a broad variety of uses when analyzing buildings and these are not necessarily similar, not even in narrowly positioned communities as the analysis developed by Ciurean et al. [24], showed for a carefully monitored region (of 247 km^2) with high earthquake risk. However, residential and storage buildings represent the largest share, higher than 50%, depending on the region.

In Europe, residential buildings represent the majority of the buildings' space, as the data in Table 1.3 outline, using as reference the EU Buildings Datamapper [38]. This sector mainly consists of apartment buildings and houses with 60% of the buildings having one single floor and less than 1% having over four floors. This structure is also the result of the regional specific and this has to be considered when assessing a community.

I. Visa et al., *Solar Energy Conversion Systems in the Built Environment*, Green Energy and Technology, https://doi.org/10.1007/978-3-030-34829-8_1

Table 1.1 Types of buildings considering their use and occupancy [57]

No.	Occupancy	Uses
1	Assembly (*A*) *A*-1 Theatres, halls for performing arts *A*-2 Restaurants (including associated commercial kitchens), banquet halls *A*-3 Recreation spaces, indoor swimming pools, libraries *A*-4 Indoor arena, tennis courts *A*-5 Stadiums	Places where people are gathering
2	Business (*B*)	Places for providing services: offices, police stations, governmental buildings, banks, barbers and beauty shops, civic administration, post offices, radio and television stations, etc.
3	Educational (*E*)	Schools, day care centres
4	Factory (*F*) *F*-1 Moderate hazards: bakeries, electricity generation plants, laundries, paper mills, woodworking, etc. *F*-2 Low hazards: ceramic products, glass products, metal products, ice, etc.	Places used for manufacturing, packaging, repairing, etc.
5	High hazard (*H*) buildings, where there are used *H*-1 Explosives, organic peroxide(s) *H*-2 Flammable gases, combustible dust, flammable liquids, pyrophoric(s), water reactive materials *H*-3 Combustible fibres, flammable solids *H*-4 Buildings embedding corrosive materials, highly toxic materials *H*-5 Semiconductors fabrication facilities	Places involving hazardous materials (manufacturing, processing, generation, storage or use) Hazardous materials stored or used on the top of the roofs and canopies are classified as outdoor storage or use and should comply with the International Fire Code
6	Institutional (*I*) *I*-1 Social rehabilitation facilities, alcohol and drug centres *I*-2 Foster care facilities, detoxification facilities, hospitals, nursing homes *I*-3 Detention centres, jails, prisons *I*-4 Adults day care centres, Children day care centres	Places where people are unable to leave without assistance, e.g. hospitals, care homes, prisons
7	Mercantile (*M*)	Commercial: shops, department stores, markets, drug stores, etc.

(continued)

Table 1.1 (continued)

No.	Occupancy	Uses
8	Residential (*R*) *R*-1 Boarding houses, hotels, motels (with more than 16 occupants) *R*-2 Apartment houses, convents, dormitories, live/work units, etc. *R*-3 Boarding houses with less than 16 occupants, care facilities for less than five persons receiving care, etc. *R*-4 Buildings for 5–16 persons receiving custodial care	Sleeping units where the occupants are transient Sleeping units or more than two dwellings where the occupants are permanent Occupancies where the occupants are permanent and are not classified in groups R-1 and R-2 Occupants that are capable to respond to an emergency and complete building evacuation
9	Storage (*S*) *S*-1 Moderate hazard storage *S*-2 Low hazard storage	Places where things are stored: aircraft hangar, buildings used for storing bags, baskets, cardboard boxes, furniture, parking places, etc. Buildings used for the storage of non-combustible materials, e.g. beverages in glass, metal or ceramic containers, cement in bags, electrical motors, frozen foods, ivory, metals, porcelain and pottery, stows, etc.
10	Utility and miscellaneous (*U*)	Accessory structures and others: agricultural buildings, grain siloes, greenhouses, private garages, towers, barns, stables, tanks, etc.

Table 1.2 Types of buildings considering the type of construction [58]

Type	Materials used
I	Building elements are made of non-combustible materials with a fire resistance of 3 h
II	Building elements are made of non-combustible materials with a fire resistance of 1 h
III	Exterior walls are made of non-combustible materials Interior elements are made of any type of material allowed by the code
IV	Exterior walls are made of non-combustible materials Interior elements are made of solid or laminated wood without concealed spaces (e.g. laminated timber, fire retardant treated wood)
V	Structural elements, exterior walls and interior walls are made of any materials allowed by the code

According to the type of inhabitants, residential buildings are divided into single-family houses and multifamily houses while, according to their architecture, these can be single-storey buildings and multi-storey buildings. Single-family houses were predominantly built up to the first half of the XX century. After the Second World War, multifamily houses started to be built and then developed as standard structures where dwellings were assembled in blocks of flats, functional but with minimal architectural value. These types of buildings were initially built with four to six storeys, further on

Table 1.3 Breakdown of building floor area in the European Union [38]

Country	Residential (%)	Non-residential (%)
Italy	89	11
Cyprus	86.2	13.8
Malta	85.3	14.7
Greece	84.2	15.8
Romania	83	17
Spain	82.7	17.3
Slovenia	81.7	18.3
Portugal	80.7	19.3
Croatia	77.7	22.3
France	76.4	23.6
United Kingdom	76.2	23.8
Estonia	75.7	24.3
Latvia	75.2	24.8
Hungary	73.1	26.9
Denmark	72.2	27.8
Bulgaria	72.1	27.9
Ireland	70.8	29.2
Germany	68.4	31.6
Finland	67.7	32.3
Belgium	67.5	32.5
Poland	67	33
Sweden	66.6	33.4
Luxembourg	66.5	33.5
Czech Republic	64.9	35.1
Lithuania	62.6	37.4
Netherlands	60.7	39.3
Slovakia	59.4	40.6
EU 28	75.8	24.2

these were extended to 10–12 floors, developing neighbourhoods that consist only of blocks of flats and additional buildings required for the community use, as, e.g. schools. The typical architecture of these buildings is highly repetitive and leaves little room on the facades for installing solar energy convertors.

Moreover, the blocks are quite narrowly positioned; thus, shading is a potential risk that has to be considered when planning the implementation of the solar energy conversion systems. This limits the use of facades for installing the solar energy convertors (photovoltaic modules or solar thermal collectors). However, these blocks of flats have usually flat roofs, as the image in Fig. 1.1 shows, that can be well used for

Fig. 1.1 Blocks of flats built during the second half of the XX century in Brasov, Romania

installing solar energy convertors, to provide clean energy directly to the building's inhabitants (e.g. thermal energy) or to the grid (e.g. electrical energy).

These types of neighbourhoods can be found in almost any city or town in countries in the Central and Eastern Europe.

Refurbishment activities on buildings of this type are targeting increased energy savings by cutting the thermal energy losses typical for these blocks of flats, usually made of concrete and with not too thick walls. Heat losses are also produced because of the windows and doors that are not fully tight and the materials used can be significantly improved without too high investments. Living in a block of flats represents the usual way of life for many people and is therefore well accepted. This is why new neighbourhoods are currently developed consisting of blocks with some improved aesthetical features and well-improved energy efficiency.

The apartments in these blocks have one, two or three rooms with a maximum overall surface of 77 m^2 at a price lower than 80,000 EUR and this can be considered affordable for many families with working members.

The lifespan of these residential buildings is usually considered to be of 70–100 years, as mentioned by Philpott [99]. This is valid for the actual building shell (made of concrete or bricks) while the pipes, the windows and the flat roofs are expected to be functional for 30 years and different other components (e.g. bathroom appliances) may last for 10–15 years. On the other hand, constructions for industrial use last usually between 30 and 40 years, when the return of investment is complete. These lifespans support the implementation, from the very beginning, of the solar energy conversion systems, with a lifetime of about 25 years, because the power output relative to the initial power of the Si-technologies PV modules reaches about 88% after 25 years with an average yearly degradation of 0.5–0.6%, as mentioned by Jordan et al. [73]. However, the monitored PV systems showed 0% degradation for the first 12.5 years and for the remaining period, an average of 1% degradation. This degradation percentage varies with the photovoltaic material in the module and with the implementation location; thus, choosing the best performing solar energy convertors represents a task that should be developed by professionals that are able to recommend the best-suited convertor for a given climatic profile, corresponding to a well-defined budget.

If the main building structure is well preserved, the building's lifespan can be surpassed allowing renovation as a possible sustainability option.

Old buildings, with historical or aesthetical merit, receive special treatment in many countries as being part of the local tradition, culture and history. One example is represented by the UK where more than 2% of the English building stock is listed by the Department for Digital Culture, Media and Sports [29] as "buildings of exceptional interest" (Grade I) or as "particularly important buildings of more than special interest" (Grade II*) or as "buildings that are of special interest, warranting every effort to preserve them" (Grade II).

Refurbishment is one common and recommended way to extend the buildings' lifetime, well adopted in Europe where 64% of the residential sector consists of individual houses among which 39% were built before 1960, 43% between 1961 and 1990 and only 17% were built from 1991 to 2010, according to Cristofori et al. [27]. This process targets the increase in the building's efficiency by improving the insulation but it can also represent a good opportunity to integrate solar energy convertors on the roof or in the facades.

As already stated, new buildings are continuously built in various parts of the Earth. The general trend is supporting the development of high buildings, giving use to the existing land in urban areas for meeting the challenges imposed by the population growth. As an example, in Shenzhen, China, there are under construction over 45 buildings of 200 m or taller as part of the general trend adopted by China that built more than half of the world skyscrapers in 2017. Further on, with 88 new skyscrapers in 2018, China occupies the first place in the world and is followed by the USA with 13 new skyscrapers, as outlined by Holland [52].

In the USA, the trend imposed for the new buildings considers the "building on height" as being more environmentally and economically efficient, a trend well considered in the eastern part during the past decades and now also adopted by the western cities. These high buildings consist of places for apartments, shops and offices, according to the needs. However, to provide an answer to the homeless problem, buildings up to ten storeys were also proposed. Implementing these strategies will increase the population density in cities and the development of sustainable buildings has to be considered as an additional prerequisite for solving the social problems.

In Europe, 11 out of the 20 tallest buildings were built in Russia, as the Lakhta Centre in Sankt Petersburg (462.5 m) finalized in 2018, or the Federation Tower (373.7 m) in Moscow finalized in 2016. Other buildings are located in London, UK: the Shard finalized in 2012 (310 m) and One Canada Square (1991, 236 m), in Frankfurt, Germany, the Commerzbank building (1997, 259 m), and the Messe Turm (1990, 257 m), in Madrid, Spain, Torre del Cristal (2008, 249 m), Torre Cepsa (2008, 248 m), and Torre PwC (2008, 236 m) and in Paris, France, Tour First (1974, 231 m). As the data show, most of these buildings (higher than 230 m) were built after 2000, when novel materials and building technologies allowed developments in a rather acceptable financial frame. There are under construction skyscrapers in more countries than those already mentioned, as in Turkey (Istanbul, Skyland Office Tower and Skyland Residential Tower—284 m) or in Poland (Warsaw, Varso Tower—310 m), as mentioned by Raza [103].

Besides landmark buildings, annually new buildings are developed and the general trend is to support the development on height as it has obvious advantages considering the available land and the structure. For very high buildings, solutions are already identified at conceptual level, focusing on the model provided by the Eiffel Tower in Paris; thus, higher buildings require a broad base and a specific architecture, as discussed by Berg [12].

The general trend is nowadays focused on decreasing the environmental impact by using sustainable materials in developing green buildings where energy efficiency is well matching the energy-saving measures and the solar energy is considered one of the key resources in the future scenarios. The green buildings market started to be significant in 2005 and reached 61 billion USD in 2011. Afterwards, the increase was slower and reached 81 billion USD in 2014, as mentioned on the Statista website [120].

A broad literature review developed by Darko et al. [28] shows that the research lines (thus also the future developments) are focused on environmental sustainability while social and economic sustainability are only marginally considered. Moreover, among the major trends in buildings development, green design and construction technologies represent a focus to create resource-efficient and environmental friendly projects.

The LEED certification represents thus a concrete target for many new buildings. Leadership in Energy and Environmental Design (LEED) is a certification system that assesses buildings considering their energy efficiency, water use, air quality and choice of building materials along with various environmental factors, as access to public transportation and responsible land use [19]. The Statista website [120] shows that by February 2018, there were 1597 LEED certified military projects in the USA, while the total number of LEED certified projects in the USA was 67,208, by September 2018. However, it is to mention that the LEED certificate is not further involved in monitoring the building during its lifetime; thus, a LEED certified building may end up using more resources as compared to its counterparts, as result of the specific practices of the building's occupants. As outlined by Roudman [108], one of the mostly cited examples is the Bank of America Tower in New York, the first skyscraper, that received LEED Platinum certification (the highest certification level), because of the novel solutions embedded as the waterless urinals, rainwater harvesting or daylight dimming control. In the USA, about 59% of the LEED certified projects correspond to corporate buildings, 10% to higher education buildings and 4% to buildings where the US Federal Government works, according to the Green Building Information Gateway [48] website.

As stated by the World Green Building Council [136], a *green building* represents a building that during its design, construction or operation reduces or eliminates the negative environmental impact and can support positive impacts on the climate and on the natural environment. This can be reached when certain features are met, as:

- Efficient use of energy, water and other (natural) resources;
- Use of renewable energies, as solar energy, to meet the building's energy demand;
- Pollution and wastes reduction along with wastes reuse and recycling (e.g. wastewater reuse);
- Good indoor environmental quality (e.g. by avoiding building materials that are responsible for indoor emissions);
- Use of materials that are non-toxic and sustainable;
- A design that supports adaptation to a changing environment;
- The environment and the occupants' life quality are considered in the design, construction and operation phases.

The development of green buildings currently represents a priority in the strategic plans of many cities. Due to the differences in the climatic conditions and in the local culture and traditions background, green buildings are not similarly/identically defined all over the world. The World Green Building Council (WGBC) has currently about 70 member states that are working for advancing the green building concept and practice, to meet environmental, economic and social goals and targets. The main barrier in implementing green buildings on a broader scale is represented by the costs that are usually higher than for regular buildings. This is why WGBC has set as primary goal for its activity to develop "green buildings for everyone, everywhere".

Currently, the *Edge Building* in The Netherlands is considered to be the greenest and the smartest office building in the world, according to the Building Research Establishment [18] that gave it in 2016 the highest sustainability score ever awarded (98.4%). As a combination of well-insulated walls and deeply recessed windows, the building uses about 30% of the electricity normally used by the typical office buildings. Moreover, as presented by Randall [102], the southern wall integrates solar PV modules (alternating with the windows) and the roof platform is hosting solar energy convertors as also some neighbouring university buildings do; thus, the building actually produces more energy than it consumes.

The green building concept focuses on the relation between the built environment and the natural environment. However, currently, there are described many demonstration green buildings raised in various large cities in the world, meaning that this concept is not well implemented yet in the architectural options of the common plans for the near future.

The term *built environment* includes buildings, human-made spaces between buildings such as parks and the infrastructure that supports the human activity as transportation and utility networks, flood defenses, telecommunications, etc.

The Construction Industry Council [25] suggests that the built environment "...encompasses all forms of buildings (housing, industrial, commercial, hospitals, schools, etc.), and civil engineering infrastructure, both above and below ground and includes the managed landscapes between and around the buildings".

The continuous increase in the world's population asks for specific steps to support sustainable development and this poses a significant pressure on the built environment. There are differences between the urban and the rural built environment, even at the definition level: in the USA, settlements with 2500 or more inhabitants are defined as urban while in Japan (with a much denser population), only settlements with 30,000 or more inhabitants are considered urban [94].

The development of a sustainable built environment mostly addresses the urban areas, with a much higher population density, and represents a key strategic direction planned for the near future. In September 2015, the 193 members of the United Nations adopted the 17 sustainable development goals (169 targets), grouped in the 2030 Agenda [127], with specific topics on climate change, economic growth, poverty eradication and urban development. The targets include synergistic effects (e.g. increasing the access to sanitation and hygiene that will reduce environmental pollution) and there also are trade-offs among the formulated targets [83], e.g. the environmental impacts of the use of renewable energy sources that were analyzed and found of limited extent, as presented by Sokka et al. [119].

The global urbanization trend also affects the rural built environment that undergoes significant changes. Ao et al. [11] showed that by the end of 2016, China's rural population represented 42.65% from the total population in the country and a decrease of this percentage (while increasing the corresponding value for the urban environment) will bring significant changes in various aspects (e.g. in the use of vehicles). Investigations on the impact of the transportation vehicles on the environmental quality of the rural area were developed and considered the distances from the village centre to the most representative places in the built environment: the nearest bus station, train station, bus stop, main road, market, school, health centre (hospital).

It is important to use a coherent system able to evaluate the sustainability of the landscapes and buildings based on jointly agreed and relevant criteria. As Loftness and Haase [82] showed, this currently represents a focus for many groups working in the development of the sustainable built environment. Ameen and Mourshed [6] analyzed the weight of the sustainability indicators in the Iraqi context, considering the state of the art in the field of sustainability indicators and the stakeholders' perception on the local urban challenges. The data were further analyzed with a group of experts and the appropriate weights were formulated. The topics that were considered relevant were: water, safety, transportation and infrastructure, local economy, jobs and business, housing, layout, waste, hazard, local culture, land use, ecology, management and construction, innovation, governance, energy, urban space, well-being. As expected, it was observed that there is a typical response for any country; this response is influenced by the country's general development level and the culture that influences the local relevance of the sustainability dimensions. The direct consequence is that global assessment methods are not fully relevant for a realistic analysis as the global tools are not fully transferable to the local context; therefore, locally relevant indicators should be investigated, formulated and applied. Thus, for the particular case considered (of Iraq), water and safety were evidenced as the two most important sustainability indicators. This is not surprising considering the geographical location and the challenges risen by this location. Important are the sub-indicators that were used: water quality, water conservation, on-site water recycling, diversity of water sources and rainwater harvesting systems for water and security by design, safety of public spaces and protection from high temperatures and sunlight for safety.

In developing a sustainable built environment, each of its components has to be carefully watched (buildings, parks, roads, etc.) as the well-cited example of the *Gardens by the Bay* in Singapore where art meets the sustainable buildings and the walking path (the Skyway) that guides the walker through supertrees at 22 m height. These man-made trees are hosting PV modules, rainwater collection systems and various air cleaning systems. This over 100 ha park also hosts the tallest indoor waterfall and a flower dome [45].

The *Merriam-Webster* dictionary defines *a community* as "a group of people with a common characteristic or interest living together within a larger society". When considering the inhabitants of the built environment, a slightly different definition may better apply, as "people with common interests living in a particular area".

Following the main topic of this book, two different types of communities can be identified: the urban and the rural communities. A strict and general definition of these community types is not available yet; however, Mondal [93] presents a review on the most important features of an urban community, compared to the rural one and these are:

- The size: in any country, in a given time period, an urban community is much larger than a rural community;
- The density of population: in a given country, the density of population in the urban area is significantly higher than in the rural communities;
- Family: in urban communities, more importance is given to the individuals than to the families;
- Marriage: love and inter-caste marriages are more often observed in urban communities, where a higher number of divorces are also reported;
- Occupation: in urban communities, the largest share of inhabitants is focused on industrial and administrative activities, with a high specialization degree;
- Class extremes: are mostly met in cities, where very poor neighbourhoods exist near the neighbourhoods of the rich people and the apartments of the middle-class population;
- Social heterogeneity: cultural and social heterogeneity is characteristic for the urban communities, where diverse peoples, of different races and cultures, are living together; in the rural communities, this diversity is more restricted. Villages are therefore considered symbols of cultural homogeneity, sharing common customs and traditions;
- Social distance: is significantly larger in the urban communities, where anonymity and heterogeneity are prevailing;
- Systems of interaction: the circles of social contact are broader in the urban community as compared to the rural ones as the social structure of urban communities is more relying on groups of interest. Therefore, more impersonal and/or casual relations characterize city life;
- Mobility: the social mobility is significant in the urban area where the social status of an individual is mostly linked to his/her merit, intelligence and perseverance;

- Materialism: the role/importance of an urban inhabitant is described based on what he/she *has* not on what he/she *is*. Thus, the status symbols are mainly related to financial assets, home appliances, etc.;
- Individualism: the inhabitants in urban communities are considering their welfare and happiness as most important and are not thinking or acting primarily for the good of the others. This trend is dominant in large urban communities where the interactions are less personalized;
- Rationality: represents one of the most important features of the peoples living in urban communities. The relationships are governed by contracts and these are closed considering the expected gain or loss;
- Anonymity: the members of the (large) urban communities are characterized by namelessness; thus, they can hardly form a unitary group, a primary group working or supporting a given idea;
- Norms and social role conflict: the urban communities are well respecting the norms. In the absence of fixed and uniform social norms, individuals or groups in the urban communities are seeking for various ends and this leads to social disorganization;
- Rapid social and cultural changes characterize the urban communities as an answer to various stimuli: changes in the population structure, culture, traditions, etc.;
- Voluntary association: because of the diversity of their inhabitants, the urban communities support the development of secondary groups (associations, societies, etc.);
- Social control: in an urban community is formal and is regulated by law courts, police and jails;
- Secularization of the outlook: as result of weaker kinship relations and the more extended economic logic.

Considering these specific features of the urban communities, it can be concluded that the society modernization will start from these communities. However, when changes at community level are provisioned, a specific study of the community profile needs to be developed to find the best-suited communication strategy and tools and, further on, to develop the best-suited instruments for implementing the change. It is also to mention that acceptance can hardly be expected from the entire community; thus, it should be gained from those groups that are relevant inside the community for the specific acceptance target.

1.2 Energy Demand in the Built Environment

1.2.1 Energy Demand at Building Level

The energy demand in buildings depends on several typical factors and, as the International Energy Agency [61] stated, these can be grouped as:

- The climate in the implementation location;
- The building's envelope characteristics;
- The building services and energy systems;
- Occupant's behaviour that can be discussed considering:
  - The building occupation and management;
  - The occupants' activities and behaviour;
  - The required indoor environmental quality.

In a building, the energy consumption is linked to specific uses and the main ones are related to heating and cooling the living space, domestic hot water supply, lighting, cooking, audio-visual and information devices, housework devices, etc. These specific uses gain different importance according to the traditions, culture, technical development level, type of building, etc. A study presented by Cao et al. [23] outlined that the energy consumed in buildings represented 33.7% of the total energy consumption in 1980 and increased at 41.1% in 2010. The first largest energy consumer in buildings was the USA and the second place corresponded to China where 27.3% of the total energy used was consumed in buildings. In EU, a share of 40% of the total energy use corresponded to the buildings in 2012 and this share remained almost unchanged in 2015 when the household share was 25.4% and the services share amounted 13.6% of the total final energy consumption in EU-28, as reported by Eurostat [42].

As stated by Caldera et al. [21], in EU, over half of the energy demand in buildings is associated to meeting the space heating needs. During the past 15 years, this energy demand registered a significant decrease in almost all the member countries, especially in those located in the Central and Eastern Europe (Latvia, Romania, Slovenia, Slovakia, Czech Republic, etc.) as result of improving the buildings' energy efficiency, by adopting better insulation solutions. Considering the climate influence, it should be outlined that rather different energy consumption values are reported in various EU countries. An average of 60–90 kWh/(m^2 year) was reported by the International Energy Agency [60] for the southern countries, with lower heating demand (e.g. Portugal, Spain, Bulgaria, Greece) while values up to 175–235 kWh/(m^2 year) are found in countries with a colder climatic profile, as Estonia, Latvia or Finland.

The properties of the buildings' envelopes are important considering the building's shape and the materials used to build it. As presented by Rodriguez et al. [106], the heat transfer coefficients (U values) of the building elements represent the most important features that have to be considered. For buildings with low U values envelopes, the energy consumption is obviously decreased, leaving room to the buildings' designers for various building forms. In moderate climates, the building shape and the window's design are expected to have little influence on the energy consumption, while for very low U values (e.g. in southern EU countries), the average energy consumption increases due to an increase in the cooling load. When considering the cold, sub-arctic climates, if the envelope is very well insulated, the building shape has a lower influence on the overall energy consumption.

All these findings have to be considered when designing energy-efficient new buildings. However, the building stock mainly consists of already built buildings; thus, the assumptions on the energy consumption should also consider the building shape. This can cover a broad variety of options and the *form factor* represents the property that should be considered when differentiating among buildings with various shapes.

As defined by National House Building Council Foundation [95], the form factor (F_F) correlates the living area of the building and the total amount of surface area that heat can escape from and is defined by

$$F_F = \frac{A_T}{A_H} \tag{1.1}$$

where

- A_T is the total heat loss area of walls, roofs, floors and openings;
- A_H is the habitable floor area of all storeys.

Obviously, lower form factor values correspond to higher building efficiency. Thus, considering five house types with the same habitable floor area (of 93 m^2), the form factors and the corresponding space heating energy demand are included in Table 1.4, where the reference value for end mid-floor apartments is close to 2000 kWh/year.

These values show the significant effect of the building shape and outline that the most important aspect to consider is the area directly exposed to the open-air environment. Different households with similar habitable floor area and various form factors are presented in Fig. 1.2.

It is important to notice that there are more architectural and building features that contribute to the thermal losses, e.g. the thermal bridges that mainly occur at the junction of the walls, floors and ceilings but their contribution is significantly less than that of the form factors.

Additional changes in the heating energy demand of the buildings result when incorporating various design elements as the recessed doors, one or multi-storey bays, 2½ storey including dormers (for a two-storey building) or the staggered terraces that all bring an increase in the building heating energy demand as increasing the exposed area to open air. The 2½ storey including roof lights, in a two-storey building,

Table 1.4 Heating energy demands for homes with different form factors

Type of home	Form factor	Additional % versus end mid-floor apartment
End mid-floor apartment	0.8	0
Mid-terrace house	1.7	56
Semi-detached house	2.1	89
Detached house	2.5	117
Bungalow	3	136

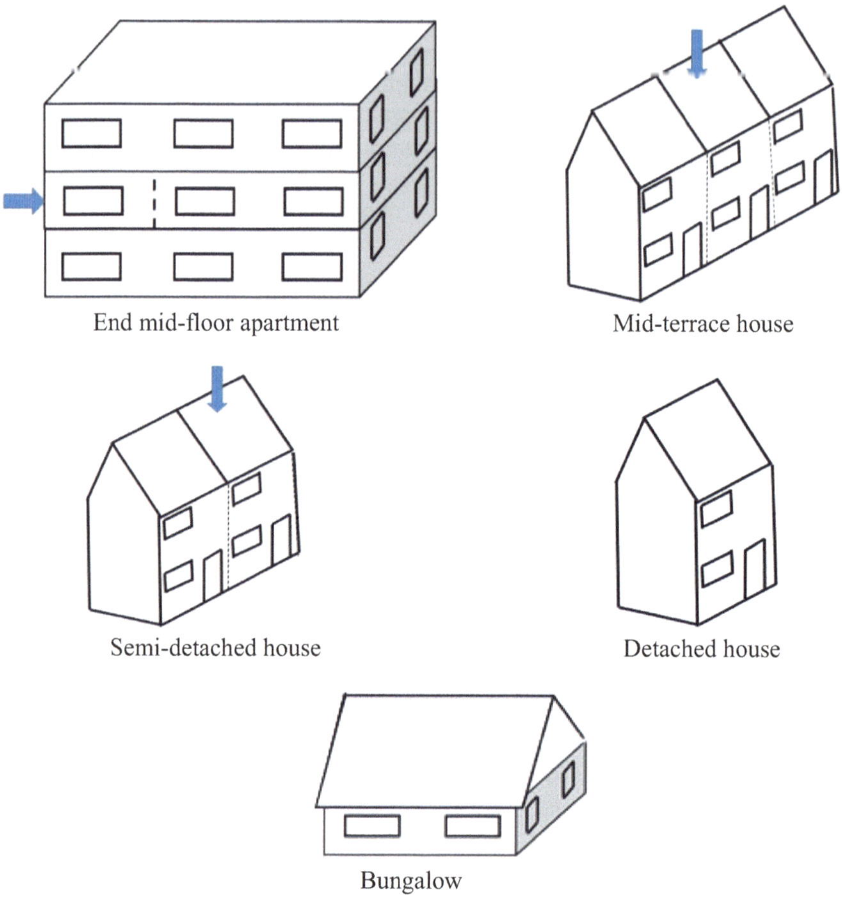

Fig. 1.2 Types of buildings with different form factors [95]

represents the only design element that reduces this demand. Moreover, integrating garages in the buildings leads to an increase in the heat energy demand of the building because when integrating a garage in a mid-terrace home, the result is a semi-detached home, with a higher form factor.

There is a general trend in designing buildings aiming at increasing daylighting. This is a well-known architectural feature, well used in the Victorian or Georgian houses where this effect was reached by raising the ceilings' height to install taller windows that allow an increased view of the sky. However, this type of architecture bares a risk considering the overheating during summertime, in warm climates. Therefore, architectural rules generally recommend a window area of about 20% of the exterior walls to get good daylighting. Moreover, shading devices can be used to limit overheating but their design needs to be done considering not to reduce the natural lighting in an unacceptable extent.

Overheating is mainly the result of "bad design, poor management or inadequate services" as concluded by McLeod and Swainson [88] when analyzing the temperature profile of a high density, low-carbon residential complex in North London. The analysis shows that overheating can be related to the lack of effective purge ventilation systems (due to the restricted windows openings) and external shading systems. Moreover, the limited insulation of the pipes and ducts contributes to the heat losses while the mechanical ventilation with heat recovery (MVHR) systems proved to be ineffective for providing cooling when needed, in all the dwellings. It is also important to mention that due to global heating, these findings have to be cautiously considered and should be periodically updated.

The period when the building was built represents another key factor to be correlated with the energy demand, because various architectural solutions focusing on energy savings were mainly developed during the past 20 years and many new construction materials are now available but were not several decades ago. As Ding et al. [32] presented in an analysis of the Chinese built environment, the heating energy consumption in 31 buildings located in Northern China varies much, between 10 and 910 MJ/(m^2 year), with an average value of 460 MJ/(m^2 year) mainly as result of upgrades to the enclosure structures and to the central heating.

A brief synthesis of the information on the energy consumption in the built environment is included in Table 1.5, where data are grouped considering the specific types of buildings:

- Residential buildings, that include all detached houses and attached dwellings, e.g. apartment complexes, blocks of flats or town houses;
- Commercial buildings including office buildings, financial buildings, special care buildings, medical buildings, multimerchandise shopping, food and beverage establishments, warehouses and other commercial structures.

The differences in the energy consumption required by different building types are also considered in the future energy scenarios involving nearly Zero Energy Buildings (nZEB) that have differently defined energy performance indicators. As analyzed by the European Commission [39], for example, in Belgium, the nZEB residential buildings in the Flemish region have to meet an energy demand of 30 kWh/(m^2 year) while for the non-residential buildings in the same region, a target of 40 kWh/(m^2 year) is formulated.

Scenarios were also developed considering the increase in the building efficiency until 2019, 2030 and 2050 for different regions with a different number of heating degree-days. The heating degree-days indicate how much (degrees) and for how long (days) the outside temperature was lower than a specific "base temperature", e.g. in UK 15.5 °C [37]. In Romania, the assumed values for new buildings were formulated by the Ministry for Regional Development and Public Administration [92] and are included in Table 1.6.

According to the United Nations [125] report *World Urbanization Prospects: the 2018 Revision*, 55.3% of the world's population was living, in 2018, in cities and this share is expected to increase, reaching 68% by 2050. This report also outlines

Table 1.5 Percentage of energy consumed in the built environment

No.	Country or region	Details	Energy share (%)	Reference
1	Worldwide	Buildings were responsible for … of the energy consumption in the world	30–40	United Nations Environment Programme [126]
2	Europe	Commercial and residential buildings account for … of the total energy consumption	38.7	Allouhi et al. [3]
3	EU	The built environment accounts for over … of the total final energy consumption in 2015	41	Krarti [78]
4	US	The built environment accounts for over … of the total final energy consumption in 2015	34	Krarti [78]
5	China	The built environment accounts for over … of the total final energy consumption in 2015	23	Krarti [78]
6	India	The built environment accounts for over … of the total final energy consumption in 2015	41	Krarti [78]
7	Sudan	The built environment accounts for over … of the total final energy consumption in 2015	57	Krarti [78]
Residential buildings				
8	US	% of total primary energy consumption in US	22	International Energy Agency [60]

(continued)

Table 1.5 (continued)

No.	Country or region	Details	Energy share (%)	Reference
9	China	The residential building sector is considered the second energy consumer (after industry) and the … share uses as reference the national energy end-use	11	Ding et al. [32]
10	Korea	The energy consumed in buildings represents about …of the total energy use	30	Jang et al. [68]
11	EU	Buildings are the largest end-use sector with …% of the total consumption	40	Cao et al. [23]
Commercial buildings				
12	US	Commercial buildings consumed about … of the building energy consumption	46	Allouhi et al. [3]
13	UK	A rate of growth in commercial energy consumption three times higher than in the domestic sector was observed during the past 25 years		Scrase [110]
14	Australia	Retail and Office buildings represent the major energy consumers in the commercial buildings sector	Retail: 35 Office: 25	Allouhi et al. [3] Council of Australia Governments [26]

that the population living in urban areas is responsible for over 75% of the non-renewable resources consumption and for about 75% of the global pollution. This is the consequence of the absence of a clear vision and strategy for the short, medium and long-term development. Therefore, professionals agreed that well-defined objectives have to be implemented by an appropriate and acceptable governance model that is formulated based on urban development, transport and infrastructure strategies

Table 1.6 Specific primary energy demand in various building types targeted in 2019, 2030 and 2050

Building type	Specific primary energy demand (kWh/(m^2 year))					
	Number of heating degree-days: 3000			Number of heating degree-days: 4250		
	2019	2030	2050	2019	2030	2050
Office and public administration (new)	48	32	10	103.76	55.81	22.1
Educational	108	50	42	190	82	65
Health and Care	48	29	8.5	116	56	24
Block of flats	97	72	48	135	100	79

and regeneration models. Beyond understanding the process, the sustainable built environment asks for novel approaches considering the needs of the inhabitants and the available resources (those already used and those existent but not commonly used, e.g. the renewable energy resources).

Various architectural solutions were identified for reducing the energy demand in the buildings implemented in various climatic profiles. Particularly, for low energy buildings, Tettey et al. [122] showed that double glazing brings the most significant energy savings, both in heating and in cooling the buildings when compared with roof insulation, insulated shutters and cavity wall insulation. However, Allen et al. [2] analyzed the glazing used for the windows implemented in low energy buildings and outlined that most of the current solutions are lacking flexibility, as the well-known example of low-e glazing. Consequently, the use of switchable glazing is discussed, designed to control the transmittance of the solar radiation, including its IR part as a possible route for decreasing the cooling energy demand during the hot seasons. The case of thermotropic windows is analyzed, considering the specific properties of this type of glazing that uses a thermochromic material that allows reversible changes in the transmittance values in response to heat. It was concluded that the use of hydroxypropyl cellulose (HPC) as membranous sandwich layer in the thermotropic glazing unit allows annual energy saving of 22% in a small building in Palermo when compared to the double glazing. This effect increases with the HPC concentration in the composite glazing and it was found that a 6% concentration offers a good compromise between minimizing the amount of heat that passes through the window (over 30% as compared to double glazing) and the transmittance during daylight. The results also showed that although the vertical mounting position does not correspond to the highest heat gain, the window is able to retain a significant heat share when exposed to the Sun, during the entire day.

The new construction technologies and materials also supported the development of energy-efficient buildings although the legal frame did not significantly change after 2006. Novel, intelligent materials were developed for decreasing the energy consumption in buildings while supporting the comfort of the inhabitants. Forzano et al. [44] reviewed the latest findings on phase-change materials (PCM) implemented in the buildings' walls to reduce the heating and the cooling loads. PCMs can be incorporated in the building walls as prefabricated panels, wallboards, bricks, etc. or can be used on existing buildings as concrete, plaster or mortar applied on the walls as presented by Al-Absi et al. [1]. Due to their high capacity for storing the latent heat, this solution can be successfully implemented both on the exterior side of the wall for decreasing the heat that can access the building or on the interior side of the wall to decrease the heat produced by various indoor appliances. The experimental tests proved that the PCMs use reduces the heating and the cooling loads. However, their use should be modelled and tested over at least one year, as a material well performing during summer can lead to a poor outcome during the winter.

As synthesized by Michael et al. [91], the refurbishment solutions for the buildings that were not designed considering the (passive) use of solar radiation have to solve several issues because, in most office buildings, 35% of the entire electricity consumption comes from artificial lighting; thus, increasing the use of daylighting can bring significant savings. When thinking about already existing buildings, the solutions that are reducing the electrical energy consumption should be acceptable; thus, these are not dramatically changing the shape and aspect of the building. These solutions include side-lighting systems mounted for reducing the unequal distribution of daylight using: glazing materials or integrated internal or external systems as light shelves, louvers and blinds. Although costly, the use of these solutions is effective and should be designed and implemented by professionals to avoid side effects, as glaring. Integrating solar energy conversion systems can be well considered through, e.g. external adaptive shading facades that can solve the issues risen by the solar protection, natural lighting and glare. Implementing renewable energy systems that support the renovation of existing buildings represents a good alternative.

Current architectural trends in the large commercial buildings design involve an atrium that supports the solar gain due to passive solar heating and natural ventilation. As Sher et al. [115] showed that the atrium became popular in the small and large commercial buildings because, if well designed, it can bring significant energy savings, particularly in temperate or cold climates. The optimal design has to consider the implementation latitude, the geometry of the atrium, its orientation towards the Sun and the light penetration, along with specific features of the glazed surface. As Hung and Chow [55] presented, the atrium position in the building can be: in the centre of the building, semi-enclosed, attached or linear. This is important when considering the main reason for including an atrium in the building design: getting significant energy savings during the warm season that compensate for the heat losses during the cold season. Overall energy savings of about 15% were reported and there is a significant increase in the use of daylight for lighting, especially when central or linear atriums are selected.

Previous architectural trends supported the development of buildings that were not particularly considering the active use of solar radiation. However, the existing building stock is periodically subject of refurbishment and this can be done by integrating solar energy conversion systems (solar thermal or PV systems) on the available places on the buildings. The most common choice is represented by the flat rooftops (in contact or at a distance over the rooftop) or on the edge of it. Bougiatioti and Michael [14] recommend this option especially for regions that have a climatic profile that benefits of large amounts of solar radiation. A recent study [123] based on satellite imagery shows that in EU, about 25% of the energy demand in the built environment can be covered using solar energy convertors installed on the building's rooftops.

Current and future trends are focusing on using the Internet of Things (IoT) for developing but also for efficiently operating the new type of buildings—the intelligent buildings. Jia et al. [70] presented a well-structured analysis on the future use of IoT for increasing the energy efficiency in the built environment while preserving the service level of building's users or occupants. Building energy management systems (BEMS) are already commercially available but research on these is ongoing for getting better correspondence between the inside and outside conditions. Examples of the use of energy only when needed are provided, as in the case of air conditioning that is turned on only when occupants are in the building. It is concluded that an IoT controlled smart grid supports the two-way communication between the utilities producers and users that enable information exchange and efficient energy transfer.

A novel concept is proposed for the future: the *living buildings* that are defined by the International Living Future Institute [65] as buildings that produce the required energy using renewable resources, capture and treat their required water, efficiently operate and show maximum beauty. This type of buildings should be developed considering the opening towards urban agriculture, thus to have a certain available soil area to cover (part of) their needs. Moreover, the support to a car-free lifestyle is highly recommended, along with the use of electrical vehicles when needed. The water used in the building should have as source the natural precipitations, the discharged water and the storm water that should be part of a treatment cycle targeting reuse. The energy used in the building should be as low as possible through a smart design that includes passive heating and cooling and natural ventilation. Renewable energy sources represent 100% of the energy production sources and this asks for an additional safe, reliable and decentralized power grid. Moreover, the living building should have a healthy interior environment allowing frequent human interactions and should be built strictly avoiding materials with negative impact on the human and ecosystem health; at least 20% of these materials should be obtained from the close vicinity of the implementation location (less than 500 km), another 30% can be obtained from a max. of 1000 km distance while 25% are allowed to be provided from maximum of 5000 km distance. Specific attention is devoted to the wastes, as the *living building* should support the wastes reduction or elimination during all the phases: design, construction, operation and end of life. Finally, yet importantly, the *living buildings* should support beauty and should be opened for visiting (at least one day per year), up to the moment when this concept is broadly accepted and replicated. Hegazy et al. [51] applied these concepts in a study for the Bibliotheca of Alexandria and they showed that the particular solutions have to be well formulated, according to the implementation location and making good use of the available natural resources.

1.2.2 Energy Demand at Community Level

The energy consumed at community level represents the sum of all the energies used by the consumers in that particular community. This includes the built environment but also additional components as transportation vehicles that use the electrical grid (railway, metro, tram, etc.), the community lighting (e.g. street lighting), the security and safety community systems, water and wastewater treatment, etc. Depending on the community type (urban or rural) and on its geographical location, this list can be enlarged and more specifically formulated.

Considering the recorded data, starting with 1950, the rural population showed a slow increase being now close to 3.4 billion and this trend is expected to reach its maximum during the next years being further followed by a slight decrease. Consequently, by 2050, a number of 3.1 billion rural inhabitants are expected, according to the United Nations [125] report *World Urbanization Prospects: the 2018 Revision*. Currently, Africa and Asia represent the home for over 90% of the rural population.

During the last century, there was a rapid growth of the urban population: there were 751 million urban inhabitants recorded in 1950, while in 2018, there were reported 4.2 billion, with 54% of the urban population living in Asia, followed by 13% in Europe and 13% Africa. The most urbanized regions are the Northern America (82% of the inhabitants living in the urban area), Latin America and the Caribbean region (81%), Europe (74%) and Oceania (68%). The urban population is expected to increase with another 2.5 billion people by 2050, with almost all the growth occurring in Asia and Africa. This supports the projections for a significant increase in the number of urban communities that should benefit of the currently formulated sustainable design concepts and on the experience gained in developing or refurbishing the built environment.

The urban population increase became significant during 1960–1979 having as direct consequence a decrease in the quality of the built environment as more constructions developed during that period replaced the traditional roof (with an insulating air chamber below it) with a concrete layer that had a much lower insulation efficiency [15]. This trend was further reversed during 1980–2006 when the thermal insulation requirements were adopted and led to buildings with higher energy efficiency.

The best locations for implementing solar energy conversion systems are usually considered in the temperate or warm climates, where a large amount of solar radiation is available. However, the study of Lobaccaro et al. [81] developed on Norwegian locations showed that good energy savings are possible when implementing a careful urban planning to give maximum use of the existing solar radiation potential. The investigated cases allowed to conclude that important are the aspect ratios of the urban canyons, the finishing materials on the façades and the ground surface when aiming at mitigating the impact of the overshadowing effect caused by the low altitude angle of the sunray at high latitudes, in urban communities. The aspect ratio represents the ratio between the average height of the buildings and the average width of the street between buildings. The results showed that the distance between the buildings has to be additionally considered when designing a community or when estimating its energy demand and its energy gain due to solar radiation. In any community, the buildings are grouped in row houses and high-raise blocks of flats that should be jointly considered in the estimations, as the effect of their interaction (e.g. by shadowing or reflections) on their energy demand is significant. In Trondheim, the reflectance values of the ground materials (most common: asphalt or grass) are $\rho = 0.2$ during the warmer periods and $\rho = 1$ due to the snow during the winter months. In this location, the Sun altitude angle at noon varies from about 50° (in summer) to 4° (in winter) and the landscape profile, with deep valleys and indented shoreline, supports overshadowing. A direct consequence is that there was not found any prevalent orientation of the buildings that clearly maximizes the total solar energy gain on the building envelopes. Small differences, of ±1% for high-rise blocks of flats and ±2% for row houses, were observed between the building's position that supports the highest received annual global solar energy and the values obtained for the other orientations. This type of infield positioning was also considered in Rjukan where three giant mirrors were installed on the top of the mountain in the neighbourhood, to track the sunrays and reflect it towards the city market square that was, before installing this solution, completely overshadowed during the most part of the year. The experimental results showed that by using bright coloured materials that have over 50% reflectance, the indirect mutual solar reflections of the surrounding surfaces are able to compensate for the overshadowing effects in the complex urban communities.

Lindroos et al. [80] report similar findings showing that the Nordic climate in the Baltic countries has not a fully negative effect on the amount of the available solar energy. Therefore, solar energy is included as a viable source for reaching the targets proposed for renewable energy use, by 2050.

The energy assessment in the building sector was broadly investigated; as synthesized by Braulio-Gonzalo et al. [15], there are mainly two approaches: the top-down one, considering the residential sector as an energy sink, mostly used for assessing macro-indicators as price, costs and climate data. The second approach is the bottom-up one that considers detailed information on buildings physics, thus allowing assessing the technological options. The variables considered in the bottom-up models are the urban morphology, the aspect ratio and the orientation of the buildings along with the building compactness (typology, shape factor) and envelope transmittance (year of construction, U values of the opaque envelope and of the windows/glazed part of the envelope). Considering an urban area located in Castellon de la Plana (East of Spain), the modelling and simulation results showed that the aspect ratio is important as it has strong influence on the solar radiation access to the building's envelope. On the other hand, the design of urban blocks of flats with internal courtyard and the South orientation of the buildings proved to have the lowest influence on the thermal energy demand for heating and cooling. The results also showed that the building's form factor and the envelop transmittance have a significant contribution, as expected, in the overall energy consumption in the residential environment. The paper also outlines that the development of new urban projects should consider energy saving that can be brought by the passive (geometrical) aspects in the buildings' and urban community's design as savings higher than 50% were reported for the investigated case study, considering a community (neighbourhood) defined with a variety of buildings.

In the past decades, research was invested for improving the energy efficiency of individual buildings but recent studies outline the importance of the entire urban configuration when assessing the total energy demand of the buildings as this can modify the solar radiation input and the convective heat transfer in buildings. As shown by Stromann-Andersen and Sattrup [121], the energy consumption in buildings increases in high-density urban areas because, in this type of communities, the passive solar gain and the daylight availability are affected. These types of urban areas are traditionally designed considering the basic buildings' geometry and height along with the distance between these buildings to regulate the access of light and solar heat. The urban canyons that create special environmental conditions represent the result and the size of these may have a long-lasting impact on the heating, cooling and lighting energy demand of the buildings. As an example, it was found that the geometry of urban canyons has a relatively high impact on the energy consumption when compared with unobstructed sites, ranging between 19% for housing up to 30% for office buildings. The direct consequence of these findings is that when building a low energy office construction in a rather low-density urban location, with an initial energy consumption of 50 kWh/(m^2 year), it may happen that the energy consumption will increase up to 70 kWh/(m^2 year) due to further building in the surroundings. The authors also investigated the contribution of the reflected light on the buildings' energy consumption, outlining that this is the main source of daylighting, available for houses and for the lower storeys in the blocks of flats in high-density urban communities. Considering the Northern European locations, it is outlined that not only daylighting but also the heat contained in the solar radiation is distributed depending on the reflectivity of the buildings' facades in the urban canyons. Thus, the buildings' facades are important considering not only the community aspect but also their contribution to the overall energy consumption.

The communities are responsible for planning the energy production or sourcing to meet the provisioned energy demand and this supports the development of the community projects that are significant for the entire energy scenario. One of the early examples is the Tvindkraft project that supported the development and implementation of the largest wind turbine in Denmark, in 1978. The idea of this project emerged among the teachers at the schools in Tvind that aimed at developing a clean and renewable energy system based on wind to supply energy to the schools and to support the quest for a clean atmosphere [124]. The wind turbine (windmill) is 53 m high and has a wingspan (rotor diameter) of 54 m, representing by far the largest wind turbine in the world at that time. This windmill is implemented in Western Jutland where the wind blows more than 300 days/year. The construction of the windmill started in 1975 and lasted for three years and the teachers in the Tvind schools covered the costs. The windmill started to work as a power plant on 26 March 1978, and up to 1980, various improvements were added. Since then, it produced over 20,000 MWh electrical energy. After 40 years, the windmill uses all the original parts except the wings and the wing bearings that were replaced in 1993.

The design and the infrastructure required to meet the energy demand are obviously more affordable at community level as compared with the dispersed version, for each building separately. This is valid for the current energy supply, mainly based on fossil or nuclear fuels and is considered to be even more recommended for the future energy scenario, involving renewable energy sources.

When speaking about energy, the "community" definition may differ depending on the type of energy (electrical or thermal), on the providers or on the consumers, etc. As example, electrical energy is mainly delivered through the grid that usually covers a very broad area (e.g. the national grid, covering an entire country) and is currently produced by large power plants based on coal or oil, or on hydrosources and, in a lower extent, on renewables as wind or solar energy. The infrastructure is complex and there was a stepwise development, considering the energy management, production, distribution and use. On the other hand, thermal energy is currently produced and distributed at district level or at individual buildings' level or, especially in less developed areas, heat is produced according to the needs in each apartment or in each room of the building. Thus, accurately assessing the thermal energy demand requires a broad range of sources.

It is important to evaluate the energy demand at community level according to the final use of these data. For example, when designing the energy system(s) required at community level, the accuracy level has to be formulated in direct correlation with the storage facilities, as the energy production facilities can be designed based on the highest energy demand identified as necessary over one year in the community. This approach is obviously a safe one; however, it leaves room for over-production and this asks for energy storage solutions for a given time period. The design should be further correlated with the climatic profile, considering how broad is the outdoor temperature range and how often the major temperature changes are expected.

Global heating and climate change may influence the energy demand in the next decades. The review developed by Andric et al. [9] shows that according to the geographical location, the cooling demand will increase while the heating demand will decrease, as expected. Less expected are the percentages of these variations, modelled by many authors for various locations. A synthesis of these results is included in Table 1.7.

The values in Table 1.7 outline that the thermal energy demand in the buildings will go through significant changes considering the time horizon 2040–2070. These changes are expected to mostly affect the locations positioned in humid, sub-tropical climates as Sydney, Australia, where a decrease in the heating energy demand is as high as 81% and an increase in the cooling energy demand of 146% can be expected [133]. Jiang et al. [71] reported similar findings for Miami, USA. However, there are slight differences among the reports for the same location as for Miami or for Sydney in the selected examples, as result of using different reference values.

Table 1.7 Impact of climate change on the building energy demand in various regions in the world

Country	City	Reference period	Targeted prediction period	Heating energy demand decrease (%)	Cooling energy demand increase (%)	References
Australia	Sydney	1990	2050	66–81	93–146	Wang et al. [133]
Australia	Sydney	2008	2070	Not available	59	Guan [50]
Australia	Melbourne	1990	2050	30–42	69–111	Wang et al. [133]
Italy	Turin	2010	2050	16	30	Waddicor et al. [131]
Sweden	Vaxjo	1996–2005	2050	13–16	39–49	Dodoo et al. [34]
Portugal	Lisbon	2010	2050	6.7–37.1	Not available	Andric et al. [8, 10]
Slovenia	Portoroz	2005	2050	6	90	Dolinar et al. [35]
United States	Miami	1991–2005	2050	79	30	Jiang et al. [71]
United States	Miami	1961–1990	2040–2069	Not available	26.6–36.4	Shen [114]
United States	Chicago	1961–1990	2040–2069	16.4–28.5	24.8–38.9	Shen [114]
Japan	Sapporo	1990–1999	2040–2050	27	23	Shibuya and Croxford [116]
United Arab Emirates	Al-Ain	2009	2050	9.5–39.2	7.3–24.1	Radhi [101]

1.3 The Energy Consumption in the Built Environment

The energy demand in the built environment is usually estimated for the post-building phase, during its occupancy. As the review paper developed by Sharma et al. [113] shows, this represents a correct assumption as over 80% of the energy used by a building all over its lifecycle is consumed during its use, for reasons the building was built. Except for this phase, energy is consumed during the construction phase and for producing the construction materials and during the end-of-life phase when the building is demolished. The same study outlines that in terms of materials, buildings consume 40% of the stone, sand and gravel, followed by wood (25%) and 40% of the energy obtained using fossil fuels. However, regardless the geographical location, the operation phase is mentioned as the highest energy consumer, as also outlined by Ortiz et al. [98] in their review paper.

According to the report published by the International Energy Agency [60], in 2010, the energy use varied in the built environment mainly depending on the building's location, as the data in Table 1.8 show.

Buildings are estimated to consume about 35% of the final overall energy consumption in the world. This demand was covered in 2010 by using coal (4%), oil (11%), natural gas (22%), electricity (28%), commercial heat (5%) and renewables (30%). These overall percentages have various specific values according to the geographical location, the type of community, the inhabitants' behaviour, etc. As an example, the residential sector in OECD countries consumed, in 2010, 30,730 PJ for the uses detailed in Table 1.8 with the highest values corresponding to North America and Europe while in the non-OECD countries, the energy consumption in residential buildings was of 56,033 PJ, with the largest share in Russia and in the African countries.

The emissions related to the final energy use in buildings grew from 2845 Mtoe CO_2 in 2010 to about 3050 Mtoe in 2018. The amount of direct CO_2 emissions was relatively steady [62] while the share of fossil fuel used in covering this energy demand decreased to about 36%. The International Energy Agency [63] shows the top five energy consumers in the residential sector in 2018, included in Table 1.9.

Yoshino and Chen [137] presented the results obtained in the project *Total energy use in the buildings: analysis and evaluation methods*. The 2016 Report mentions that there is a broad variety of values that describe the energy use in office buildings; for example, similar values are reported for the heating energy required in Austria, Belgium, Northern China and Norway and values ranging between 40 and 60 kWh/(m^2 year) for electricity. In terms of electricity use, large differences were outlined among countries with rather similar climatic profile as, for example, the energy demand in

Table 1.8 Buildings end-use energy consumption in 2010 [60]

Energy use	% in cold climate countries (60 EJ)	% in moderate and warm climate countries (57 EJ)
Space heating	45	13
Water heating	15	30
Space cooling	5	3
Lighting	7	5
Cooking	3	38
Appliances and other equipment	25	11

Table 1.9 Top five countries by total final energy consumption in the residential sector [63]

Country	China	United States	India	Russia	Japan
Total final consumption (Mtoe)	325	247	278	115	44

office buildings in France was close to 110 kWh/(m^2 year) with the most significant differences in the electrical energy consumption registered for ventilation and air conditioning along with the office appliances. This can be the results of different types of HVAC systems and their operation modes. Moreover, the report shows that large office buildings, with the floor area above 30,000 m^2, have a specific energy consumption significantly higher when compared to smaller office buildings. Further on, based on the analysis of 80 residential buildings, it could be concluded that the most relevant data for estimating the energy demand in buildings are those related to the number of heating degree-days (relative importance: 27%), family members (relative importance: 22%), heat loss coefficient (relative importance: 19%) and the building age (relative importance: 16%). Factors as the floor area, the number of cooling degree-days, or the equivalent area of interstices have a relative importance lower than 10%. It is to mention that the occupants' behaviour represents one major factor that has to be considered, both in estimating the energy demand but also in training on energy saving in the built environment. It was observed, based on the analysis done on buildings in Japan, that changing the lifestyle can bring more significant energy savings than an increased thermal insulation, with the most significant contribution in heating and cooling and in the use of domestic hot water.

The International Energy Agency [62] confirms the decision towards sustainability and, although the overall electrical energy consumption did not decrease, in 2017, about 35% of the electrical energy was produced using low-carbon sources, with the share of renewables reaching 25% and the decline of the nuclear energy down to 10%. As this report outlines, the carbon intensity of the power generation declined since 2010, but remained steady in 2017 at 491 gCO_2/kWh, mainly as result of the global use of coal, that increased by 3% (280 TWh) in 2017. The future projections for 2030 are targeting a share of energy generation using low-carbon technologies as high as 63% with 50% of this growth being the result of using PV and wind systems on a larger scale. However, to meet the Sustainable Development Scenario for 2030 (220 gCO_2/kWh), the carbon intensity has to decline much faster, with a 5.6% share per year, while for meeting the 2040 target (80 gCO_2/kWh), the decline rate has to be significantly higher.

The International Energy Agency [62] outlines the constant average building energy use per person since 1990 at 5 MWh/(person year). However, this value has to decrease down to 4.5 MWh/(person year) up to 2025 to be in line with the sustainable development targets.

The electricity consumption represents up to 50% of the global final energy consumption [64]. The building types, as presented in Table 1.10, well fit this percentage, except the buildings included in "other sectors" in non-OECD countries, with a higher share, possibly due to a larger (and uncontrolled) energy consumption of the heating, ventilation and air conditioning (HVAC) systems.

On average, the annual energy production growth between 2000 and 2010 was 1.1% in OECD countries and 6.4% in non-OECD countries.

The electrical energy production used various sources as outlined in Fig. 1.3. As the report shows [64], in 2010, in the world, the electrical energy was mainly produced in power plants, using coal (46.35%) and natural gas (21.87%). Until 2016, there was

Table 1.10 Electricity use in buildings as share of the total energy consumption. Comparative analysis in 1990 and 2010 [62]

Type of buildings	Share in OECD countries (%)		Share in non-OECD countries (%)	
	1990	2010	1990	2010
Residential buildings	31	32	17	23
Service buildings	26	32	10	14
Other sectors buildings	43	36	73	63

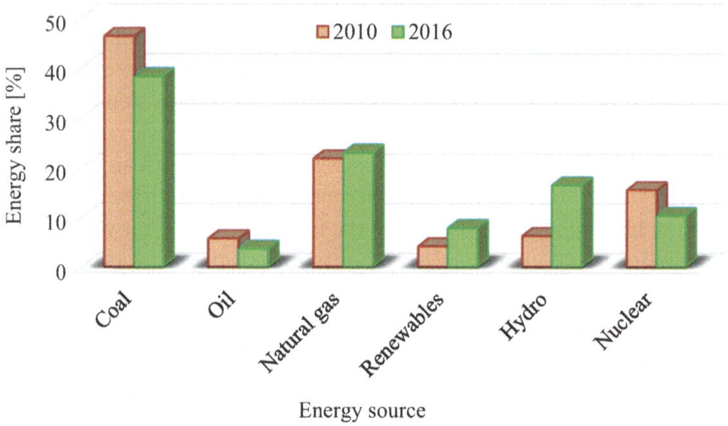

Fig. 1.3 Primary energy sources used for electricity generation [64]

a decrease in the use of fossil fuels (coal and oil) for electricity production and a slight increase in the use of natural gas, with an overall decrease in the use of raw fossil fuel, supporting the decrease of the CO_2 emissions. As already mentioned, the use of nuclear energy also decreased supporting the quest for more secure energy supply. The decrease in the fossil fuels use was compensated by the use of renewables (PV systems along with wind, geothermal, hydroenergy and biomass systems). The increase of wind in the electricity production amounted 71 TWh and the increase due to the PV systems was 50 TWh. A small increase due to the use of biomass and wastes was also recorded while a decrease was registered in 2016 for the use of natural gas, coal and oil with values of 47, 32 and 18 TWh, respectively.

A significant increase in the electricity production was recorded for hydroenergy. Additionally, a substantial electricity trade occurred in the OECD countries in Europe, where electricity imports grew at an average annual rate of 4.1% during the 1974–2017 periods. In these countries, the major energy sources for electricity production remained, in 2016, coal and natural gas (about 27% for each). The annual growth rate in the electricity consumption in the residential environment was 2%, for the period 2000–2010, and −0.4% during 2010–2016, while for commercial and service buildings these rates were 2.1 and −0.5%.

The top ten countries considering the electrical energy consumption in 2016 are included in Fig. 1.4, based on the data published by the International Energy Agency [64]. As the results show, large electrical energy consumption may be mainly the result of a large population in the country (as for China, Brazil or India) or of a significant economic development (as for Germany, Korea or Japan) or of both as for the USA.

The thermal energy consumption also varies with the geographical location of the buildings and with their construction year, considering that efficient thermal insulation solutions were consistently developed during the past two decades.

In EU, over 50% of the thermal energy demand is due to space heating, with values varying from country to country, according to the climatic profile, the buildings' energy performance and the users' habits [23]. A significant decrease in the yearly energy demand in almost all EU countries was observed, mainly as result of refurbishment supported by various national strategies and as consequence of building new houses with better energy performance.

Following the Eurostat [42] synthesis values, the final European energy consumption in the residential sector was split among space heating (64.1%), water heating (14.3%), lighting and appliances (14.4%), cooking (5.6%), space cooling (0.3%) and various other uses. There are quite broad differences in the thermal energy shares (space heating, space cooling and water heating) corresponding to the EU countries, as detailed in Table 1.11. These differences are mainly the result of the buildings' performances as the highest share for space heating mainly corresponds to countries

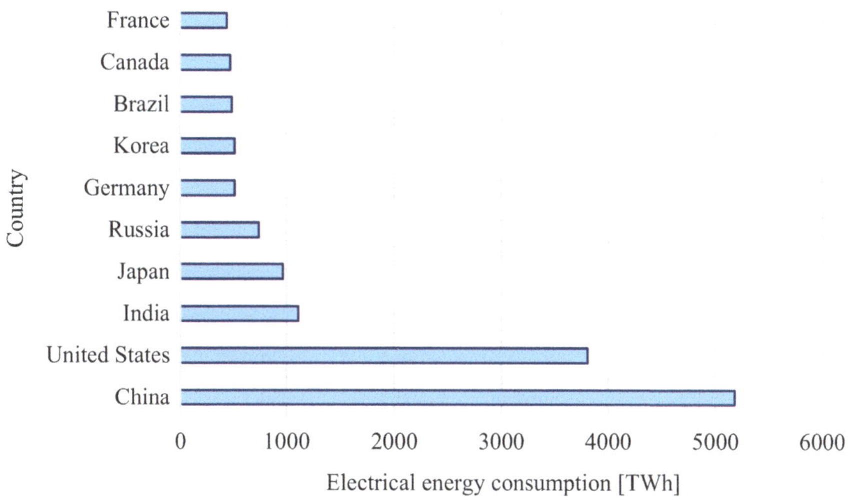

Fig. 1.4 Top ten countries with the highest electrical energy consumption in 2016 [64]

Table 1.11 Space heating, space cooling and water heating shares of the final energy consumption in the residential sector in 2017 [43]

Country	Space heating share (%)	Space cooling share (%)	Water heating share (%)
Albania	31.7	5.5	21.4
Austria	69.6	0	14.9
Bulgaria	54.3	0.5	17.2
Belgium	73.8	0.1	11.4
Croatia	68.7	1.8	10.0
Czech Republic	69.0	0.1	16.2
Cyprus	Not available	Not available	Not available
Denmark	62.5	0	21.2
Estonia	Not available	Not available	Not available
Finland	65.8	0.1	14.9
France	66.1	0.2	11.1
Germany	67.1	0.2	16.1
Georgia	58.8	0.3	11.3
Greece	56.2	4.4	13.5
Hungary	74.0	0.1	12.0
Ireland	58.9	0	19.8
Italy	67.5	0.7	11.9
Kosovo	71.3	3.5	6.5
Latvia	65.6	Not available	18.6
Lithuania	70.3	0	9.2
Luxembourg	79.3	0.2	7.1
Malta	15.0	8.1	9.8
Moldova	70.7	0.1	10.0
Netherlands	63.6	0.2	16.7
Norway	43.8	0	14.2
Poland	66.0	0	16.1
Portugal	21.1	0.7	19.1
Romania	63.4	0.3	13.4
Serbia	60.2	0.5	14.4
Slovakia	68.3	0.1	14.3
Slovenia	63.7	0.5	16.0
Spain	43.4	0.9	19.1
Sweden	54.5	0	13.6
United Kingdom	62.1	0	17.2
Average Europe	64.1	0.3	14.8

that have not fully implemented a refurbishing strategy (e.g. Moldova) or to countries where energy saving is not a stringent priority yet. The lowest space heating shares correspond to countries with a rather warm climate as Malta, Portugal and Spain. The trend for meeting the heating energy demand is focused on district heating considering the "urban transition" that characterizes the development predicted by the European Commission up to 2050. The main advantage of the district heating systems is the lower heating costs when the fossil fuel prices are high. As Werner [134] presented, these costs are low in dense urban areas with concentrated heating demands. Thus, the main disadvantages are related to the high distribution costs in sub-urban or in rural areas and to the lower competitiveness when the price of the fossil fuel is low. Major district heating systems are reported in cities as Moscow, St. Petersburg, Beijing, New York, Kiev, Seoul, Warsaw, Berlin, Hamburg, Helsinki, Stockholm, Copenhagen, Paris, Prague, Sofia, Bucharest, Vienna, and Milan, while the total number of systems estimated all over the world is 80,000 out of which 6000 in Europe [134]. In the EU, the heat supplied with district heating systems reached the highest value of about 2.7 EJ/year in 2010, with the highest shares corresponding to natural gas, followed by coal and coal products and to biofuels and wastes. A similar trend is observed when analyzing the heat sources supplied to the district heating systems all over the world.

It is to mention that the amounts of resources used for district heating decreased after 2010 in EU, reaching in 2014 about 2.3 EJ/year, while all over the world, this amount increased from about 14.2 EJ/year in 2010 up to 14.8 EJ/year in 2014, followed by a continuous decrease.

These results allowed formulating the conclusion that there still is a high share of fossil fuels used in district heating in EU. Moreover, about 90% of the heat produced in district heating systems all over the world relies on fossil fuels resources, the most important energy sources used in combined heat and power plants (CHPs) and in boiler plants, with the frontrunners being Russia (using natural gas) and China using coal as main fuels.

To reach the sustainability targets, these shares have to be decreased by replacing the fossil fuel sources with non-fossil fuel ones. Moreover, the use of biomass for meeting the energy demand in the built environment declined because this source will be more broadly used in the transportation sector; thus, other renewables are required to be employed.

A study presented by Zhao et al. [138] showed that solar hybrid heating systems might represent a feasible solution for the near future to insure the thermal performance, particularly in buildings in the rural area. The results showed that each additional 2 m^2 to the air collector area may raise the maximum indoor temperature with approximatively 2 °C but, as expected, the insulation of the exterior walls has to be carefully considered. It was also observed during the infield testing that, under extremely cold weather conditions, an additional heating source is required.

District heating systems are also reported to use the excess heat resulted from the high electricity consumers, as the large data centres that are providing cloud services. Heat recovery is also mentioned from the nuclear reactors and 26 PJ heat was supplied in 2014 in Russia, Bulgaria, Czech and Slovak Republics, Hungary, Switzerland and

Ukraine using this source. Moreover, heat recycling from wastes incineration is also considered as in Denmark and Sweden where all wastes incineration plants are "waste to energy" installations connected to district heating systems for heat recycling [134].

The use of renewables is also mentioned [134] for meeting the energy demand in the built environment but the amounts are still low. As Mauthner et al. [87] mentioned, the global solar thermal capacity of unglazed and glazed water collectors in operation was 435 GW in 2015. Geothermal energy is used in district heating systems for many years in Iceland and in France and is in the beginning in Germany, Hungary, Italy, Romania, Belgium, etc. In the entire world, about 220 TJ/year of heat were supplied using geothermal heat pumps [135]. The use of biofuels was estimated in 2014 at 524 PJ, out of which 405 PJ in the EU, more often reported in Sweden, Finland, Denmark and Austria [134].

The developments expected in the urban areas require specific measures for avoiding or at least reducing the environmental and energy supply challenges, as stated by Buffa et al. [16]. This paper outlines that in 2017 there were about 6000 district heating networks in EU, that were able to cover about 12% of the heat demand while 115 district cooling networks were registered only, covering a significantly lower share (2%) of the total cooling energy demand. The traditional district heating (DH) networks consist of power stations that feed hot water or steam into the piping system to distribute the heat into the urban areas. These systems have rather high installation costs and are responsible for heat losses, especially during summer when only domestic hot water (DHW) is required and the heat losses may reach 30% as result of a higher retention time of the thermal fluid (water) in the network. These limitations are addressed by the fourth and fifth generations of district heating and cooling (DHC) networks, aiming at reaching high efficiencies by operating at lower temperatures. The systems developed in the frame of the fourth generation are not able to simultaneously provide heating and cooling to different buildings and this is the main focus of the fifth-generation systems. The paper shows that there are currently 40 such networks in Europe, most of them implemented in Italy, Germany, Switzerland and The Netherlands; such systems are also implemented in Belgium, Norway and UK (one in each country). Most of these systems use as heat source groundwater, seawater, the ground or multisources that are available in the implementation locations.

As already mentioned, district heating represents a feasible alternative for urban areas but it is not fully suitable considering the less dense communities, as for example, the rural ones. The rural built environment has several characteristic features, as outlined by Zheng and Bu [139] for China but that are valid for almost all the underdevelopment regions. One typical feature is the self-sufficiency of the houses due to the rather low density at community level. A second feature characterizes the building's energy efficiency that is less good as compared with the urban area houses mainly because there is a poorer form factor that leads to increased energy consumption. Most of the buildings in the rural communities are detached houses or are of bungalow type. Moreover, these houses are built using less proper materials, with rather poor insulation properties. Consequently, the heating energy demand is higher for these houses.

To satisfy the energy demand, the rural population is mostly relying on traditional sources as, for example, the wooden biomass in certain regions of Romania (where there are large forest surfaces in the mountain areas). In China, the mostly used resources in the rural area are the traditional biomass and coal, for winter heating and the study developed by Duan et al. [36] outlined a negative correlation between the share of households using solid fossil fuels and their income levels. The *National Improved Stove Programme* run in China between 1980 and 1990 aiming at providing fuel-efficient devices to almost one billion rural inhabitants. The analysis of the results showed that improved cooking stoves are not enough [118] and all the household energy need to be carefully approached and intervention programmes have to be developed for all the intended purposes [72]. The review developed by Kerimray et al. [75] shows that in Mongolia, the coal and wood burning is essential in the traditional Mongolian nomadic tent-like dwelling (*gers*) with over 98% of the surveyed households using these sources for heating. Further, as Meirmans [89] reported, about 29% of the heating demand in rural regions in the Czech Republic is met using coal.

As the discussions so far presented showed, the energy consumption in buildings is covered using various fuel types; a synthesis is presented in Table 1.12 for the USA, China and the European Union in 2010 [23].

As the results in Table 1.12 show, in the USA and in EU, the largest share of energy used in buildings relies on electricity or is obtained using natural gas. In the USA, electrical energy represents the largest share and is mainly used for heating, cooling, lighting and other appliances.

However, natural gas represents the main energy source for domestic hot water preparation. The results also show that in China, the energy obtained using biomass and wastes holds the highest share in the total energy consumption, mainly used for heating and cooking. The data also outline that in China, the buildings in the residential sector consume about five times the energy consumed in the buildings in

Table 1.12 Buildings final energy consumption by fuel/source type in the United States, China and European Union in 2010 [23, 60]

Energy source	Country		
	United States (%)	China (%)	European Union (%)
Coal	0.3	14.3	2.7
Oil	7.7	11.6	13.6
Natural gas	38.4	7.4	36.7
Electricity	50.3	15.2	31.1
Commercial heat	0.3	4.1	7.0
Biomass and wastes	2.6	47.1	8.5
Other renewables	0.3	0.4	0.4

the service sector; therefore, changing from biomass to other green fuels represents a strategic decision especially for the rural areas where this consumption is high.

Using district heating systems represents a trend that is well accepted in the EU countries. As presented by Tettey et al. [122], in Sweden, multi-storey apartment buildings are using the energy provided by biomass-based district heating systems consisting of a combined heat and power (CHP) plant that uses wooden chips or wood powder.

The projections for 2050 and 2100 show a significant increase in the residential building energy demand for heating and for cooling, as outlined by Isaac and Van Vuuren [67]. The results show that the heating and cooling energy demand will increase mainly in cooling-dominant and less developed regions, e.g. in India, where the cooling energy is expected to reach more than 3000 PJ in 2050 and 17,000 PJ in 2100. The scenarios show that the energy demand for cooling will register the most significant increase all over the world as result of global heating and increase in the population number, thus in the built environment units. This is why there is a certain need for improving the design of the buildings' envelopes by including air-sealing systems and well-insulated windows for reducing the energy demand and increasing the sustainability.

To cover this energy demand without increasing the pressure of global heating, the implementation of renewable energy systems in the buildings or nearby, in the built environment, is required. The future energy scenarios target the reduction in the use of fossil fuels as the EU strategy states or the one developed by the Australian government that plans to reach 100% of renewable energy generation in the gross final energy consumption in the built environment. As Walker et al. [132] showed, this target can be reached, considering that for EU-28, the share of renewables grew from 8.5% (2004) to 16.7% (2015) and the three major actions supporting this target are already implemented: energy saving in the energy consumption processes in the building, efficiency increase in the energy production systems and the use of renewables. This implies that energy neutral and CO_2 neutral buildings represent the targeted options in the future, especially for the new buildings. Moreover, introducing community renewable energy systems and community storage systems may represent the viable path as presented by Polly et al. [100].

Finally, yet importantly, it has to be mentioned that there is a need for specific education and/or training for implementing the actions that lead to a sustainable built environment. The measures considered to reduce the energy consumption in the built environment should be well planned and supported by the buildings' users, while the direct environmental benefits should be well presented and perceived. As reported by Al-Marri et al. [4] a group of Qatari students that were studying in the UK was subject to testing the behavioural changes and the causes that supported these. It could be observed that 91.2% of the students were switching off the lights and the heating and cooling devices when living home in the UK while only 45.2% of the students were doing the same things when in Qatar. The results represent a clear indication of changing behaviour and are considered the result of the high costs of energy in the UK (representing up to 12% of the students' income), when compared to Qatar that has very low costs associated to the energy consumption. Moreover, the analysis showed that before studying in the UK, over 50% of the students were not aware of the environmental damage resulted from the use of fossil fuels and that raising the awareness led to significantly improved behaviour towards energy saving, thus towards sustainability.

This type of differences is experienced also among groups of experts from different geographical regions, as outlined by Si and Marjanovic-Halburd [117]. When comparing the selection criteria of optimal green technologies in building retrofit developed by the Chinese and the UK professionals, it was found that the English architects are emphasizing the economic criteria (e.g. cost or financial incentives) while the Chinese ones are more focused on technical criteria. Obviously, these are the consequences of what the clients (the building's users) are asking for: in the UK, the main requests are related to the reduction in the operation costs and to the increase in the building's value while in China, the main request is related to the building's safety and security. However, both groups of professionals outlined the reduction in the energy used at building level and in its water consumption as important. This shows that there still is much work to be done, to get a more unitary way of discussing the need for sustainability and the concrete steps to be followed.

Moreover, the assessment of the feasibility applied in green energy production and in green technologies in the built environment has to consider the stakeholders priorities and to harmonize them; the mostly relevant stakeholders in this respect are the architects, the civil engineers and the building's users.

1.4 Indicators for Buildings Efficiency and Sustainability

According to the United Nations [125] latest report *World Urbanization Prospects: the 2018 Revision*, over half of the world's population is living in cities, with the highest share in more developed countries (78.7%) and in high income countries (81.5%) and the lowest share in least developed countries (33.6%) and low income countries (32.2%). These percentages are expected to increase due to the strong urbanization trend, reaching in 2050 a share of urban population in the total number of inhabitants estimated at 68%. Based on the analyzed data and on modelled predictions, in 2015, the world population was of about 7.3 billion and is expected to reach 9.7 billion inhabitants in 2050; thus, the number of citizens will increase with more than 2 billion persons. This increase has predictable consequences on the energy consumption and on the greenhouse gases emissions. Already ten years ago, the International Energy Agency [59] recognized cities as major contributors to CO_2 emissions (over 70%) while over two-thirds of the final energy use was associated with the urban areas [49]; therefore, the transition towards zero energy buildings is currently part of many local development strategies [61].

As reported by Cajota et al. [20], cities, where about 40% of the energy use is in the buildings sector, have to go through a changing process focusing on energy planning to reach the sustainability targets. These changes have to focus on increasing the buildings' energy performance but also to address the environmental, social and economic impacts of the constructions, as mentioned by Mattonia et al. [86].

The recast version of the Energy Performance of the Buildings Directive formulated by the European Parliament and Council [41] outlined the general framework for the *nearly Zero Energy Building* (nZEB) as "a building that has a very high energy performance [...] the amount of energy required should be covered to a very significant extent by energy from renewable sources, including energy from renewable energy sources produced on-site or nearby". This definition leaves room for the member states to define numerical evaluation criteria, depending on the specific climate and on the potential of increasing the efficiency of the energy use. The directive requires any new building to be *nZEB* by the end of 2020 and any new public building by the end of 2018. Moreover, the USA, Japan and Korea have rather recently formulated nZEB strategies and goals.

The nZEB concept is related to the *Zero Energy Building* [66] concept that describes an energy-efficient structure that balances its energy consumption by using the energy produced by on-site renewable energy systems, as described by Marszal et al. [85]. Another related concept is the *Net Zero Energy Building* (NZEB) representing, according to Kaewunruen et al. [74], "a building that can meet its entire energy demand using renewable energy systems". The definition according to EU [40] states that a NZEB represents a highly energy-efficient building, having thus a low annual energy demand covered using renewable energy systems on-site installed. This can be reached through reducing the energy use of the buildings by extending the daylight use combined with increased insulation, passive solar heating or natural ventilation associated with certain features, e.g. using high performance windows, efficient light sources and well-managed HVAC systems.

This European characterization mostly applies to new buildings. When considering the older or the refurbished buildings, most commonly, two terms are used: *green buildings* and *sustainable buildings*.

According to the US Environmental Protection Agency [128], a *green building* represents "a structure that is environmentally friendly and resource-efficient over the entire lifecycle". According to Robichaud and Anantatmula [105], green buildings are following four main targets: minimizing the environmental impact, enhancing the health conditions of the building's users, supporting the economical returns to investors and local community and reaching acceptable lifecycle impacts during the planning, development and operational phases.

Following the review reported by Alwisy et al. [5], the most significant *green building* design criteria are represented by:

- *The indoor environment quality* directly influences the health of the occupants; the thermal comfort, the lighting comfort, the air quality and the acoustic comfort represent important measurable properties that are usually included in the evaluation tools of *green buildings*. The indoor quality can be supported by using novel phase-change materials in the building's envelope to avoid overheating in the super-insulated residential buildings. Lighting is also linked to the occupants' health and satisfaction and solutions involving advanced lighting control and the use of LED sources are recognized to improve the building's efficiency. Moreover, the acoustic level in the building has a significant influence on the productivity and comfort of the building's users. Last but not least, air quality is considered along with the HVAC systems and the most important issues outlined are those related to the ventilation systems in the building, that are able to provide fresh air (without any contaminants) and to keep a certain level of the air humidity if coupled with specific monitoring systems;

- *The energy use* is considered the most influential criterion when describing a building as *green*. Energy-saving measures are part of the *green building* design, considering the shape, orientation and the type and size of the windows as the most influential factors. Moreover, Ghahramani et al. [46] reported the control of the heating and cooling programmes as reducing over 30% of the energy consumed in the buildings. Lee et al. [79] proved that over 47% of savings in lighting and HVAC energy consumption could be reached when implementing specific control strategies. This energy is further reduced by using efficient systems, particularly considering the HVAC ones. Moreover, the use of renewables implemented on or nearby the building represents a viable alternative as, when analyzing two buildings located in different climatic profiles (Finland and Brazil), it was observed that PV systems could cover between 20% and 40% of the total electrical energy demand of the analyzed constructions;
- *The water use* represents a key issue in the *green buildings* design considering that only 3% of the Earth's water supply is potable, the rest requiring treatment processes before use. This is why the following aspects related to water are important in the *green building*'s design: water savings, leakage monitoring systems, water-efficient equipment and the use of rainwater and grey water. Collecting, storing and using the rainwater (for local external or internal non-potable uses) was investigated by Campisano and Modica [22] who reported savings between 25 and 50% in the investigated households (located in the Patti town, part of the Metropolitan region of Messina in Sicily, Italy), during a two weeks monitoring period;
- *Materials used* in developing *green buildings* has to focus on recycled materials and on the limited use of non-renewable materials (as plastics or polymeric foams). To get an improved materials efficiency and sustainability, the lifecycle impact of various common materials was comparatively analyzed with that of eco-materials to assess the primary energy demand, the CO_2 emissions potential and the water demand in obtaining and using them by Anastasiou et al. [7]. They reported the successful use of common wastes, as fly ash and steel slag, in producing concrete for the buildings' foundation;
- *Wastes* resulted during the construction phase of the building and *pollution prevention systems* represent important features when assessing a *green building*. Considering pollution, most important seems to be the air and light pollution in buildings, and studies were developed outlining that HVAC systems can be mostly responsible for the deterioration of the indoor air quality due to inadequate ventilation or ducts improper cleaning. Light pollution is mainly associated with improper artificial lighting and with the limited use of the daylight resources;

- *Land use* represents another criterion recommended in the assessment of the *green buildings* and two sub-criteria were found to be mostly relevant: rainwater management and heat island effect reduction. A study developed by Shafique and Kim [111] showed that storing the rainwater on the roof can decrease the roof temperature up to 9 °C due to evaporation, thus limiting the heat island effect.

As outlined by Doan et al. [33], currently, the *green building* term is mainly linked to the environmental improvements while *sustainable building* has the main focus on the four pillars of sustainable development (environmental, social, economic and institutional pillars) and more pillars can/will be added to define the *sustainable building* concept, e.g. culture or epistemology. Nguyen and Altan [96] showed that there were 382 software tools in 2011 for assessing the energy efficiency, renewable energy use and energy performance of the buildings. However, the sustainability metrics are far from being uniform and seldom address all the sustainability aspects, as defined. This is the reason why the two terms are also grouped, as in the review developed by Shan and Hwang [112] on green buildings rating systems, i.e. on the framework developed by building and construction authorities, international organizations and/or consultancy companies for assessing and verifying "the sustainability and the greenness of the buildings". Moreover, the *green building* concept was defined by the US Environmental Protection Agency [128] as being also known as "green construction" or "sustainable building", as presented by Ding et al. [31]. Moreover, although *Green Buildings* and *Sustainable Buildings* are not inter-changeable terms, many times they are mentioned as they were. Rating tools for assessing the building's sustainability are developed outlining the capacity for further development towards sustainability [104]. Table 1.13 gathers the mostly used buildings rating systems, considering their green features and sustainability.

These most common tools are described either as "green building rating systems" [107] or "rating system tools for evaluating buildings sustainability" [84]. They also concluded that most of these tools are not equally addressing all the sustainability issues and are mostly focusing on the environmental impacts ignoring the social or economic impacts.

BREEAM or *Building Research Establishment Environmental Assessment Method* represents a sustainability assessment method launched in 1990 by the Building Research Establishment [18]. According to the information on the company website, the *BREEAM* scheme provides third-party certification to assess the sustainability performance of individual buildings, communities and infrastructure projects. The certification can cover various stages in the built environment lifecycle, including the design, the construction, the building's operation and its refurbishment. The categories that are considered in the certification system are related to energy, health and well-being, innovation, land use, materials, management, pollution, transport, wastes and water. Each of these categories is sub-divided into assessment topics that are awarded a set of credit points.

The final assessment considers the sum of these credits and a certain category weighting. Specific standards were developed for various building types, as

Table 1.13 Green building rating systems (selection)

Rating system	Developer	Year of the first/last version	Website/References
BREEAM	Building Research Establishment Ltd., UK	1990/2016	https://www.breeam.com Shan and Hwang [112]
LEED	US Green Building Council (USGBC)	1994/2016	https://new.usgbc.org/leed Shan and Hwang [112]
HK BEAM/BEAM Plus	BEAM Society Limited	1996/2010	https://www.hkgbc.org.hk/eng/BEAMPlus.aspx Nguyen and Altan [96]
Green Star	Green Building Council Australia	2003/2015	https://new.gbca.org.au/green-star/ Nguyen and Altan [96]
CASBEE	Ministry of Land, Infrastructure, Transport and Tourism of Japan	2004/2016	http://www.ibec.or.jp/CASBEE/english/ Nguyen and Altan [96]
DGNB	German Sustainable Building Council	2008/2018	https://www.dgnb.de/en/index.php Marjaba and Chidiac [84]
HQE	HQE Association	1997/2016	http://hqe.publishingcenter.fr/hqe-in-the-world/list-of-projects HQE Association [54]

office, retail, industrial, data centres, education, healthcare, residential mixed use and various other building types.

The *BREEAM* rating system has six levels, as presented in Table 1.14. Specific standards are formulated for different development lifecycle stages: communities, infrastructure, new construction (home and commercial buildings), in-use (commercial buildings), refurbishment and fit-out (home and commercial buildings). The

Table 1.14 The *BREEAM* rating [17]

Rating	Acceptable	Pass	Good	Very good	Excellent	Outstanding
No. of stars	*	**	***	****	*****	******

BREEAM system was used in more than 80 countries all over the world. Among these, countries with a significant number of awarded certificates are: UK (10,699 certificates), Austria (3063), Spain (2233) and Sweden (1295); moreover, there are also confirmed as countries with *BREEAM* certified buildings, The Netherlands, Norway, Germany, Switzerland, China, Turkey, Saudi Arabia, Russia, etc. Currently, there are over 568,000 issued certificates and more than 2.2 million registered buildings [18].

LEED or *Leadership in Energy and Environmental Design* was developed by the US Green Buildings Council (USGBC) and is, according to the information on its website, the most widely used green building rating system in the world [129]. According to the website, "*LEED* is for all building types and all building phases including new construction, interior fit outs, operations and maintenance, and core and shell. There's a LEED for every type of building project". The latest version of the *LEED* rating system (v4.1) was developed for the following groups: *Building Design and Construction* (BD + C), *Interior Design and Construction* (ID + C), *Building Operation and Management* (O + M), *Neighbourhood Development* (ND), *Homes*, *Cities and Communities*, *LEED Recertification* and *LEED Zero* developed for projects aiming at net zero goals in carbon and/or resources.

Third-party technical reviewers (part of the *Green Business Certification Inc.*) assess the projects that are subject to *LEED* certification. They are allocating points to the following main categories: location and transportation, sustainable sites, water efficiency, energy and atmosphere, materials and resources, indoor environmental quality and innovation.

Following the number of points that are awarded, the project may get into one of the four rating levels, as the data in Table 1.15 show.

Until 2016, the *LEED* rating system was applied in more than 160 countries, on over 93,000 projects. During 2017–2018, various *LEED* certificates were awarded to buildings in the USA (Houston, Honolulu, Chicago, Baton Rouge, Sacramento, Austin, Long Beach, Los Angeles, etc.), Spain (Barcelona, BD + C), Japan (Kyushu, O + M), China (Guangzhou, O + M), Mexico (Guadalajara, BD + C), Italy (Milano, BD + C), France (Grenoble, O + C), Turkey (Istanbul, O + M), Brazil (Sao Paolo, BD + C), etc.

HK BEAM stands for *Hong Kong Building Environment Assessment Method*. This tool was developed by the *BEAM Society Limited* (BSL) to assess green buildings and to offer training to BEAM professionals and affiliates. In 2010, the *Hong Kong Green Building Council* certified the latest version of the tool, called Beam Plus, and in 2012, its V1.2 version was released. This system contains the following assessment options: *BEAM Plus New Buildings*, *BEAM Plus Existing Buildings*, *BEAM Plus Interiors* and *BEAM Plus Neighbourhood*. According to the information provided by the *BEAM Plus* brochure [53], the main topics that are subject of evaluation

Table 1.15 The LEED rating system [109]

Rating	Certified	Silver	Gold	Platinum
No. of points	40–49	50–59	60–79	80+

Table 1.16 The Green Star rating system [47]

No. of stars	1 *	2 **	3 ***	4 ****	5 *****	6 ******
Rating	Minimum practice	Average practice	Good practice	Best practice	Australian excellence	World leadership
Rating tool	Performance					
				Design and as built		
				Interiors		
				Communities		

are: *community aspects* (CA), *site aspects* (SA), *green building attributes* (GBA), *management* (MAN), *materials and wastes aspects* (MWA), *energy use* (EU), *water use* (WU), *outdoor/indoor environment quality* (OEQ/IEQ) and *innovations* (IA).

The rating system has four levels: bronze, silver, gold and platinum. A rating level can be awarded to a project after completing the provisional assessment (*A*) or the final assessment (FA). Using this assessment system, over 18.6 million m^2 of green buildings were certified, in Hong Kong, Beijing, Shanghai, Shenzhen, etc.

Green Star represents a building rating system developed by the *Green Building Council in Australia* [47]. There are currently four rating tools to certify the design, construction and operation of buildings and of communities. These are *Green Stars—Communities, Green Stars Design—As built, Green Stars—Interiors* and *Green Stars—Performance*. The *performance* category focuses on nine topics aiming at assessing the building's sustainability, as *management, indoor environmental quality, energy, transport, water, materials, land use and ecology, emissions and innovation*. This category recently included new carbon neutral standards for buildings' certification. This tool applies to any building type except single-detached dwellings (single houses).

The *Green Star* rating scale has six categories that are differently available, according to the rating tools, as presented in Table 1.16. Moreover, the *Green Star* website includes a calculator that allows a preliminary assessment of the building's quality, based on specific indicators related to the indoor comfort (air quality, lighting, thermal and acoustic comfort), greenhouse gases emissions, peak electrical energy demand, potable water, wastes during operation and refurbishment, the use of refrigerants, etc. A specific Renewables and Offsets *Green Star Guide* was additionally released in 2018 to support the calculators and to raise the awareness of these resources.

According to the information in Table 1.16, projects on buildings and community design and construction will get certified only if their rating is at least at the *best practice* level (four stars). Buildings assessed based on the *Green Star—Performance* tool gets certified at any of the 1–6 stars levels.

CASBEE stands for *Comprehensive Assessment System for Built Environment Efficiency*. According to the information on the CASBEE website [69], a committee established in 2001, joining representatives from academia, industry and national and local governments, formed the Japan Sustainable Building Consortium (JSBC) under the auspice of the Ministry of Land, Infrastructure, Transport and Tourism and further on developed the *CASBEE* system.

The *CASBEE* concept considers the *Built Environment Efficiency* (BEE) defined as the ratio of the *Environmental Quality of the Building* and the *Environmental Load of the Building*. These two main assessment categories are defined as follows:

- *Q* (Quality) Built environment quality evaluates "improvement in living amenity for the building users, within the virtual enclosed space (the private property)";
- *L* (Load) Built environment load evaluates "negative aspects of environmental impact which go beyond the virtual enclosed space to the outside (the public property)". The environmental impact is evaluated considering the emissions of pollutants in the air, of greenhouse gases, heat, noise, etc., the resources consumption and other various categories.

The assessment system has tools formulated according to the building type or to the assessment purpose and the latest versions are the following:

- CASBEE—NC, for Buildings (New Construction) 2014 edition*
- CASBEE—EB, for Buildings (Existing Buildings) 2014 edition
- CASBEE—RN, for Buildings (Renovation) 2014 edition
- CASBEE for Real Estate 2014 edition*
- CASBEE for Commercial Interiors 2014 edition
- CASBEE—TC, for Temporary Construction 2007 edition
- CASBEE—HI, for Heat Island 2010 edition
- CASBEE—UD, for Urban Development 2014 edition*
- CASBEE for Cities 2013 edition*
- CASBEE for Cities—Pilot version 2015 edition*
- CASBEE for Detached Houses (New Construction) 2014 edition
- CASBEE for Dwelling Unit (New Construction) 2014 edition
- CASBEE Health Checklist
- CASBEE for Housing Renovation Checklist
- CASBEE Community Health Checklist

CASBEE focuses on four topics: *Energy Efficiency, Resources Efficiency, Local Environment* and *Interior Environment*. These topics were more specifically reformulated, to allow a coherent calculation of the *BEE* indicator.

The current assessment sub-categories are three for the Q indicator (*Q1 Indoor environment, Q2 Quality of services* and *Q3 Outdoor environment*) and three for the L indicator. The environmental load reduction has the following sub-components: *LR1 Energy, LR2 Resources and Materials* and *LR3 Off-site environment*. The rating scale of the *CASBEE* system has five levels, as presented in Table 1.17.

Since 2008, the *CASBEE* system includes a CO_2 lifecycle impact assessment tool (LCCO2) that evaluates the CO_2 emissions from the building phase, through the operation phases, up to the building's demolition and disposal. Moreover, the LCCO2 performance is also evaluated based on a maximum five stars rating system using as reference the emissions of a building that satisfies the standard defined by the Energy Conservation Law.

Table 1.17 The CASBEE rating system [56]

Rating	C (poor)	B− (fairly poor)	B+ (good)	A (very good)	S (excellent)
Number of stars	*	**	***	****	*****
BEE and Q values	0–0.5	0.5–1	1–1.5	1.5–3 or BEE = 3 if Q < 50	BEE > 3 and Q > 50

The assessment results represent the % of CO_2 emissions of the building considering 100% the emissions of the reference building. This assessment tool focuses on improving the energy efficiency of the building, on the use of ecological materials, on extending the building's lifetime, on on-site solar power generation that supports the procurement of a green power certificate or of carbon credits, etc.

Up to March 2016, about 500 buildings were assessed in Japan and over 14,000 CASBEE accredited professionals were trained for implementing the system, according to the users' needs.

DGNB is the acronym of the *Deutsches Gesellschaft fur Nachhaltiges Bauen (German Sustainable Building Council)* that was founded in 2007 and has now over 1200 members representing industry and professionals in buildings construction and in the real estate value chain. In 2009, the *DGNB* system was developed and, according to the information provided on the dedicated website, it represents a tool aiming at evaluating the buildings and urban districts. In 2018, the latest version of the certification system for new buildings was launched, strengthening the focus on sustainability criteria.

Considering the broad variety of buildings, the *DGNB* system is developed to certify office and administrative buildings, retail buildings, industrial buildings, hotels, residential buildings, mixed-use buildings and educational facilities. Currently, more than 20 schemes consider four categories:

- Existing buildings: buildings in use and renovated buildings;
- New constructions: educational buildings, offices, healthcare buildings, retail buildings, hotels, industrial buildings, (small) apartment buildings, laboratory buildings, buildings for mixed use, multi-storey car parks, sport halls, buildings used for meetings/gatherings;
- Interiors: office, retail, restaurants, hotels;
- Districts: urban districts, office and business districts, industrial sites, events areas, resorts, vertical cities.

The assessment criteria consider six topics:

- Environmental quality: building lifecycle assessment, local environmental impact, sustainable resource extraction, potable water demand and wastewater volume, land use and biodiversity at the site;
- Economic quality: lifecycle cost, flexibility, adaptability and commercial viability;

Table 1.18 DGNB rating system [30]

Rating	Bronze (%)	Silver (%)	Gold (%)	Platinum (%)
Total performance index	35–50	50–65	65–80	>80
Minimum Performance index	–	35	50	65

- Socio-cultural and functional quality: thermal comfort, indoor air quality, acoustic comfort, visual comfort, user control, quality of the indoor and outdoor spaces, safety and security along with design for all;
- Technical quality: sound insulation, quality of the building envelope, use and integration of the building technology, ease of cleaning the building components, ease of recovery and recycling, immissions control and mobility infrastructure;
- Site quality: local environment, influence on the district, transport access and access to amenities;
- Process quality.

The score for these topics is calculated based on the combination of evaluation points (allocated for each specific criterion) and the relevance weighting. There were more than 2800 *DGNB* certifications granted by the end of 2017, most of them obtained starting with 2015 when there were close to 1000 emitted certificates. Starting with 2015, the rating system has four levels: bronze, silver, gold and platinum that are awarded according to the scores in Table 1.18. The *DGNB* website also declares the "diamond" rating as being the highest level that was awarded so far to only five buildings, while the platinum certificate was awarded so far to 225 projects and the gold certificate to 831 projects. It is also to mention that bronze certificates can be awarded only to existing buildings; thus, new buildings have to comply with at least the silver certificate requirements.

A precertification system supports the building's developers and planners to reach an optimum, sustainable solution for their particular building.

The *DGNB* certification system was implemented in many countries all over the world, as Germany, Austria, China, Denmark, Canada, Luxembourg, Poland, Romania, Russia, Switzerland, Slovakia, Spain, Thailand, Czech Republic, Turkey, The Netherlands, Ukraine, Hungary, UK, Vietnam, etc.

HQE stands for *Haute Qualité Environnementale* (environmental high quality) and represents an assessment tool developed in France by the HQE Association [54]. The tool contains 15 topics that can be fulfilled at three performance levels: "high performance", "performance" or "base", directly referring to environmental aspects as: energy saving, resources saving, recyclability, indoor air quality, etc. According to the HQE website, the HQE certification was awarded to over 500 projects located in France, Brazil, Luxembourg, Poland, Belgium, Germany, Italy, Monaco, Morocco, Gabon, Algeria, Vietnam, etc. The final rating has five levels: pass, good, very good, excellent and exceptional.

The *HQE* certification system can be applied to four categories:

- Residential buildings, when the environmental performance is assessed for four themes: (1) energy and savings with the sub-divisions: energy, water and maintenance; (2) environment (site, components, worksite, waste); (3) health and safety (space quality, air quality and water quality); (4) comfort (hydrothermal comfort, acoustic comfort, visual comfort and olfactory comfort);
- Non-residential buildings that are assessed using the same four themes;
- Non-residential buildings in operation that are assessed considering three categories: sustainable buildings, sustainable management or sustainable use, according to the type of requester (landlord, property manager or building user);
- Sustainable planning as a management tool that can be applied at district level.

Besides these rating systems, worldwide used and having, each, over 500 certified projects, there are plenty of national assessment tools, as outlined by Bernardi et al. [13] in their synthesis material. Based on a large number of certification schemes assessed, they concluded that there are certain categories that are mostly used in the assessment tools, as the energy performance and the solid waste management. Moreover, many tools are also including categories related to the building materials, water and wastewater management, ecology and environmental quality. The categories that are only seldom formulated are resistance against natural disasters (considered only by *CASBEE, DGNB* and *HQE* tools) while *the building information and users guide* is only part of the *BREEAM* tools and of some sub-schemes of *LEED, HQE* and *DGNB*.

All the rating systems in Table 1.13 have specific tools for the evaluation and certification of residential buildings, office buildings, commercial buildings, educational buildings. For industrial buildings, some of these systems have not particularly developed a specific rating tool (e.g. *LEED*). Moreover, the assessment tools can be applied to all the building's development stages: predesign and design, construction, post-construction and use/maintenance as, for example, the *BREEAM, CASBEE, DGNB* or *HQE* or for only part of these (as for *LEED* that does not apply for the first stage). The tools are covering new buildings, existing buildings or buildings subject of refurbishment. According to the information in Table 1.13, most of the rating tools are developed by governmental structures or professional councils and this supports the assumption that all over the world, the green buildings represent a coherent part of the future development scenarios.

Shan and Hwang [112] published a synthesis of some relevant green buildings rating systems in 2018. The paper outlines the most relevant evaluation criteria for 15 rating systems that include BREEAM, LEED, Green Star and CASBEE. They found out that, among 29 criteria, the ones detailed in Table 1.19 are the most important and these are mentioned in direct relation with the assessment tools above presented.

The most important criterion considered by all the assessment tools is *energy*. According to the review, this criterion addresses the peak load reduction, the monitoring of the energy consumption, energy efficiency of various appliances and equipment installed in buildings and the use of renewable energy systems, with a particular

Table 1.19 Weightings of the essential evaluation criteria for the mostly used green building rating systems [112]

Rating system	Energy (%)	Site (%)	Indoor environment (%)	Land and outdoor environment (%)	Materials (%)	Water (%)	Innovation (%)
BREEAM	19.97	–	–	7.69	12.5	6.73	10
LEED	28	12	14	–	16	9	4
Green Star	25	–	16	10	14	10	9
CASBEE	20	–	20	15	12	3	–
HQ BEAM	35	25	20	–	8	12	–

focus on the use of solar energy. Either this is considered as passive use, according to the classic sustainable design principles as orientation or solar gain, as outlined by O'Malley et al. [97], or the active use is assessed, considering the solar energy conversion systems installed on the building or nearby. The *site* criterion pays attention to the sustainability issues during the building construction. The *indoor environment* criteria evaluate a wide range of building's features, responsible with the sound environment, thermal comfort, lighting and illumination, visual and acoustic comfort, air quality, etc. This criterion along with the *land and outdoor environment* criterion has the lowest standard deviation of the weights in various assessment tools, proving that these are uniformly considered all over the world. On the other hand, the *water* criterion has a much broader standard deviation from the average weighting (6.1%) as result of how its calculation is included in various rating systems as, for example, the lowest weighting corresponds to CASBEE that considers water quality and water savings as part of other categories. The *water* criterion is more highly weighted in arid regions, where water-efficient use is important (e.g. in the Abu Dhabi EPRS rating system). The weighting of these different criteria may vary from one type of building to another and from one group of respondents to another, e.g. *indoor environment* was considered to be the most important criterion for healthcare buildings, as outlined by Kim and Osmond [76]. The *DGNB* buildings rating system considers the sustainability aspects related to the environment as divided into *environmental quality* and *technical quality*, as outlined by Marjaba and Chidiac [84]. In their paper, a comparative analysis of the *LEED, BREEAM* and *DGNB* rating systems is developed and the results show that the environmental aspects have the lowest weight in the *DGNB* tools (31%) while the other two rating tools have a significantly larger share dedicated to this topic: 72% (*LEED*) and 69% (*BREEAM*). The economic and social criteria are much better represented in the *DGNB* rating system, at 28% and 31%, respectively, when compared to *BREEAM* (2.5 and 12.5%) or to *LEED* (3.5 and 18%).

Table 1.20 Rating tools for sustainable communities

Rating system	Rating tool for communities	Date when it was launched	Number of certified or registered projects
BREEAM	Communities	2009 Updated in 2012	20 certified projects Over 35 registered projects
LEED	Cities and Communities	2019	
Green Star	Communities	2015 v.1.0 2016 v.1.1.	
CASBEE	CASBEE City (low-carbon type)	Under development	

During the past decade, almost all certification systems developed for buildings were extended with module(s) dedicated to *urban planning*, focusing on the development of sustainable communities. As the information synthesized in Table 1.20 shows, for the analyzed systems, these tools are further presented.

The *BREEAM Communities* tool [17] is an international certification scheme that integrates sustainable design in the development of the communities' master plan and allows assessing the economic, social and environmental sustainability of the developments (e.g. new or refurbished ones). As outlined by the Building Research Establishment [17], the certification scheme supports the integration of the sustainability opportunities in the design stages, resulting not only in sustainable solutions but also in financial savings. Considering the entire community, the advantages include the possibility of developing site-wide energy solutions or drainage systems. The well-defined set of criteria developed for the UK can be adapted to the local context, considering the climatic profile, the local standards and regulation and other issues resulted from the cultural specific features of the location. The mandatory *BREEAM Communities* standards consist of five core categories (governance, land use and ecology, resources and energy, social and economic well-being, transport and movement) and a sixth category called innovation, in forty individual assessment issues that are awarded a number of credits. Table 1.21 gives an overview of this certification scheme.

The rating benchmark for the *BREEAM Communities* has six levels: unclassified (less than 30%), pass (higher or equal to 30%), good (higher or equal to 45%), very good (higher or equal to 55%), excellent (higher or equal to 70%) and outstanding (higher or equal to 85%).

This certification tool was applied in various countries in Europe (Island, UK, Norway, Sweden, Finland, Germany, Hungary, Bulgaria, Turkey, etc.) and Asia (China, South Korea, etc.). *BREEAM* certified communities are registered also in the USA and in the Middle East.

The *LEED Cities and Communities* programme [130] started on 1 April 2019 aiming at implementing the assessment scheme for urban sustainability and standards of living.

Table 1.21 BREEAM Communities rating system (Building Research Establishment 2017)

Technical category	Assessment issues	Weighting (%)
Governance	G01 Consultation plan G02 Consultation and engagement G03 Design review G04 Community management of facilities	9.3
Social and economic well-being	SE01 Economic impact SE02 Demographic needs and priorities SE03 Flood risk assessment SE04 Noise pollution SE05 Housing provision SE06 Delivery of services, facilities and amenities SE07 Public realm SE08 Microclimate SE09 Utilities SE10 Adapting to climate change SE11 Green infrastructure SE12 Local parking SE13 Flood risk management SE14 Local vernacular SE15 Inclusive design SE16 Light pollution SE 17 Training and skills	42.7 Sub-categories Local economy (SE01, SE17): 8.4 Environmental conditions (SE03, SE04, SE08, SE10, Se13, SE16): 7.8 Social well-being (SE02, SE05, SE06, SE07, SE09, SE11, SE12, SE14, SE15): 12.3
Resources and energy	RE01 Energy strategy RE02 Existing buildings and infrastructure RE03 Water strategy RE04 Sustainable buildings RE05 Low impact materials RE06 Resource efficiency RE07 Transport carbon emissions	21.6
Land use and ecology	LE01 Ecology strategy LE02 Land use LE03 Water pollution LE04 Enhancement of ecological value LE05 Landscape LE06 Rainwater harvesting	12.6

(continued)

Table 1.21 (continued)

Technical category	Assessment issues	Weighting (%)
Transport and movement	TM01 Transport assessment TM02 Safe and appealing streets TM03 Cycling network TM04 Access to public transport TM05 Cycling facilities TM06 Public transport facilities	13.8

It is expected that about 14 cities will receive a package value of approximately 25,000 USD. According to the US Green Building Council, the certification scheme relies on the categories outlined in Table 1.22.

This recent version replaces the former *LEED for Cities* and *LEED for Communities* with an expanded form that further includes projects' assessment in the design and planning phases and contains a tool for assessing existing cities.

The main categories included in the *Green Star—Communities* assessment [47] are:

- *Governance*: maximum 28 points can be awarded on criteria dealing with the design review, engagement, adaptation and resilience, corporate responsibility, sustainability awareness, community participation and governance and environmental management;
- *Liveability*: maximum 20 points shared considering the following: healthy and active living, community development, sustainable buildings, culture, heritage and identity, walkable access to amenities, access to fresh food, safe places;
- *Economic prosperity*: maximum 21 points for: community investments, affordability, employment and economic resilience, education and skills development, return of investments, incentive programmes, digital infrastructure and peak electricity demand reduction;
- *Environment*: maximum 29 points for: integrated water cycle, GHG strategy, materials, sustainable transport and movement, sustainable sites, ecological value, waste management, heat island effect, light pollution;
- *Innovation* (10 points available).

The *CASBEE City* (low-carbon cities) represents an assessment tool that is similarly developed as the one corresponding to buildings and relies on the CASBEE UD (urban development). As outlined by Klemes [77], the 80 assessment criteria are part of two main categories:

- Performance, with criteria formulated to assess the natural environment, the quality of services and the contribution to the natural environment;
- Environmental load, that focuses on the impact on the local environment, on the social infrastructure, and on the energy and materials consumption.

Table 1.22 LEED cities and communities rating system [130]

Category	Assessment issues	Maximum score
Integrative process (IP)	Integrative planning and leadership Green building policies and incentives	5
Natural systems and ecology	Ecosystem assessment Green spaces Natural resources conservation and restoration Light pollution reduction Resilience planning	9
Transportation and land use	Transportation performance Compact, mixed use and transit-oriented development Access to quality-oriented transit Alternative fuel vehicles Smart mobility and transportation policy High-priority site	15
Water efficiency	Water access and quality Water performance Integrated water management Storm water management Smart water systems	11
Energy and greenhouse gases emissions	Power access, reliability and resiliency Energy and GHG emissions performance Energy efficiency Renewable energy Low-carbon economy Grid harmonization	30
Materials and resources	Solid waste management Waste performance Special waste streams management Responsible sourcing and infrastructure Material recovery Smart waste management systems	10
Quality of life	Demographic assessment Quality of life performance Trend improvement Distributional equity Environmental justice Housing and transportation affordability Civic and community engagement Civil and human rights	20
Innovation	Innovation	6
Regional priority	Regional priority	4

The system also promotes the participation of the local stakeholders in choosing the weighting coefficients assigned to different criteria.

Currently, almost all the certification systems recognize the need for developing tools to assess the built environment at community level; however, the assessment criteria cover a very broad range of topics. The most important are the technical ones (e.g. materials, energy, water) that are directly responsible for the environmental quality in the analyzed community. Other criteria consider the land use, the sustainable building and restauration processes along with social issues. Many tools include specific points (up to 10% of the score) dedicated to innovation, that reward novel actions developed at community level that are supporting sustainability.

It is to outline that none of these certification schemes are compulsory at national level; thus, the choice to develop a construction project according to the sustainability criteria included in a certification tool is the community's volunteer choice, considering the sustainable development and additional positive consequences of certification. Among these consequences, one can mention the fact that each tool supports the community to sustainable development and provides a clear discussion topic between the community planners and the community inhabitants.

References

1. Al-Absi ZAAS, Isa MHM, Ismail M (2019) Application of phase change materials (PCMs) in building walls: a review. Lecture notes in civil engineering: the advances in civil engineering materials, vol 19. Springer Nature Singapore, pp 73–82
2. Allen K, Connelly K, Rutherford P, Wu Y (2017) Smart windows—dynamic control of building energy performance. Energy Build 139:535–546
3. Allouhi A, El Fouih Y, Kousksou T, Jamil A, Zeraouli Y, Mourad Y (2015) Energy consumption and efficiency in buildings: current status and future trends. J Clean Prod 109:118–130
4. Al-Marri W, Al-Habaibeh A, Abdo H (2017) Exploring the relationship between energy cost and people's consumption behaviour. Enrgy Procedia 105:3464–3470
5. Alwisy A, BuHamdan S, Gül M (2018) Criteria-based ranking of green building design factors according to leading rating systems. Energy Build 178:347–359
6. Ameen RFM, Mourshed M (2019) Urban sustainability assessment framework development: the ranking and weighting of sustainability indicators using analytic hierarchy process. Sustain City Soc 44:356–366
7. Anastasiou E, Filikas KG, Stefanidou M (2014) Utilization of fine recycled aggregates in concrete with fly ash and steel slag. Constr Build Mater 50:154–161
8. Andric I, Gomes N, Pina A, Ferrao P, Fournier J, Lacarriere B, Le Corre O (2016) Modeling the long-term effect of climate change on building heat demand: case study on a district level. Energy Build 126:77–93
9. Andric I, Koc M, Al-Ghamdi SG (2019) A review of climate change implications for built environment: impacts, mitigation measures and associated challenges in developed and developing countries. J Clean Prod 211:83–102
10. Andric I, Pina A, Ferrao P, Lacarriere B, Le Corre O (2017) On the performance of district heating systems in urban environment: an energy approach. J Clean Prod 142(1):109–120
11. Ao Y, Chen C, Yang D, Wang Y (2018) Relationship between rural built environment and household vehicle ownership: an empirical analysis in Rural Sichuan, China. Sustainability 10(5):1566. https://doi.org/10.3390/su10051566

12. Berg N (2012) Is there a limit to how tall buildings can get? https://www.citylab.com/design/2012/08/there-limit-how-tall-buildings-can-get/2963/. Accessed on 16.04.2019
13. Bernardi E, Carlucci S, Cornaro C, Bohne RA (2017) An analysis of the most adopted rating systems for assessing the environmental impact of buildings. Sustainability 9(1226):1–27
14. Bougiatioti F, Michael A (2015) The architectural integration of active solar systems. Building applications in the Eastern Mediterranean region. Renew Sustain Energy Rev 47:966–982
15. Braulio-Gonzalo M, Bovea MD, Rua MJ, Juan P (2016) A methodology for predicting the energy performance and indoor thermal comfort of residential stocks on the neighbourhood and city scales. A case study in Spain. J Clean Prod 139:646–665
16. Buffa S, Cozzini M, D'Antoni M, Baratieri M, Fedrizzi R (2019) 5th generation district heating and cooling systems: a review of existing cases in Europe. Renew Sustain Energy Rev 104:504–522
17. Building Research Establishment (2018) Achieving Sustainable Masterplans, BREEAM Communities. https://www.breeam.com/discover/technical-standards/communities/. Accessed 24 Apr 2019
18. Building Research Establishment (2019) The edge, Amsterdam. https://www.breeam.com/case-studies/offices/the-edge-amsterdam/. Accessed 24 Apr 2019
19. Burger R (2018) What are the benefits of LEED certification. https://www.thebalancesmb.com/what-are-the-benefits-of-leed-certification-845365. Accessed 03 May 2019
20. Cajota S, Peterb M, Bahub J-M, Guignetc F, Koch A, Maréchal F (2017) Obstacles in energy planning at the urban scale. Sustain Cities Soc 30:223–236
21. Caldera M, Corgnati SP, Filippi M (2008) Energy demand for space heating through a statistical approach: application to residential buildings. Energy Build 40(10):1972–1983
22. Campisano A, Modica C (2014) Selecting time scale resolution to evaluate water saving and retention potential of rainwater harvesting tanks. Procedia Eng 70:218–227
23. Cao X, Dai X, Liu J (2016) Building energy-consumption status worldwide and the state-of-the-art technologies for zero-energy buildings during the past decade. Energy Build 128:198–213
24. Ciurean R, Chen L, Hussin H, van Weste CJ, Glade T, Frigerio S, Pasuto A (2014) Quantitative vulnerability assessment of buildings to debris—flows in Fella River Basin using run-out modelling and damage data from the 29th of August 2003 event. In: Conference proceedings: analysis and management of changing risks for natural hazards, Padua, Italy, 18–19 Nov 2014
25. Construction Industry Council (2018) APPG for excellence in the Built Environment. http://cic.org.uk/about-us/. Accessed on 03 Jun 2019
26. Council of Australian Governments (2012) National strategy on energy efficiency, baseline energy consumption and greenhouse Gas emissions. In: Commercial Buildings in Australia Part 1-Report, Nov 2012
27. Cristofori C, Carutasiu MB, Canalleti JL, Norvainsiene J, Motte F, Notton G (2019) Building integration of solar thermal systems-example of a refurbishment of a church rectory. Renew Energ 137:67–81
28. Darko A, Chan APC, Huo X, Owusu-Manu DG (2019) A scientometric analysis and visualization of global green building research. Build Environ 149:501–511
29. Department for Digital, Culture, Media & Sport, UK (2018) Principles of selection for listed buildings. https://assets.publishing.service.gov.uk/government/uploads/system/uploads/attachment_data/file/757054/Revised_Principles_of_Selection_2018.pdf. Accessed 13 June 2019
30. DGNB Deutsches Gesellschaft fur Nachhaltiges Bauen (2018) https://www.dgnb.de/en/index.php. Accessed 03 June 2019
31. Ding Z, Fan Z, Tam VWY, Bian Y, Li S, Chethan IM, Illankoon S, Moon S (2018) Green building evaluation system implementation. Build Environ 133:32–40
32. Ding Z, Zhu H, Wang Y, Ge X (2017) Study and analysis of office building energy consumption performance in severe cold and cold region, China. Adv Mech Environ 9(11):1–21
33. Doan DT, Ghaffarianhoseini A, Naismith N, Zhang T, Ghaffarianhoseini A, Tookey J (2017) A critical comparison of green building rating systems. Build Environ 123:243–260

34. Dodoo A, Gustavsson L, Bonakdar F (2014) Effects of future climate change scenarios on overheating risk and primary energy use for Swedish residential, buildings. Energy Procedia 61:1179–1182
35. Dolinar M, Vidrih B, Kajfez-Bogataj L, Medved S (2010) Predicted changes in energy demands for heating and cooling due to climate change. Phys Chem Earth Parts A/B/C 35:100–106
36. Duan X, Jiang Y, Wang B, Zhao X, Shen G, Cao S (2014) Household fuel use for cooking and heating in China: results from the first Chinese environmental exposure-related human activity patterns survey (CEERHAPS). Appl Energy 136:692–703
37. Energy Lens (2019) Degree Days—handle with care. https://www.energylens.com/articles/degree-days#hdd. Accessed 16 Sept 2019
38. EU Buildings Datamapper (2013) https://ec.europa.eu/energy/en/eu-buildings-datamapper. Accessed 18 May 2019
39. European Commission (2013) National plans for nearly zero energy buildings. http://ec.europa.eu/energy/efficiency/buildings/implementation_en.htm. Accessed 08 June 2019
40. European Parliament (2009) On the proposal for a directive of the European Parliament and of the council on the energy performance of buildings (recast). http://www.europarl.europa.eu/doceo/document/A-8-2017-0314_EN.html. Accessed 08 June 2019
41. European Parliament and Council (2010) Directive 2010/31/EU on the energy performances of buildings (recast). http://www.buildup.eu/en/practices/publications/directive-201031eu-energy-performance-buildings-recast-19-may-2010. Accessed 08 June 2019
42. Eurostat (2018) Consumption of energy. https://ec.europa.eu/eurostat/statistics-explained/index.php?title=Archive:Consumption_of_energy#End-users. Accessed 18 May 2019
43. Eurostat (2019) Energy consumption in households. Statistics explained. https://ec.europa.eu/eurostat/statistics-explained/pdfscache/58200.pdf. Accessed 28 July 2019
44. Forzano C, Baggio P, Buonomano A, Palombo A (2019) Building integrating phase change materials: a dynamic hydrothermal simulation model for system analysis. J Sustain Dev Energy Water Environ Syst 7(2):325–342
45. Gardens by the Bay (2017) https://www.gardensbythebay.com.sg/en/plan-your-visit/getting-here-and-parking.html. Accessed 22 Apr 2019
46. Ghahramani A, Karvigh SA, Becerik-Gerber B (2017) HVAC system energy optimization using an adaptive hybrid metaheuristic. Energy Build 152:149–161
47. Green Building Council Australia (2019) Green star rating system https://new.gbca.org.au/green-star/rating-system/. Accessed 21 Mar 2019
48. Green Building Information Gateway (2018) http://www.gbig.org/. Accessed 26 Apr 2019
49. Gruber A, Bai X, Buettner T, Dhakal S, Fisk DJ, Ichinose T (2012) Urban energy systems. Global energy assessment (GEA) toward a sustainable future. Cambridge University Press, Cambridge, UK and New York, NY, USA and the International Institute for Applied Systems Analysis, Laxenburg, Austria, pp 1307–1400
50. Guan L (2009) Implication of global warming on air-conditioned office buildings in Australia. Build Res Inf 37:43–54
51. Hegazy I, Seddik W, Ibrahim H (2017) The living building: integrating the built environment with nature evaluating the Bibliotheca of Alexandria according to the challenge imperatives. Int J Low-Carbon Technol 12:244–255
52. Holland O (2018) China built more skyscrapers in 2018 than ever before. https://edition.cnn.com/style/article/skyscraper-china-ctbuh-2018/index.html. Accessed 23 Apr 2019
53. Hong Kong Green Building Council (2018) BEAM PLUS procedures manual, Version 4.0. https://www.beamsociety.org.hk/files/BEAM-Plus-Procedures-Manual_v4.0.pdf. Accessed 13 June 2019
54. HQE Association (2016) HQE Certification. http://hqe.publishingcenter.fr/cerway/essentials. Accessed 24 Apr 2019
55. Hung W, Chow W (2001) A review on architectural aspects of atrium buildings. Archit Sci Rev 44(3):285–295

56. Institute for building environment and energy conservation (2016) Comprehensive assessment system for built environment efficiency, CASBEE. http://www.ibec.or.jp/CASBEE/english/downloadE.htm. Accessed 23 Mar 2019
57. International Code Council (2015) International building code (Chap. 3) Use and occupancy classification. https://codes.iccsafe.org/content/IBC2015/chapter-3-use-and-occupancy-classification. Accessed 18 May 2019
58. International Code Council (2018) International Building code (Chap. 6) Types of constructions. https://codes.iccsafe.org/content/IBC2018/chapter-6-types-of-construction. Accessed 03 Jun 2019
59. International Energy Agency (2008) World Energy Outlook 2008, Paris, France. https://www.iea.org/media/weowebsite/2008-1994/WEO2008.pdf. Accessed 08 Jun 2019
60. International Energy Agency (2013) Transition to sustainable buildings: strategies and opportunities to 2050. http://www.iea.org/etp/buildings/. Accessed 19 Apr 2019
61. International Energy Agency (2016) Energy Technology Perspectives (Executive Summary) IEA 371. http://dx.doi.org/10.1787/energy_tech-2014-en
62. International Energy Agency (2019a) Buildings. Tracking clean energy progress. https://www.iea.org/tcep/buildings/. Accessed 24 Apr 2019
63. International Energy Agency (2019b) https://www.iea.org/statistics/kwes/consumption/. Accessed 24 Apr 2019
64. International Energy Agency (2019c) Electricity information overview. https://www.iea.org/statistics/electricity/. Accessed 24 Apr 2019
65. International Living Future Institute (2014) Living building challenge 3.0: a visionary path to a regenerative future. https://living-future.org/wp-content/uploads/2016/12/Living-Building-Challenge-3.0-Standard.pdf. Accessed 26 Apr 2019
66. International Partnership for Energy Efficiency Cooperation (2018) Zero energy building definitions and policy activity—an international review. https://ipeec.org/upload/publication_related_language/pdf/766.pdf. Accessed 23 May 2019
67. Isaac M, Van Vuuren DP (2009) Modelling global residential sector energy demand for heating and air conditioning in the context of climate change. Energy Policy 37(2):507–521
68. Jang J, Baek J, Leigh S-B (2019) Prediction of optimum heating timing based on artificial neural network by utilizing BEMS data. J Build Eng 22:66–74
69. Japan Sustainable Building Consortium (2019) Comprehensive assessment system for built environment efficiency. http://www.ibec.or.jp/CASBEE/english/. Accessed 20 July 2019
70. Jia M, Komeily A, Wang Y, Srinivasan RS (2019) Adopting Internet of Things for the development of smart buildings: a review of enabling technologies and applications. Autom Constr 101:111–126
71. Jiang A, Zhu Y, Elsafty A, Tumeo M (2017) Effects of global climate change on building energy consumption and its implications in Florida. Int J Constr Educ Res 14:1–24
72. Jin Y, Ma X, Chen X, Cheng Y, Barisc E, Ezzati M (2006) Exposure to indoor air pollution from household energy use in rural China: the interactions of technology, behaviour, and knowledge in health risk management. Soc Sci Med 62:3161–3176
73. Jordan DC, Kurtz SR, VanSant K, Newmiller J (2016) Compendium of photovoltaic degradation rates. Prog Photovolt Res Appl 24:978–989
74. Kaewunruen S, Rungskunroch P, Welsh J (2019) A digital-twin, evaluation of net zero energy building for existing buildings. Sustainability 11(1):159. https://doi.org/10.3390/su11010159
75. Kerimray A, Rojas-Solorzano L, Trokmahalleh MA, Hopke PK, O'Gallachoir BP (2017) Coal use for residential heating: patterns, health implications and lessons learned. Energy Sustain Dev 40:19–30
76. Kim S, Osmond P (2014) Analyzing green building rating tools for healthcare buildings from the building user's perspective. Indoor Built Environ 23:757–766
77. Klemes JJ (2015) Assessing and measuring environmental impact and sustainability. Butterworth-Heinemann, Elsevier
78. Krarti M (2018) Optimal design and retrofit of energy efficient buildings, communities and urban centers. Butterworth-Heinemann, Elsevier

79. Lee D, Cha G, Park S (2016) A study on data visualization of embedded sensors for building energy monitoring using BIM. Int J Precis Eng Manuf 17:807–814
80. Lindroos TJ, Lehtila A, Kaljonen T (2018) Baltic energy technology scenarios, Nordic Council of Ministers. Nordisk Ministerråd, Copenhagen. http://www.diva-portal.org/smash/get/diva2:1195548/FULLTEXT01.pdf. Accessed 05 Jun 2019
81. Lobaccaro G, Carlucci S, Croce S, Paparella R, Finocchiaro L (2017) Boosting solar accessibility and potential of urban districts in the Nordic climate: a case study in Trondheim. Sol Energy 149:347–369
82. Loftness V, Haase D (2013) Sustainable built environments. Springer Science + Business Media, New York
83. Maes MJ, Jones KE, Tolledano MB, Milligan B (2019) Mapping synergies and trade-offs between urban ecosystems and the sustainable development goals. Environ Sci Policy 93:181–188
84. Marjaba GE, Chidiac SE (2016) Sustainability and resiliency metrics for buildings—critical review. Build Environ 101:116–125
85. Marszal AJ, Heiselberg P, Bourrelle JS, Musall E, Voss K, Sartori I (2011) Zero energy building—a review of definitions and calculation methodologies. Energy Build 43:971–979
86. Mattonia B, Guattari C, Evangelisti L, Bisegna F, Gori P, Asdrubali F (2018) Critical review and methodological approach to evaluate the differences among international green building rating tools. Renew Sustain Energy Rev 82:950–960
87. Mauthner F, Weiss W, Spork-Durr M (2016) Solar heat worldwide. Markets and contribution to the energy supply 2014. IEA Solar Heating & Cooling Programme. https://www.aee-intec.at/0uploads/dateien1105.pdf. Accessed 05 Apr 2019
88. McLeod RS, Swainson M (2017) Chronic overheating in low carbon urban developments in a temperate Climate. Renew Sustain Energy Rev 74:201–220
89. Meirmans K (2013) Household direct energy consumption and CO_2 emissions in European countries. Master programme energy and environmental sciences. University of Groningen
90. Merriam-Webster dictionary (2019) Community. https://www.merriam-webster.com/dictionary/community. Accessed 03 June 2019
91. Michael A, Gregoriu S, Kalogirou S (2018) Environmental assessment of an integrated adaptive system for the improvement of indoor visual comfort of existing buildings. Renew Energ 115:620–633
92. Ministry for Regional Development and Public Administration, Romania (2013) Plan for increasing the Nearly Zero Energy Buildings. https://ec.europa.eu/energy/en/topics/energy-efficiency/energy-performance-of-buildings/nearly-zero-energy-buildings. Accessed 23 Apr 2019
93. Mondal P (2015) 20 Important characteristics of urban community. http://www.yourarticlelibrary.com/sociology/20-important-characteristics-of-urban-community-sociology/4873/. Accessed 19 Apr 2019
94. National Geographic (2018) Urban Area. https://www.nationalgeographic.org/encyclopedia/urban-area/. Accessed 03 June 2019
95. National House Building Council Foundation (2016) The challenge of shape and form. https://www.nhbcfoundation.org/wp-content/uploads/2016/10/NF-72-NHBC-Foundation_Shape-and-Form.pdf. Accessed 13 June 2019
96. Nguyen B, Altan H (2011) Comparative review of five sustainable rating systems. Procedia Eng 21:376–386
97. O'Malley C, Piroozfar PAE, Farr ERP, Gates J (2014) Evaluating the efficacy of BREEAM code for sustainable homes (CSH): a cross-sectional study. Enrgy Procedia 62:210–219
98. Ortiz O, Castells F, Sonnemann G (2009) Sustainability in the construction industry: a review of recent developments based on LCA. Constr Build Mater 23:28–39
99. Philpott L (2018) What is the average lifespan of modern buildings? https://www.quora.com/What-is-the-average-life-span-of-modern-buildings. Accessed 23 Apr 2019
100. Polly B, Kutscher C, Macumber D, Schott M, Pless S, Livingood B (2017) Zero energy buildings to zero energy districts. @2016 ACEEE Summer Study Energy Efficient Buildings,

NREL. https://aceee.org/files/proceedings/2016/data/papers/10_566.pdf. Accessed 03 June 2019
101. Radhi H (2009) Evaluating the potential impact of global warming on the UAE residential buildings—a contribution to reduce the CO_2 emissions. Build Environ 44:2451–2462
102. Randall T (2015) The smartest building in the world. Bloomberg business week. https://www.bloomberg.com/features/2015-the-edge-the-worlds-greenest-building/. Accessed 23 Apr 2019
103. Raza A (2016) The 10 tallest buildings in Europe. https://theculturetrip.com/europe/articles/the-10-tallest-buildings-in-europe/. Accessed 22 Apr 2019
104. Reed R, Bilos A, Wilkinson S, Schulte KW (2009) International comparison of sustainable rating tools. J Sustain Real Estate 1(1):1–22
105. Robichaud LB, Anantatmula VS (2010) Greening project management practices for sustainable construction. J Manag Eng 27(1):48–57
106. Rodriguez E, Fernandes MS, Soares N, Gomes A, Rodrigues Gaspar A, Costa JJ (2018) The potential impact of low thermal transmittance construction on the European design guidelines of residential buildings. Energy Build 178:378–390
107. Roostaie A, Nawari N, Kibert CJ (2019) Sustainability and resilience: a review of definitions, relationships, and their integration into a combined building assessment framework. Build Environ 154:132–144
108. Roudman S (2013) Bank of America's Toxic Tower. The New Republic, Digital Edition. https://newrepublic.com/article/113942/bank-america-tower-and-leed-ratings-racket. Accessed 22 Apr 2019
109. Samarasekera R (2017) LEED Credits prerequisites and points: how are they different? https://www.usgbc.org/articles/whats-difference-between-leed-credit-leed-prerequisite-and-leed-point. Accessed 01 May 2019
110. Scrase I (2000) White Collar CO_2. Energy consumption in the service sector. https://www.ukace.org/wp-content/uploads/2012/11/ACE-Research-2000-08-White-collar-CO2-Energy-consumption-in-the-service-sector.pdf. Accessed 03 June 2019
111. Shafique M, Kim R (2017) Application of green blue roof to mitigate heat island phenomena and resilient to climate change in urban areas: a case study from Seoul, Korea. J Water Land Dev 33:165–170
112. Shan M, Hwang B (2018) Green building rating systems: global reviews of practices and research efforts. Sustain Cities Soc 39:172–180
113. Sharma A, Saxena A, Sethi M, Shree V, Varun M (2011) Life cycle assessment of buildings: a review. Renew Sust Energy Rev 15:871–875
114. Shen P (2017) Impacts of climate change on U.S. building energy use by using downscaled hourly future weather data. Energy Build 134:61–70
115. Sher F, Kawai A, Gulec F, Sadiq H (2019) Sustainable energy saving alternatives in small buildings. Sustain Energy Technol Assess 32:92–99
116. Shibuya T, Croxford B (2016) The effect of climate change on office building energy consumption in Japan. Energy Build 117:1–11
117. Si J, Marjanovic-Halburd L (2018) Criteria weighting for green technology selection as part of retrofit decision making process for existing non-domestic buildings. Sustain Cities Soc 41:625–638
118. Sinton JE, Smith KR, Peabody JW, Yaping L, Xiliang Z, Edwards R, Quan G (2004) An assessment of programs to promote improved household stoves in China. Energy Sustain Dev l VIII(3):33–52
119. Sokka L, Sinkko T, Holma A, Manninen K, Pasanen K, Rantala M, Leskinen P (2016) Environmental impacts of the national renewable energy targets—a case study from Finland. Renew Sust Energy Rev 59:1599–1610
120. Statista (2018) Size of Green Buildings Market in the US from 2005 to 2014. https://www.statista.com/statistics/248060/value-of-us-green-building-market/. Accessed 22 Apr 2019
121. Stromann-Andersen J, Sattrup PA (2011) The urban canyon and building energy use: urban density versus daylight and passive solar gains. Energy Build 43:2011–2020

122. Tettey UYA, Dodoo A, Gustavsson L (2019) Design strategies and measures to minimise operation energy use for passive houses under different climate scenarios. Energy Effic 12:299–313
123. Tirone J (2019) Study shows rooftop solar could power 25 percent of Europe, Renew Energy World. https://www.renewableenergyworld.com/articles/2019/09/study-shows-rooftop-solar-could-power-25-percent-of-europe.html?cmpid=&utm_source=enl&utm_medium=email&utm_campaign=solar_energy_news&utm_content=2019-09-04&eid=288395584&bid=2513472. Accessed 05 Sept 2019
124. Tvindkraft (2019) The Tvindkraft project. https://www.tvindkraft.dk/en/. Accessed 13 Jun 2019
125. United Nations (2018) World Urbanization Prospects: the 2018 Revision https://population.un.org/wup/Publications/Files/WUP2018-KeyFacts.pdf. Accessed 13 June 2019
126. United Nations Environment Program (2007) Buildings and climate change, UNEP Publications. http://www.unep.fr/shared/publications/pdf/DTIx0916xPA-BuildingsClimate.pdf. Accessed 13 Jun 2019
127. United Nations General Assembly (2015) Transforming our world: the 2030 Agenda for Sustainable Development, Resolution A/RES/70/1, New York. https://sustainabledevelopment.un.org/post2015/transformingourworld. Accessed on 22 May 2019
128. United States Environmental Protection Agency (2009) Green building. http://www.epa.gov. Accessed 13 Jun 2019
129. United States Green Building Council (2019a) LEED 4.1. https://new.usgbc.org/leed-v41. Accessed 01 May 2019
130. United States Green Building Council (2019b) LEED 4.1. Cities and Communities: Existing brochure. https://www.usgbc.org/resources/2019-leed-cities-and-communities-grant-program-guide. Accessed 16 May 2019
131. Waddicor DA, Fuentes E, Siso L, Salom J, Favre B, Jimenez C, Azar M (2016) Climate change and building ageing impact on building energy performance and mitigation measures application: a case study. Built Environ 102:13–25
132. Walker S, Labeodan T, Boxem G, Maasen W, Zeiler W (2018) An assessment methodology of sustainable energy transition scenarios for realizing energy neutral neighbourhoods. Appl Energy 228:2346–2360
133. Wang X, Chen D, Ren Z (2010) Assessment of climate change impact on residential building heating and cooling energy requirement in Australia. Build Environ 45:1663–1682
134. Werner S (2017) International review of district heating and cooling. Energy 137:617–631
135. World Energy Council (2013) World energy resources: geothermal. https://www.worldenergy.org/wp-content/uploads/2013/09/WER_2013_9_Geothermal.pdf. Accessed 21 Mar 2019
136. World Green Building Council (2016) What is a green building? https://www.worldgbc.org/what-green-building. Accessed 22 Apr 2019
137. Yoshino H, Chen S (2016) Total energy use in buildings: Analysis and evaluation methods, Project summary report. International Energy Agency. http://www.iea-ebc.org/Data/publications/EBC_PSR_Annex53.pdf. Accessed 22 Apr 2019
138. Zhao D, Ji J, Yu H, Zhao X (2019) A study on thermal characteristic and sleeping comfort of a hybrid solar heating system applied in cold rural areas. Energy Build 182:242–250
139. Zheng G, Bu W (2018) Review of heating methods for rural houses in China. Energies 11(12):3402. https://doi.org/10.3390/en11123402

Chapter 2
Renewable Energy Sources and Systems

2.1 Solar Energy in the Built Environment

2.1.1 Solar Radiation

Any substance emits electromagnetic waves called radiation if its temperature is above 0 K. The solar radiation represents the electromagnetic radiation generated by the Sun during the fusion process involving hydrogen atoms to form helium. The *irradiance* represents the power of the solar radiation received over 1 m^2, expressed in W/m^2. The energy of the solar radiation received on a 1 m^2 surface during a specific time interval can be thus calculated by integrating the irradiance over that time interval; the result is called *solar energy* (also called solar irradiation or insolation) and is expressed usually in Wh/m^2.

Although neither the internal processes in the Sun that generate the solar radiation are reversible nor the Sun recovers its burned matter (the Sun actually consumes itself while generating radiation), the solar radiation can be considered a *renewable energy source* at the timescale of human history.

Solar energy is the most abundant renewable energy on which all the other renewable energy sources rely on. As an example, the wind is generated by the movement of air masses differently heated by the ground that received different amounts of solar radiation while the water flows have their source in the vaporization of the water heated by the solar radiation in lakes, seas, oceans. Moreover, the shallow geothermal energy would exhaust after short time intervals of heat pump use if no contribution would be added by the solar radiation.

The solar radiation is uniformly sent by the Sun in space with an irradiance of about 63.3×10^6 W/m^2 [31]. However, because of the huge distance between the Sun and the Earth (close to 150×10^6 km), only about 1367 W/m^2 reaches the Earth atmosphere; thus, over the year, the entire Earth system (land surfaces, oceans and atmosphere) absorbs an average of 340 W/m^2 of solar power [103].

I. Visa et al., *Solar Energy Conversion Systems in the Built Environment*, Green Energy and Technology, https://doi.org/10.1007/978-3-030-34829-8_2

The annual amount of solar energy received by the Earth is estimated at 10^{18} kWh/year [29]. This represents between 6000 times [97] to 10,000 times [29] of the annual energy consumed by the entire world population. This means that about 0.01% of the solar energy would be enough to cover the entire energy demand of the humans.

2.1.1.1 The Extraterrestrial Solar Radiation

The irradiance represents the solar power that a unit area can receive if normally positioned to the sunrays' direction. It depends on the distance from the Sun and on the time during the nearly periodic 11-year solar cycle. Over one year, the distance between Sun and Earth is varying (Fig. 2.1) from approx. 147.1×10^6 km at perihelion (3 January) to 152.1×10^6 km at aphelion (4 July). The irradiance of the extraterrestrial solar radiation at perihelion is about 1412 W/m^2 and at aphelion 1325 W/m^2 [29, 82].

According to the American Society for Testing Materials that defined the AM 0 reference solar spectrum (ASTM E-490), at an average distance between Earth and Sun the irradiance of the extraterrestrial solar radiation is 1366.1 W/m^2 and it is called *solar constant* (I_0) [107].

During any day, the extraterrestrial solar irradiance (B_0) can be calculated as proposed by Duffie and Beckman [33], using

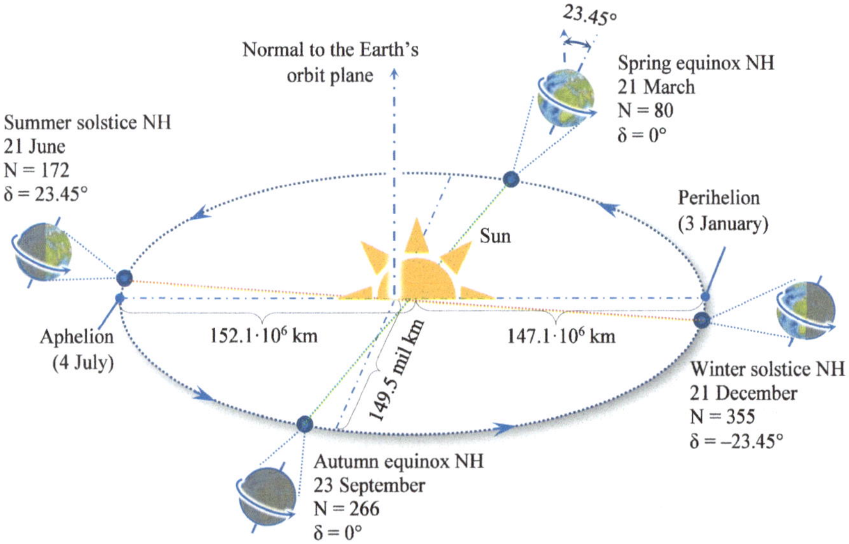

Fig. 2.1 Elliptical trajectory of the Earth around the Sun, the particular positions (solstices and equinoxes, aphelion and perihelion) and the angle of the Earth's axis with the Earth orbit plane (where N is the day number in the year, δ is the solar declination angle, and NH represents the northern hemisphere)

$$B_0 = I_0 \cdot \left(1 + 0.033 \cdot \cos \frac{360 \cdot N}{365}\right) \tag{2.1}$$

where N is the day number in the year.

2.1.1.2 Atmospheric Attenuation of the Solar Radiation

The extraterrestrial solar radiation spectrum (AM 0) is similar to the spectrum of a standardized black body, heated at 5250 °C as presented in Fig. 2.2 using the yellow surface (1) and black line (3). When passing through the Earth atmosphere, the solar radiation is reduced as a result of scattering, back reflection to the space and absorption and its spectrum is modified according to:

- The composition and the thickness of the atmospheric layers (especially the ozone layer);
- The distance the solar radiation is passing through the atmosphere up to the Earth surface;
- The impurities type and amount in air (e.g. water vapours, dust particles, various pollutants);
- The type of the cloud cover.

The ionosphere located at high altitudes (60–1000 km) around the Earth is able to absorb very short wavelength radiation such as the X and gamma rays. The ozone (O_3) layer located at 15–35 km above the Earth surface absorbs radiation with longer wavelengths, usually within the ultraviolet (UV) range. The lower layers of the atmosphere absorb an important part of the radiation, with longer wavelengths within

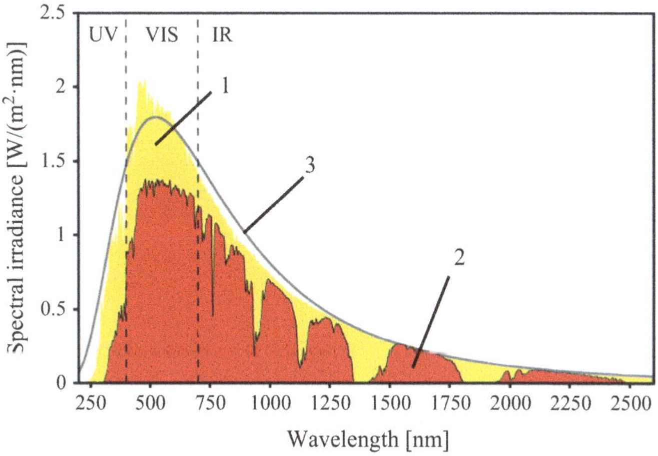

Fig. 2.2 Spectrum of the solar radiation above the atmosphere (1) and at sea level (2) along with the spectrum of the radiation generated by a blackbody at 5250 °C (3) [131]

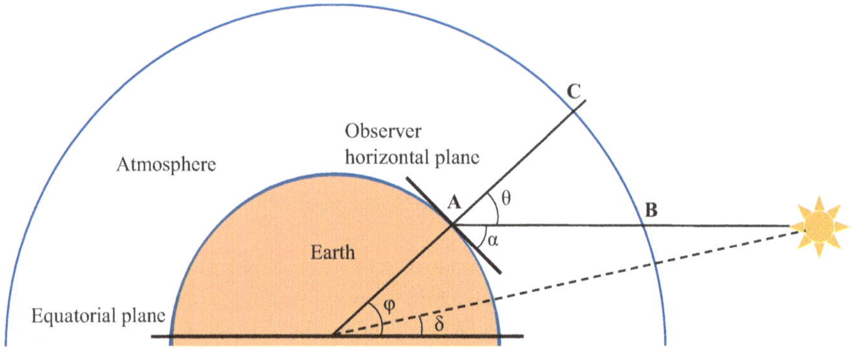

Fig. 2.3 Distance passed through the atmosphere by the sunrays and the relevant angles: altitude angle (α), zenith angle (θ), latitude (φ), declination angle (δ)

the visible (VIS) and infrared (IR) spectral range, because of the significant oxygen (O_2), water vapours (H_2O) and carbon dioxide (CO_2) content.

The solar radiation absorption in the atmosphere is generally proportional to the distance covered by the solar radiation on its way to the ground. This distance mainly depends on the Sun position in the sky, relative to the observer location. It is minimal when the Sun is directly above the observer ($\alpha = 90°$), and it increases when the Sun is closer to the observer horizontal plane, as outlined in Fig. 2.3.

The air mass (AM) coefficient characterizes the distance passed by the sunray through the atmosphere and the spectrum of the solar radiation for that distance. Above the atmosphere AM = 0, while at ground level the air mass coefficient can be estimated [82] using

$$\mathrm{AM} = \frac{AB}{AC} \simeq \frac{1}{\cos\theta} = \frac{1}{\sin\alpha} \tag{2.2}$$

where

- α is the altitude angle of the Sun.
- θ is the zenith angle of the Sun (the complement of the altitude angle α) (Fig. 2.3).

According to Visa et al. [155], the altitude angle can be calculated using

$$\alpha = \arcsin(\cos\omega \cdot \cos\delta \cdot \cos\varphi + \sin\delta \cdot \sin\varphi) \tag{2.3}$$

where

- φ is the latitude of the considered location.
- δ is the declination angle, formed by the sunrays direction and the equatorial plane of the Earth (Fig. 2.3)

$$\delta = 23.45 \cdot \sin\frac{360 \cdot (N - 80)}{365} \tag{2.4}$$

– ω is the hour angle that changes with 15°/h

$$\omega = 15 \cdot (12 - t_s) \tag{2.5}$$

where t_s is the solar time expressed in hours.

Further information on the sunray angles is detailed in Sect. 3.3.2.

Air mass AM = 1 indicates that the Sun is exactly overhead the observer ($\alpha = 90°$); this situation occurs between Tropics twice each year. If the Sun altitude angle is $\alpha = 30°$, the air mass coefficient AM = 2, etc.

Since the air mass coefficient has different values during a day and during the year, an average value of AM = 1.5 is usually considered when testing solar energy conversion systems. The corresponding solar radiation that reaches the sea level has the red spectrum presented in Fig. 2.2.

2.1.1.3 Solar Radiation at the Ground Level

The solar radiation that is able to pass through the atmosphere and reaches the ground level has two main components according to the direction where these come from, as schematically presented in Fig. 2.4: the *direct (beam)* and the *diffuse* solar radiation. A third component, *the reflected/Albedo* solar radiation may also occur under certain conditions, further described. Their sum represents the *global (total)* solar radiation, and its irradiance can be calculated using

$$G = B + D + R \tag{2.6}$$

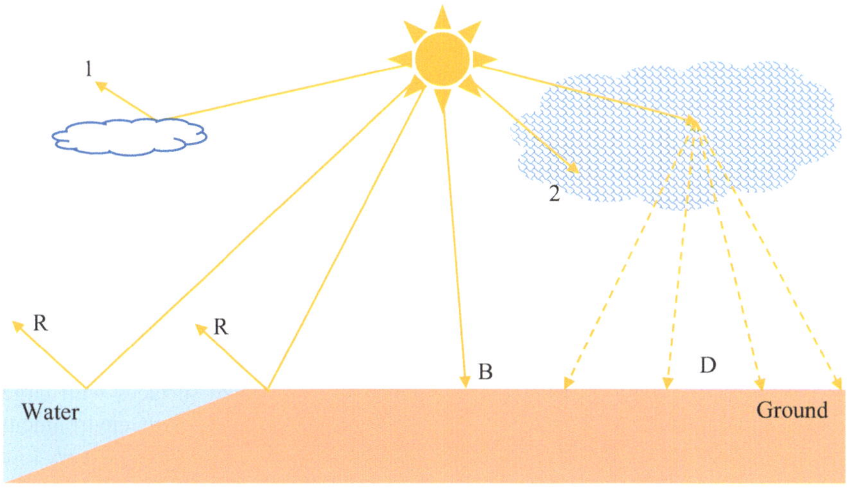

Fig. 2.4 Three components (B, D, R) of the solar radiation at ground level as it passes through the atmosphere where part of it is: reflected by the clouds (1) or absorbed by the clouds (2)

where

- G is the global solar irradiance.
- B is the direct solar irradiance.
- D is the diffuse solar irradiance.
- R is the reflected solar irradiance.

Several parameters influence the variation in time of the solar radiation at ground level. Among these, the most important are as follows:

- The geographical coordinates of the location, given by the latitude and the longitude on the Earth;
- The topography of the location: irregular landforms such as mountains and hills that can obstruct solar radiation during certain time intervals while large water- or snow-covered surfaces can increase the reflected component;
- The climate in the location with its specific humidity, temperature and dust/pollutant particles in the air that support scattering, thus decreasing the solar irradiance.

The most important component, considering the input to renewable energy systems, is the *direct solar radiation* that is coming from the Sun as parallel sunrays without being obstructed by non-transparent objects (e.g. clouds) and is reduced only by the various transparent components in the atmospheric layers. The direct solar radiation generates dark shadows of the lighted objects.

The second component which is always part of the solar radiation is the *diffuse solar radiation*; this represents the radiation that was scattered by components in the atmosphere, especially by clouds, vapours, chemicals or any other particles suspended in the atmosphere. The diffuse radiation comes from the entire sky vault, without any particular direction, and gives the colour of the sky; therefore, it is also known in the literature as skylight [78]. The diffuse radiation accounts for different shares of the available solar radiation depending on how cloudy the sky is: during sunny days only 10–20%, whereas during cloudy days up to 100% [104]. For Brasov, Romania, a mountain located in the temperate climate, the variation of the direct and diffuse solar radiation during one year is presented in Fig. 2.5, based on the values obtained using the Meteonorm software interpolated from nearby locations. According to this data source, in the horizontal plane the ratio between the yearly direct and diffuse solar energy is roughly unitary, summing up over one year approximately 1200 kWh/m^2.

A third component is the *reflected solar radiation* that is also received from multiple directions; this is why the reflected radiation is usually considered as part of the diffuse solar radiation and is measured using the same devices.

The reflected solar radiation is significant mainly on tilted surfaces with large angles to the horizontal plane, for areas with high ground or built environment reflectance (Albedo); otherwise, on horizontal surfaces, this component is very low and can be neglected [104]. Depending on the surface type on the ground and on the existing buildings, the reflectance Albedo coefficient (ρ) can take values between 0 and 1; higher values indicate a higher reflection of the solar radiation and a lighter

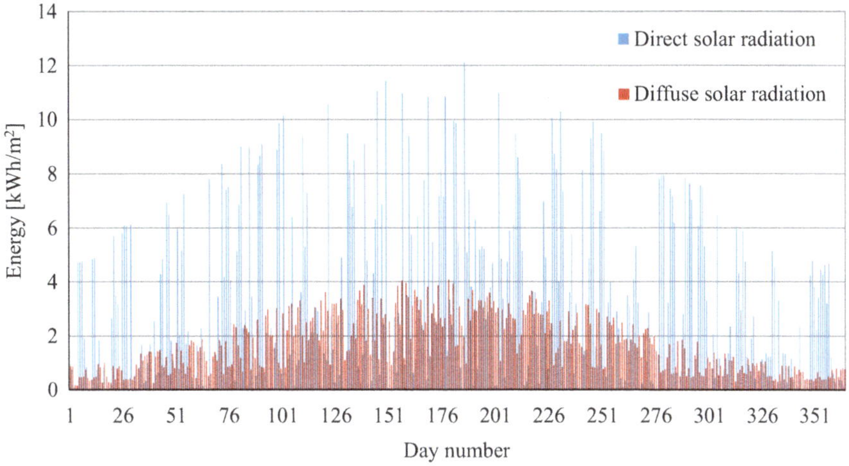

Fig. 2.5 Daily energy over the year for the direct and diffuse solar radiation on the horizontal plane, in Brasov, Romania (data extracted using the Meteonorm software)

surrounding area with higher diffuse solar irradiance. Accepted Albedo reflectance values for various environments are presented in Sect. 2.1.2. If the existing environment is not analysed with high accuracy, an average Albedo value of 0.2 can be used to calculate this component.

2.1.1.4 Equipment used for Measuring the Solar Radiation at the Ground Level

Before the direct and the diffuse solar radiation components could be individually measured at small time intervals, using accurate electronic devices, a good indication of the available solar radiation was offered by *the sunshine duration*. This parameter indicates the number of hours when the Sun is visible, uncovered, on the sky vault.

In the early days of solar monitoring, the most common method used to measure the sunshine duration over a long period was the use of the Campbell–Stokes instrument, invented in 1853. It consists of a glass or quartz sphere that concentrates the sunlight and burns lines on a graded, non-flammable paper, positioned under the sphere. Any break in the burned line during daytime indicates that during that interval, the sky was covered with clouds [29].

Currently, the instrument used to measure the *global solar irradiance* is called *pyranometer* (from the Greek words *pyros*—fire, *ano*—sky, *metron*—measure). The pyranometers can be connected to digital autonomous data loggers and have two operation principles with different accuracies.

High-precision pyranometers use *thermopile-based sensors* that contain a grid/array of thermocouples able to generate a small electric voltage proportional

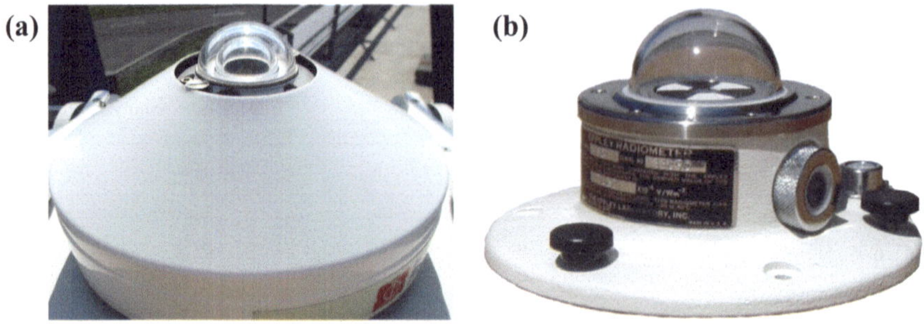

Fig. 2.6 Pyranometers with thermopile sensors: **a** Kipp & Zonen model CMP21 for global solar irradiance; **b** Eppley Laboratory model 8-48*A* mainly used to measure the diffuse solar irradiance [44]

to the solar irradiance received on the sensors' surface. These pyranometers provide accurate measurements across a wide range of wavelengths, but are the most expensive (Fig. 2.6).

The second type of pyranometers is based on small *photovoltaic cells* or *photodiodes* (Fig. 2.7). Their operating principle is similar to that of the thermopile and is based on the generation of a variable electrical current that is proportional to the incident solar radiation on the sensor's surface. They are less accurate in terms of absolute values and are able to measure a much narrower solar radiation spectrum that corresponds to the type of the photoelectric material used. However, they are less expensive compared to the first type and are suited to most measurements required in the renewable energy sector as estimating the solar energy potential, system control, etc. [34].

Both pyranometer types are able to measure the global solar irradiance available in the hemispherical field of view of the planar sensor used within the instrument. Besides using them for estimating the solar energy potential in a particular location, they can be also used to measure the global solar irradiance received on the flat surfaces of the solar energy convertors, either photovoltaic or solar thermal. In this case,

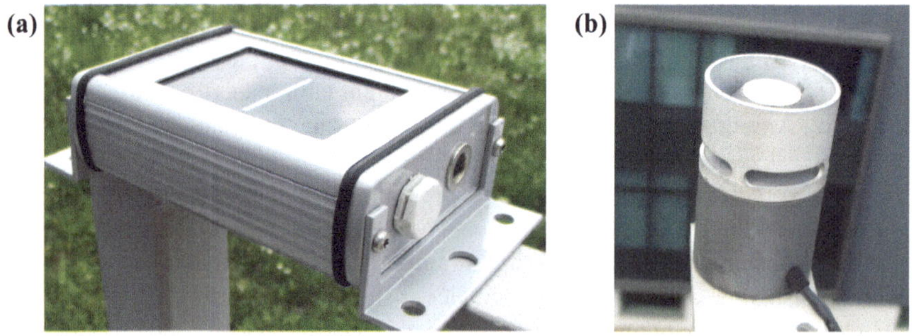

Fig. 2.7 Lower accuracy pyranometers based on: **a** photovoltaic cells; **b** photodiode

the pyranometer is mounted in the same plane as the solar energy convertor, allowing the user to monitor its operation and to quickly identify possible malfunctions.

The *diffuse solar irradiance* along with the *reflected* one can also be measured using pyranometers if shading elements (disc, ball or ring) are used to block the direct solar component. Considering the Sun's path on the sky vault, these additional devices must have an adjustable position. For the variant with a ring (Fig. 2.8a), the position can be automatically or manually adjusted on a daily basis, or once every few days. The other two variants, with disc (Fig. 2.8b) or ball (Fig. 2.8c), require an automatic positioning every few minutes. Considering the three variants, the one with the ring shading element is the least accurate because it occupies a larger part

(a) (b) (c)

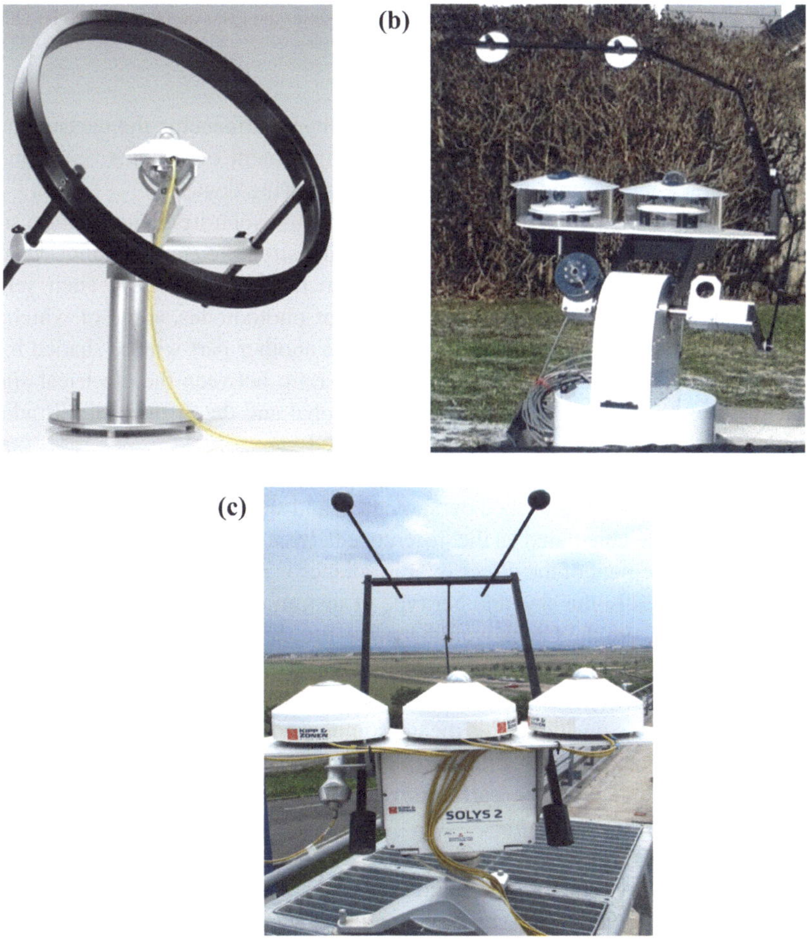

Fig. 2.8 Variants of shaded pyranometers for measuring the diffuse solar irradiance: **a** *Kipp & Zonen* ring model CM 121B/C [85]; **b** *Eppley Laboratory* disc model [43]; **c** Kipp & Zonen SOLYS2 ball model

(a) (b)

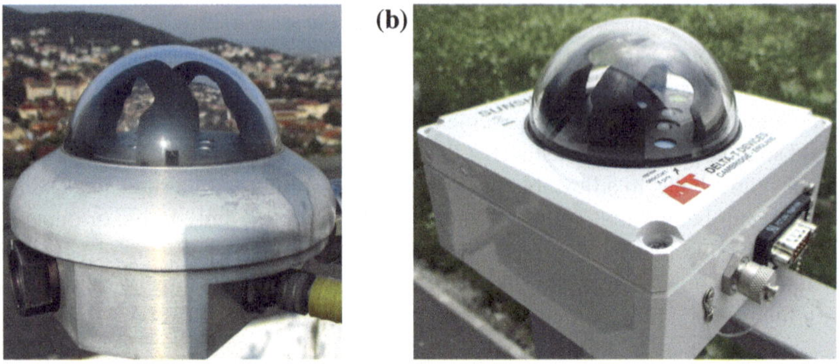

Fig. 2.9 Pyranometers with shading element included under the glass dome, provided by Delta-T: **a** SPN1 model; **b** BF5 model

of the sky vault than the disc or ball, thus negatively influencing the measurements. However, it has the lowest cost (single-axis movement of the ring, compared to dual-axis tracking required for disc and ball shadowing devices).

An option for measuring the global and the diffuse solar irradiance using a single instrument is a mixed pyranometer with a shading structure enclosed under its glass dome, e.g. the SPN1 or the BF5 models from Delta-T (Fig. 2.9). Their sensing element usually contains several thermopiles or photodiodes, some of which will be irradiated by the direct solar radiation while another part will be shaded by the included shading structure. Depending on the ratio between the electrical signals generated by the two types of elements, the global and the diffuse solar irradiance available in the horizontal plane can be estimated.

If the global and the diffuse solar irradiance are measured in the same plane, using one of the above-described pyranometers and shading devices, the *direct solar irradiance* can be calculated as the difference between these two components.

If the measurement of the direct solar irradiance is required in the direction where the Sun is located in the sky, another type of instrument has to be used. It is called *pyrheliometer* (from the Greek words: *pyros*—fire, *helios*—Sun, *metron*—measure, Fig. 2.10) and has a very narrow field of view that contains only the solar disc (about 5°). To properly function, the pyrheliometers have to be continuously oriented, following the Sun's position. This means complex additional equipment required to accurately track the position of the Sun along with instruments that allow the correct positioning of the shading disc/ball required for measuring the diffuse solar irradiance (Fig. 2.8b, c).

This additional equipment consists of two components: the two-axis tracking system and the control system of the angular displacement. The latter usually calculates the position of the Sun in the sky based on the exact location of the equipment and the exact time, both taken from the GPS satellite network.

An alternative to measure the global solar irradiance is the use of *reference cells* (Fig. 2.11). These are particularly employed for calibrating the flash or pulse solar simulators required to measure the photovoltaic output in laboratory conditions. A

Fig. 2.10 Pyrheliometer installed on an accurate solar tracking system [77]

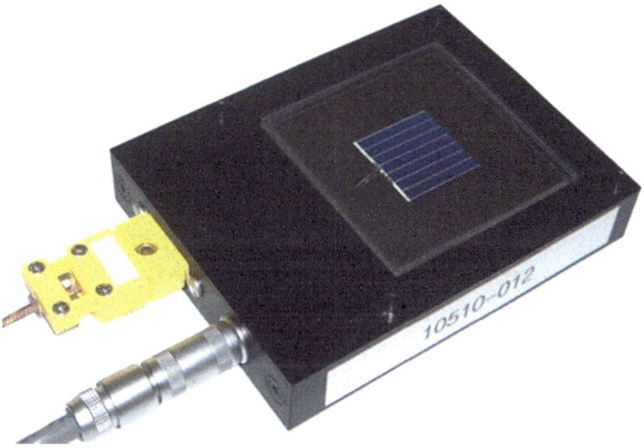

Fig. 2.11 Photovoltaic reference cell Copyright MKS Instruments, Inc. Used with permission [108]

reference cell is actually a photovoltaic cell encapsulated in an aluminium block covered with a layer of optical glass and most frequently includes a thermocouple (temperature sensor) on the backside of the cell for correcting its output based on the cell temperature. It is similar to a pyranometer based on photovoltaic cells, but uses more accurate sensing materials; moreover, its encapsulation does not allow the outdoor installation. These devices have the advantage of a very small response time compared to thermopile-based pyranometers. A reference cell produces a known amount of electrical energy that varies according to the received solar irradiance; this means that the output current can be used to indirectly measure the irradiance. The conversion factor of the reference cell depends on the air mass, AM, and is usually expressed in mA/(kW m^2) at AM = 1.5 and 25 °C.

For analysing the *spectral distribution* of the solar radiation, *spectroradiometers* are used, able to measure the power of the incident radiation for each individual wavelength. These instruments are rather complex and expensive, while their operation and maintenance are also difficult. This is why, in contrast to pyranometers and pyrheliometers, these are not outdoor used for long-term solar monitoring but rather in laboratories for testing purposes or in order to generate/validate models able to predict the spectral distribution of sunlight according to the meteorological data and Sun's position.

All the instruments measuring the solar irradiance are directly or indirectly calibrated according to the World Radiological Reference (WRR) using an absolute cavity radiometer owned by the World Meteorological Organization (WMO). During calibration, the instruments are compared under standard reference conditions, while the sensor's output will be scaled using an appropriate factor. High-precision instruments are calibrated directly according to the WRR, while instruments with lower precision are tested in natural sunlight and compared to already calibrated, high-precision instruments.

Daily measurements of the direct, diffuse and global solar irradiance are further presented, and specific aspects for each solar radiation type are discussed. The experimental data were recorded every minute, using a Kipp & Zonen SOLYS2 Sun-tracking systems with two secondary standard CMP22 pyranometers installed in the horizontal plane (for the global and the diffuse radiation) and a first-class CHP1 pyrheliometer continuously oriented towards the Sun (for the direct solar radiation). These devices are installed in the Renewable Energy Systems and Recycling (RESREC) R&D Centre, in the Transilvania University of Brasov, Romania (45.6°N and 25.5°E), and are presented in Fig. 2.8c.

Three months were selected for the analysis: December (winter solstice), March (spring equinox), June (summer solstice), and during each month, three types of days (cloudy, sunny and mixed) were analysed.

During fully cloudy days (Fig. 2.12), regardless of the month, all three components of the solar radiation have very low irradiance, a bit higher during the spring and summer months. As expected, the direct solar radiation is almost continuously absent, because the sky is covered with clouds and the global solar radiation has the same irradiance as the diffuse one.

During sunny days (Fig. 2.13), the following aspects can be noticed:

- The different sunrise and sunset hours indicate shorter winter days and longer days during the summer;
- During the winter days, at noon, the maximum global solar irradiance measured in the horizontal plane (Fig. 2.13a) is smaller than the direct irradiance (Fig. 2.14a);

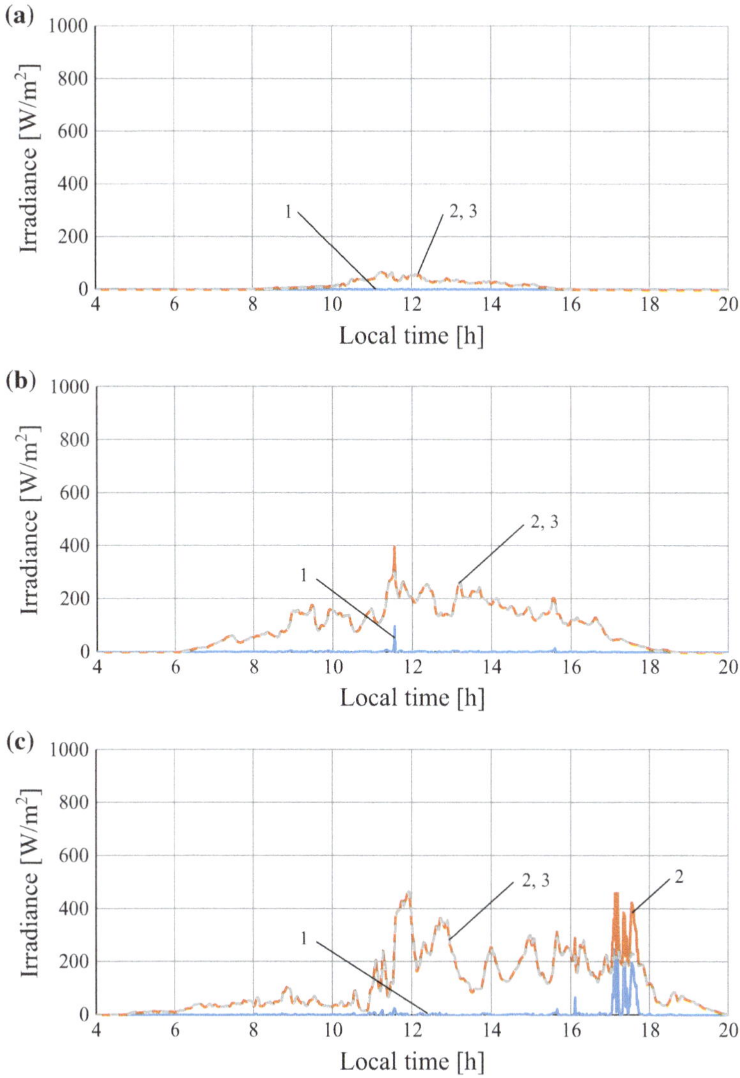

Fig. 2.12 Irradiance in the horizontal plane, during cloudy days, of the direct solar radiation (1), global solar radiation (2) and diffuse solar radiation (3) in: **a** December; **b** March; **c** June

this occurs because the Sun is positioned closer to the horizon line, and therefore, a smaller portion of the solar radiation reaches the horizontally positioned global radiation pyranometer; the same is valid for all other winter months, with a decreasing effect as the days come closer to the equinoxes;

- During the rest of the year, the maximum global solar irradiance is similar to the one of the direct solar radiation.

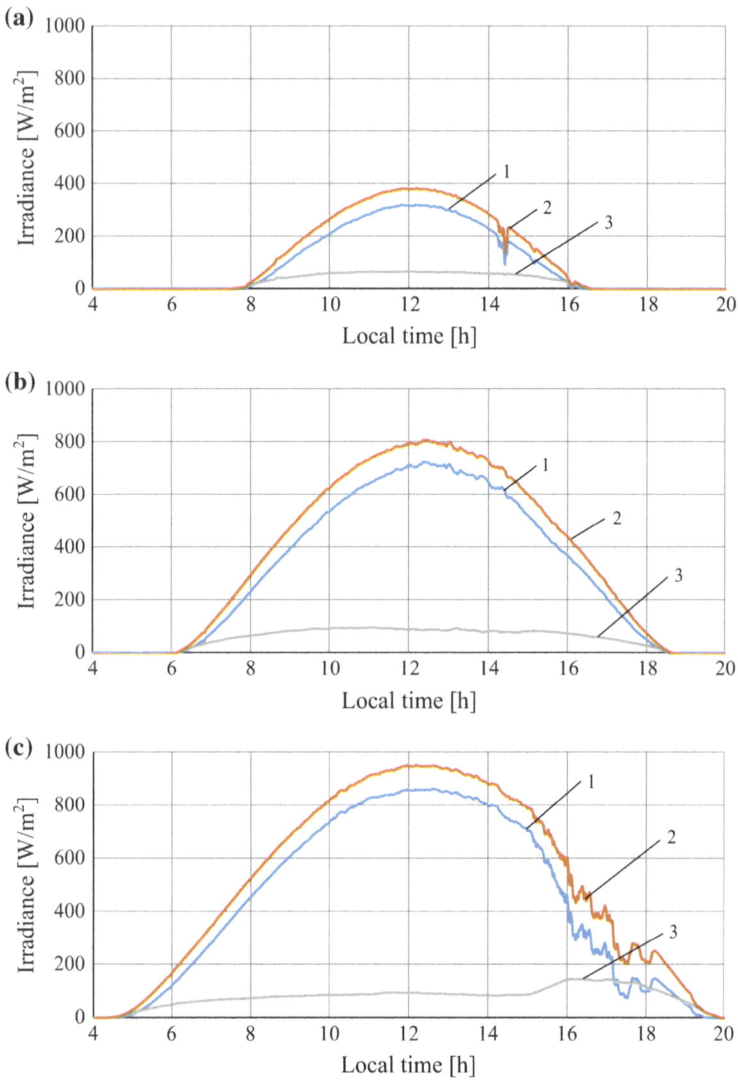

Fig. 2.13 Irradiance in the horizontal plane, during sunny days, of the direct solar radiation (1), global solar radiation (2) and diffuse solar radiation (3) in **a** December; **b** March; **c** June

Figure 2.14 outlines the differences between the direct solar irradiance received during a sunny and a cloudy day in the same month. Despite the different length of the winter/summer days, the received direct irradiance reaches similar maximum values over the year, of 800–900 W/m^2.

Figure 2.15 presents a mixed day (2 July 2018) in comparison with a roughly clear, sunny day (19 June 2018). These two days were selected in the same summer

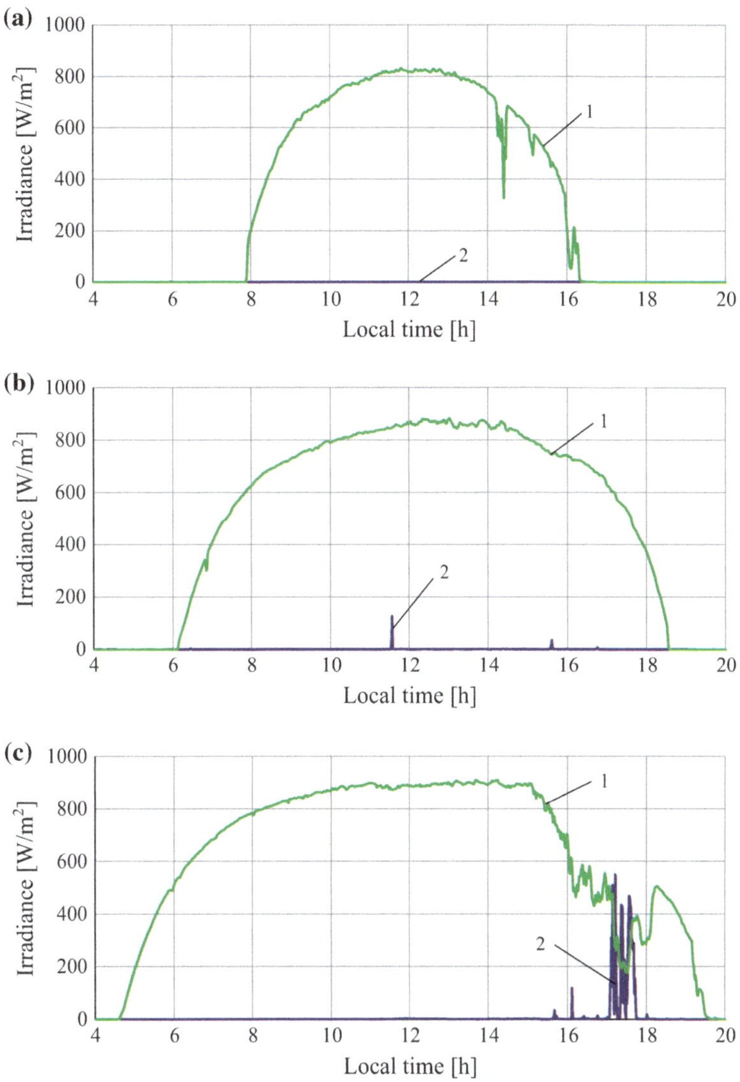

Fig. 2.14 Irradiance of the direct solar radiation during a sunny (1) and a cloudy day (2) in: **a** December; **b** March; **c** June

interval to avoid the possible influence of different atmospheric conditions or of the seasonal variations of the Sun position in the sky and outline the following aspects:

- The maximum global solar irradiance during the mixed day is higher compared to that during the sunny day (roughly 1350 W/m^2 versus 1000 W/m^2) although the direct solar irradiance is smaller during these days. This is the consequence of

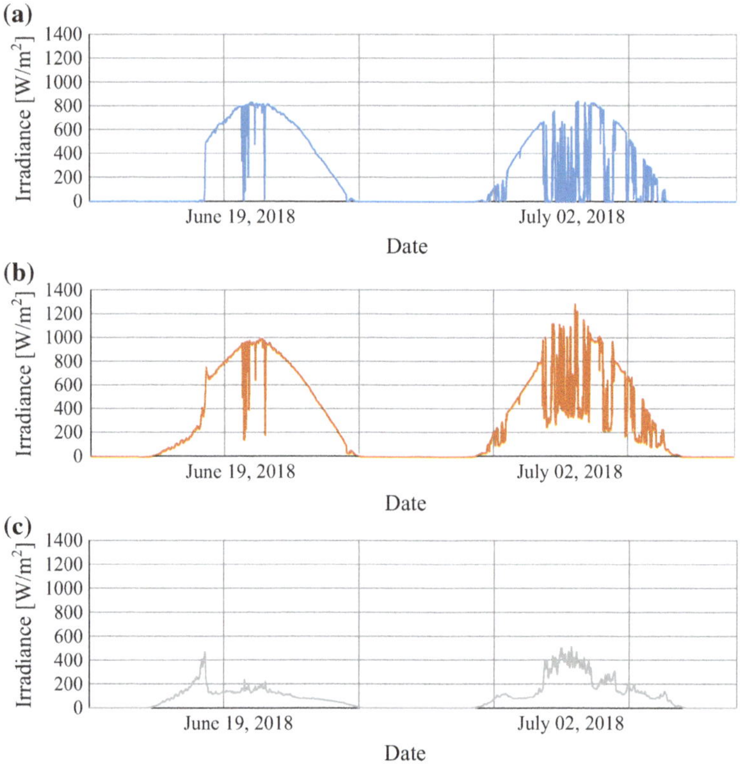

Fig. 2.15 Solar irradiance in the horizontal plane on an almost sunny day (19 June 2018) and a mixed day (02 July 2018) during the summer season: **a** direct solar irradiance variation; **b** global solar irradiance variation; **c** diffuse solar irradiance variation

the fact that the direct solar radiation also includes the reflected component by the clouds on the horizontal plane, where the global solar radiation is measured;

- The reflection phenomenon can be responsible for global solar irradiance close to the solar constant (1366.1 W/m^2) and is important for the correct sizing of the conversion system (inverter or charge controller) that manages the electrical power produced by the PV system.

2.1.1.5 Evaluating the Solar Energy Potential

The solar energy potential represents an important parameter that influences all life aspects: the energy consumption and production, the renewable energy systems' operation, the animals or plants growing, etc. According to the requirements, the solar energy potential can be estimated for longer time intervals (years, months) or shorter ones (weeks or days), or even instantaneous values can be modelled.

Experimentally, it can be measured as the energy received on a surface during the analysed time interval (Wh/m^2) or as the average power on that surface (W/m^2).

There are different ways to evaluate the solar energy potential in a given location using:

- Locally measured data;
- Images of the sky;
- Mathematical models;
- Data provided by geostationary satellites;
- Solar radiation maps and online databases.

Locally measured data, using solar radiation stations with pyranometers and other instruments as previously presented, installed in the implementation location, represent the best solution for identifying the solar energy potential. However, for a valid conclusion, local measurements have to cover at least an entire year. Moreover, two or three years of recorded data are required, to mitigate the seasonal differences during consecutive years. There are three types of stations for measuring the global solar radiation [104]:

- The most complex type includes instruments for measuring the global and the diffuse solar irradiance in the horizontal plane and the direct solar irradiance on the Sun direction, during each moment of the day. So, any two of the three components can be used to validate the third component. These stations may also contain equipment for measuring the solar radiation (direct, diffuse and/or global) on tilted or on tracked surfaces;
- Common stations are able to measure only two components, while the third one is calculated. The usual solution is to measure only the global and the diffuse solar irradiance in the horizontal plane and to calculate the direct solar irradiance. This supports a significant simplification of the station, because it eliminates the solar tracking mechanism required for measuring the direct solar irradiance. However, this also implies a lower accuracy in estimating the direct solar irradiance;
- Low-cost stations, that are mostly common nowadays, are only measuring the global solar irradiance in the horizontal plane. The other two components are calculated with a lower accuracy.

Images of the sky taken from the ground at small and constant time intervals enable to sense the cloud coverage; based on these, the estimation of the solar irradiance can be done. Figure 2.16 outlines the general idea of the estimation process: based on the initial image of the sky, three types of pixels are identified (blue/yellow/cyan), corresponding to Sun/clear sky and clouds; various ratios of the number of pixels can be used to identify the value of the solar radiation components [24].

Mathematical models have an even lower accuracy but are the only available when the solar irradiance was not previously measured in the location of interest. Many mathematical models are reported in the literature, ranging from general to more specific ones that may consider special conditions in the implementation location (ground, water surfaces, etc.). These may also consider cloudiness data observed by trained meteorologists, sunshine duration, atmospheric turbidity, etc. [8, 12, 19].

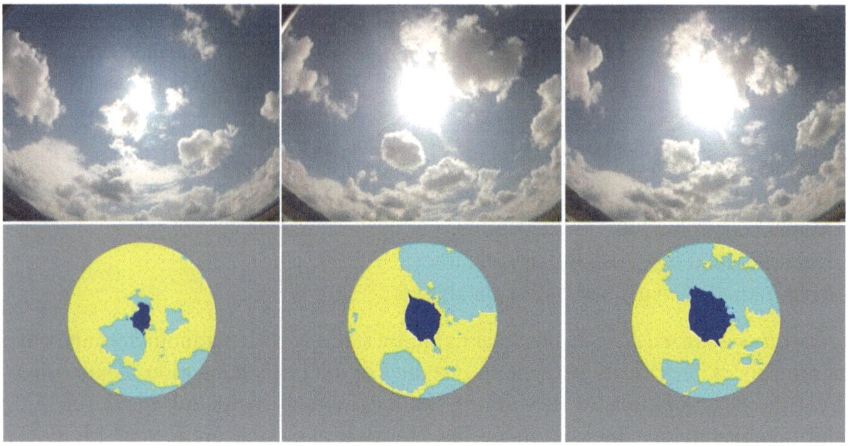

Fig. 2.16 Processed sky images used to estimate the solar irradiance components [24]

Various methods for modelling the global, direct and diffuse components of the solar radiation based on parametric and decomposition models are described in the literature [117]. Empirical methods represent another alternative [115, 116], while artificial neural network methods are also accepted as they are able to synthesize complex estimation problems [143, 148].

As an example, a rather simple model for *ground surfaces* was developed by Meliss [98] for clear sky days. The direct solar irradiance (B) is time-dependent and can be calculated using

$$B = B_0 \cdot e^{-\frac{T_R}{0.9+9.4 \cdot \sin \alpha}} \tag{2.7}$$

where

- B_0 is the extraterrestrial solar irradiance (Eq. 2.1).
- T_R is the turbidity factor.
- α is the altitude angle of the Sun (Eq. 2.3).

The turbidity factor is correlated with the attenuation of the direct solar radiation during clear sky conditions; it varies with the optical path length AM [70] and with the temperature, pollution, humidity of the atmospheric layers crossed by the sunrays [14].

In usual atmospheric conditions, typical T_R values are given in Table 2.1.

Following the sketch in Fig. 2.17, the direct solar irradiance received in the horizontal plane (B_h) can be calculated using

$$B_h = B \cdot \sin \alpha \tag{2.8}$$

Table 2.1 Values of the turbidity factor [14]

Atmospheric features	Turbidity factor
Ideal pure atmosphere	1
Arctic clean and cold air	2
Clean and warm air	3
Warm and wet air	4–6
Polluted atmosphere	8

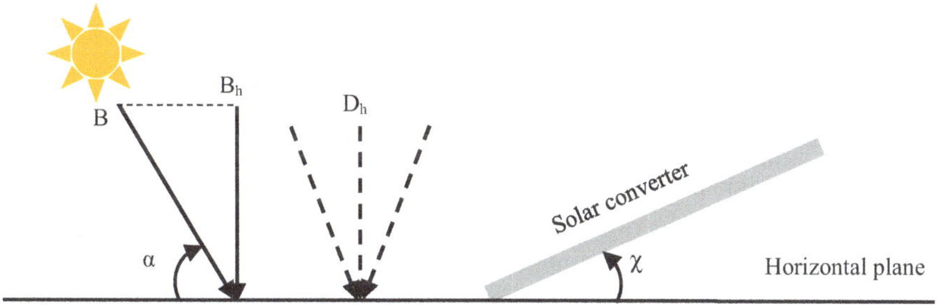

Fig. 2.17 Solar radiation components and angles required for estimating the solar irradiance

The diffuse solar irradiance (D) can be calculated on the direction of the Sun using

$$D = \frac{D_\mathrm{h} \cdot (1 + \sin \alpha)}{2} \tag{2.9}$$

where D_h is the maximum diffuse solar irradiance received on the horizontal plane (Fig. 2.17) and can be calculated according to Meliss [98] using

$$D_\mathrm{h} = \frac{1}{3} \cdot (B_0 - B) \cdot \sin \alpha \tag{2.10}$$

The irradiance of the ground reflected solar radiation (R_n) on a tilted surface at an angle χ from the horizontal plane (Fig. 2.17) can be calculated according to Kalogirou [82] using

$$R_n = \rho \cdot (B_\mathrm{h} + D_\mathrm{h}) \cdot \frac{1 - \cos \chi}{2} \tag{2.11}$$

where ρ is the Albedo reflectance factor of the surrounding ground.

The global solar energy (E_G) available during a day can be calculated by summing up the integrals over the daylight time interval (between sunrise t_sr and sunset t_ss) of the direct, diffuse and reflected irradiances, using

$$E_\mathrm{G} = E_\mathrm{B} + E_\mathrm{D} + E_\mathrm{R} = \int B\mathrm{d}t + \int D\mathrm{d}t + \int R\mathrm{d}t \tag{2.12}$$

where

- E_B is the direct solar energy.
- E_D is the diffuse solar energy.
- E_R is the reflected solar energy.

In real sky conditions, when clouds are crossing the sky, the direct solar radiation is negatively influenced by the humidity and clouds within the atmospheric layers. According to Diaconescu et al. [30], the monthly energy of the direct solar radiation received on the ground (E_{Bm}) in real sky conditions can be roughly estimated using

$$E_{Bm} = E_B \cdot F_{CC} \cdot N_m \tag{2.13}$$

where

- N_m is the number of days in the month.
- F_{CC} is the monthly direct radiation factor of clouds crossing; it represents the ratio between the monthly energy of the direct solar radiation measured at ground level and the direct solar energy estimated at ground level considering that all days in the month are clear sky days.

Similar statements are valid for the monthly energy of the diffuse solar radiation (E_{Dm}) in real sky conditions that can be estimated using a monthly diffuse radiation factor (C_D) with

$$E_{Dm} = 3 \cdot E_D \cdot C_D \cdot N_m \tag{2.14}$$

The values of these two factors, calculated for Brasov, Romania, are included in Table 2.2.

The solar irradiance on *large water surfaces* can be estimated using empirical methods as reviewed by Cotorcea [27]; most of these were developed and tested for the Pacific and Atlantic Oceans [9, 127]. In addition, models were developed by Bason [14, 15] and tested during the Danish Galathea III expedition carried out between August 2006 and April 2007.

Geostationary satellites collecting data on the Earth's atmosphere and the cloud coverage degree with high spatial and temporal resolution can also be used to estimate the solar energy potential. Along with local solar data sets and meteorological parameters recorded at the ground level, several models were developed to estimate the solar radiation components at the Earth surface based on satellite imagery [123, 128]. These models are grouped into physical and statistical (empirical) models; the latter is less complex because they do not require much information about the atmosphere and are mainly based on satellite data processing and ground measurements. Examples are the *Perez model* [122] and the *Heliosat model* developed for Meteosat satellites [129, 130] in the frame of the European Solar Radiation Atlas (ESRA) programme. Studies based on satellite data for water surfaces were also developed by Gautier [50] who assessed the energy flow in the Indian Ocean.

Table 2.2 Monthly and annual values of the factor of clouds crossing (F_{CC}) and of the diffuse radiation factor (C_D) in Brasov, Romania [30]

Month	January	February	March	April	May	June	July	August	September	October	November	December	Annual
F_{CC}	0.30	0.30	0.35	0.35	0.35	0.41	0.45	0.38	0.35	0.35	0.35	0.25	0.35
C_D	0.35	0.45	0.45	0.5	0.55	0.55	0.55	0.5	0.45	0.45	0.35	0.3	0.45

Solar radiation maps and online databases represent another option to calculate the solar energy potential in a given location. These can be obtained either using satellite data [55] or by interpolating the data recorded by infield solar radiation or meteorological stations, as an example: SolarGIS [145] in Fig. 2.18 or PVGIS [125] in Fig. 2.19 or SoDa [144], and the latest is used for estimating the solar radiation on large water surfaces.

Offline databases and estimation software (e.g. Meteonorm, RETScreen) represent another source for estimating the solar energy potential in a given location. These tools are usually based on measured and interpolated data, as e.g. the Meteonorm database that uses data recorded by worldwide meteorological stations. Moreover, for locations where no measured values exist in the database, the software interpolates measurements from the nearest five stations. This may sometimes generate values that do not fully match the reality because the closest stations may be located in areas with different topographies and climatic profiles as the location of interest.

The solar radiation maps and the satellite images have a low spatial resolution as each pixel may represent tens to hundreds of square kilometres and the estimated result is an average solar energy for the entire surface. Each location on that surface has a different potential, influenced by the local parameters, e.g. the irregular relief forms or different meteorological conditions; therefore, such resolutions may not be accurate enough to estimate the energy produced by the renewable energy systems.

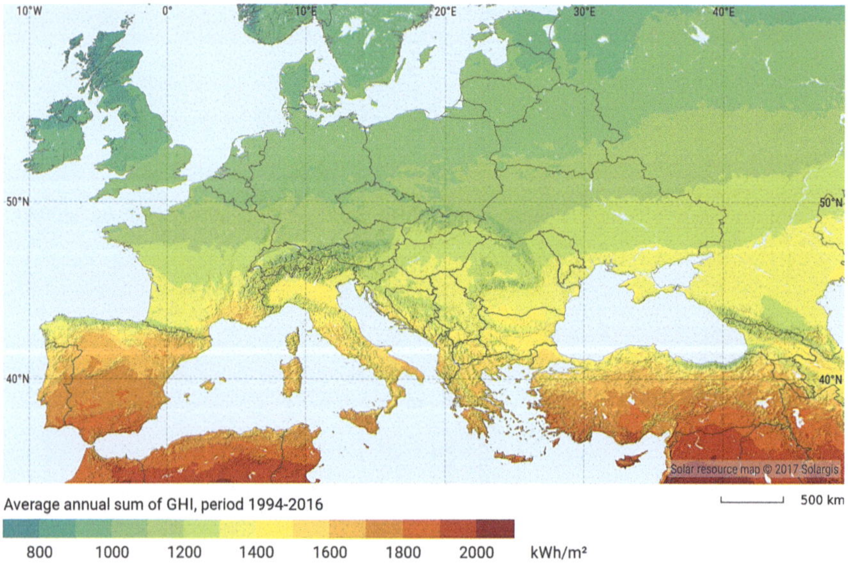

Fig. 2.18 Average annual solar energy map of global horizontal radiation in Europe, courtesy of SolarGIS [145]

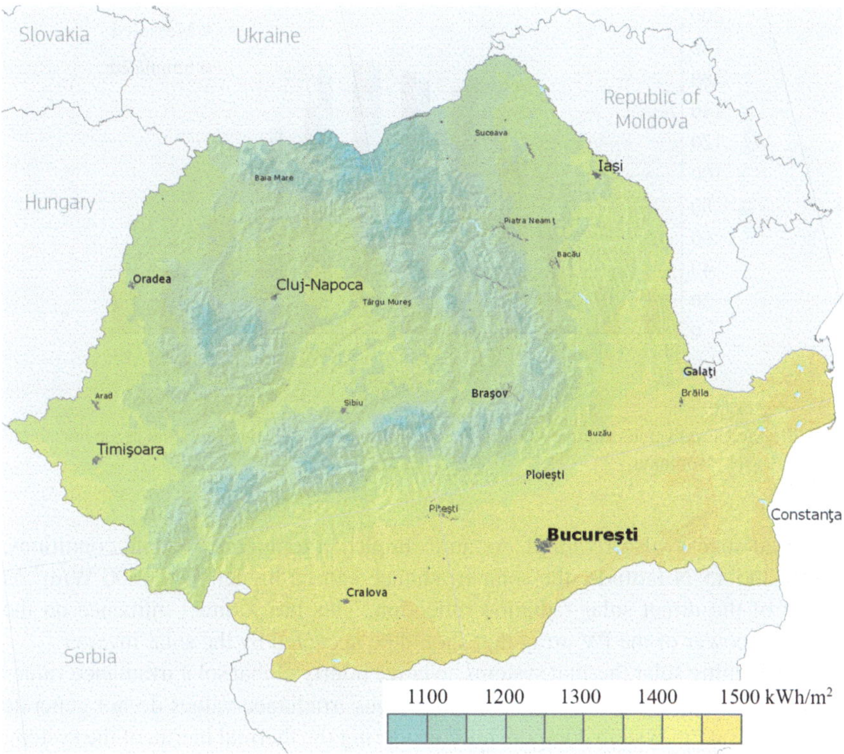

Fig. 2.19 Annual solar energy map of the global horizontal solar radiation estimated for Romania [125]

Depending on the location and on the specific parameters used in each estimation, these solutions are useful in estimating the monthly or yearly solar energy potential (Fig. 2.20).

If the solar irradiance needs to be estimated for shorter time intervals (weeks, days, hours or even minutes), these solutions are not able to provide fully reliable data. For these situations, one of the following options can be selected:

- Local solar measurements;
- Sky images taken from the ground;
- Live images from geostationary satellites combined with neuronal networks that are able to comparatively analyse current images and similar ones recorded in the past and decide on the current values of the solar radiation components.

When designing photovoltaic systems, the solar energy potential is necessary and sufficient for sizing the PV array on an annual basis, according to the local load. However, to correctly size all the electric components of a PV system, the maximum

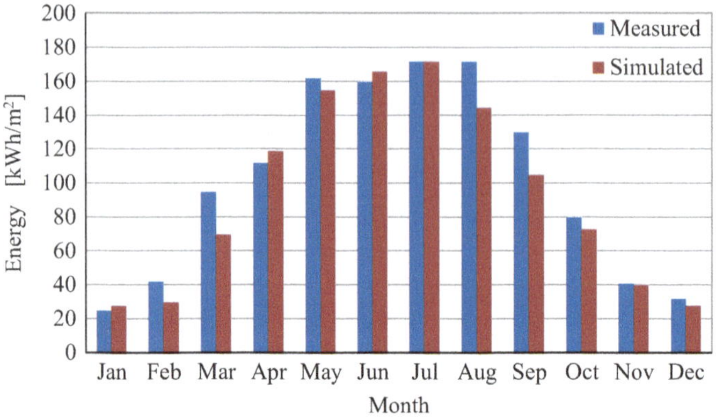

Fig. 2.20 Measured versus simulated monthly energy of the global solar radiation in the horizontal plane in Brasov, Romania

solar irradiance is also required. As an example, in temperate climate conditions, close to the 45°N latitude, the solar irradiance can be higher than 1300 W/m^2 as a result of the direct solar radiation reflection. This has a direct influence on the maximum power of the PV array that should be accepted by the solar inverter.

For designing solar thermal systems, average hourly global solar irradiance values are enough because the maximum instantaneous irradiance values do not generate an overload on the system components considering the thermal inertia of the system.

2.1.2 Solar Energy Available in the Built Environment

Besides the factors that influence the solar energy potential in any location, the solar energy available in the built environment is additionally influenced by the following factors:

- *The orientation of the building's surfaces* considered for installing the solar energy convertors (solar thermal collectors and/or photovoltaic modules);
- *The shading* due to the neighbouring obstacles (buildings, trees, etc.);
- *The Albedo effect* generated by the surrounding surfaces reflecting the solar radiation;
- *The heat island effect* consisting of higher air temperatures in the built areas than in the natural ones;
- *The street canyon effect* occurring when tall buildings continuously limit both sides of a street, influencing the micro-climate in the built area.

2.1.2.1 The Orientation of the Building's Surfaces

The orientation of the building's surfaces considered for installing solar energy convertors is usually described using as reference the horizontal plane and the South direction.

Relative to the horizontal plane (Fig. 2.21), the position of any surface can be defined through the tilt angle (χ) or its complement—the altitude angle (α_n). The tilt angle represents the angle between the surface and the horizontal plane, and the altitude angle (α_n) is defined between the normal line to the surface (n) and the horizontal plane.

Relative to the South direction (Fig. 2.22), the position of the surface can be defined through the azimuth angle (ψ_n) that is the angle between the projection (n_{H}) in the horizontal plane of the surface's normal line (n) and the local South direction. In this book, the azimuth angle is measured from the South direction, with positive values eastward and negative values westward. Thus, for a solar energy convertor installed in the plane of a South-oriented roof, its azimuth angle is $\psi_n = 0°$; when the roof is eastward or westward oriented, the azimuth angle of the solar convertor is $\psi_n = +90°$ and $\psi_n = -90°$, respectively, while for a North orientation, the azimuth angle is $\psi_n = 180°$.

In other studies, the North direction is considered as reference, and in this case, the azimuth angle of the surface has values ranging from 0° (North-oriented surface) to 90° (East), 180° (South) and 270° (West).

The tilt and the azimuth angles are usually called surface angles and are required to calculate the incidence angle of the sunray (v), defined as the angle between the sunray and the normal line (n) to the solar energy convertor installed on the available surface, as detailed in Fig. 2.23. Its values range from 0° (when the sunray is perpendicular on the front side of the receiving surface) to 180° (when the sunray is perpendicular on the backside of the receiving surface). According to Visa et al.

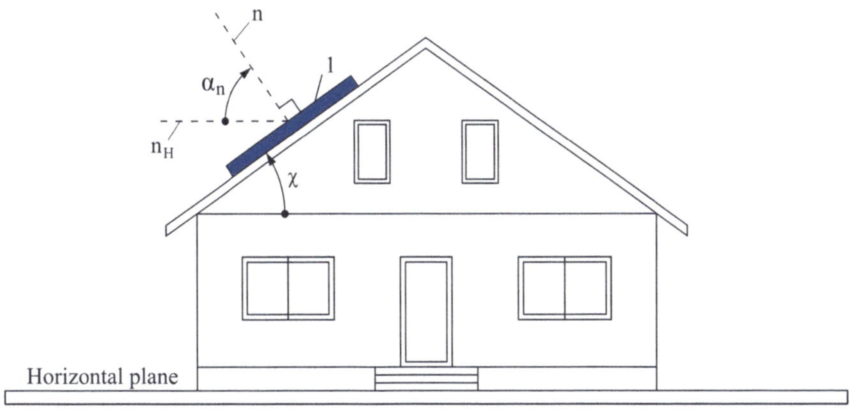

Fig. 2.21 Orientation of the solar energy convertor (1) relative to the horizontal plane, where n_{H} is the projection in the horizontal plane of the normal line (n) to the solar energy convertor (1)

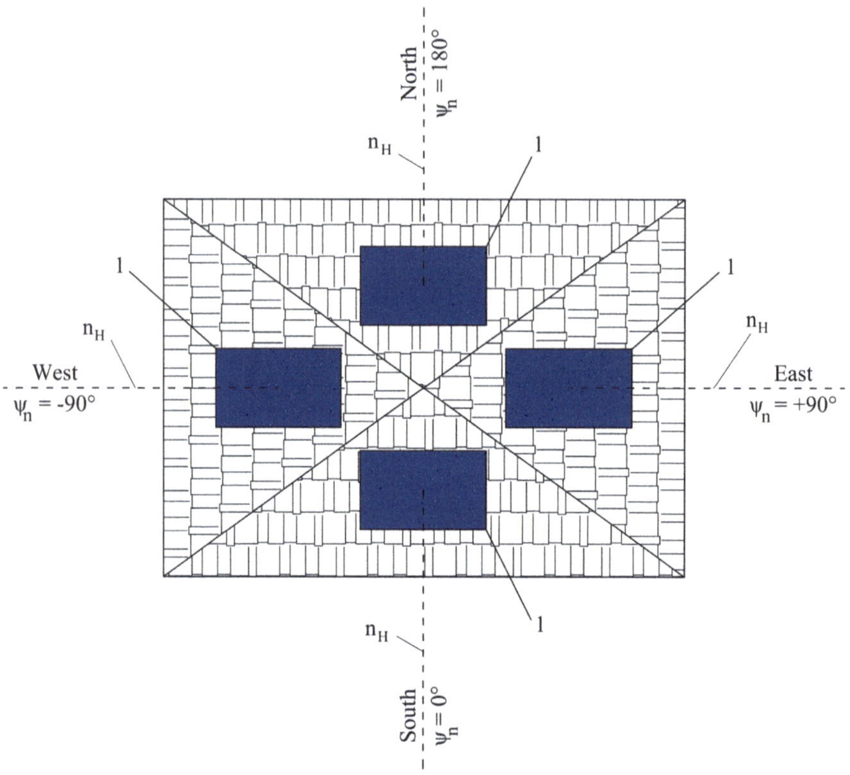

Fig. 2.22 Top view of the roof and the azimuth angle of the solar energy convertor (1) according to its position on the building's roof

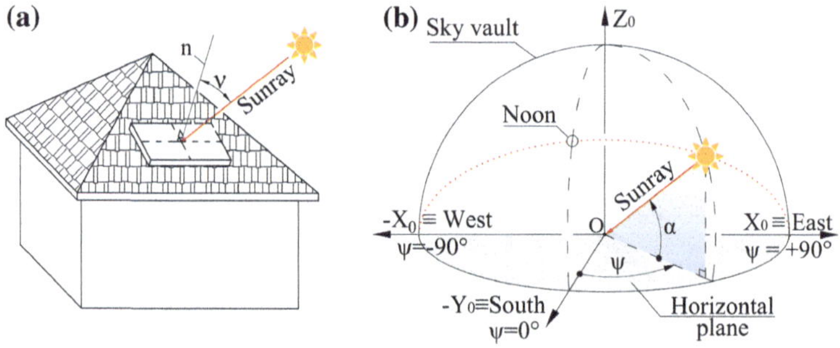

Fig. 2.23 **a** Incidence angle (ν) between sunray and the normal line (n) to the solar energy convertor surface; **b** the solar altitude (α) and azimuth (ψ) angles

[155], the incidence angle can be calculated using

$$\nu = \arccos(\cos\alpha \cdot \cos\alpha_n \cdot \cos(\psi - \psi_n) + \sin\alpha \cdot \sin\alpha_n) \tag{2.15}$$

where

- α is the solar altitude angle, Eq. (2.3).
- ψ is the solar azimuth angle calculated using

$$\psi = \mathrm{sgn}\omega \cdot \arccos\left(\frac{\cos\delta \cdot \cos\omega \cdot \sin\varphi - \sin\delta \cdot \cos\varphi}{\cos\alpha}\right) \tag{2.16}$$

where

- ω is the hour angle, Eq. (2.5).
- δ is the declination angle, Eq. (2.4).
- φ is the latitude of the implementation location.

Knowing that the tilt angle (χ) of the receiving surface is complementary to the altitude angle (α_n), Eq. (2.15) can be written as follows:

$$\nu = \arccos(\cos\alpha \cdot \sin\chi \cdot \cos(\psi - \psi_n) + \sin\alpha \cdot \cos\chi) \tag{2.17}$$

Moreover, Eq. (2.15) can get more simple forms for particular positions of the receiving surface. For the horizontally mounted solar energy convertors ($\chi = 0°$ and $\alpha_n = 90°$), the azimuth angle has no longer influence on the amount of the received direct solar radiation; thus, Eqs. (2.15) and (2.17) become

$$\nu = 90 - \alpha \tag{2.18}$$

For vertically mounted solar energy convertors ($\chi = 90°$ and $\alpha_n = 0°$), Eqs. (2.15) and (2.17) will get the following particular form

$$\nu = \arccos(\cos\alpha \cdot \cos(\psi - \psi_n)) \tag{2.19}$$

For South- or North-oriented vertical surfaces ($\psi_n = 0°$ or $\psi_n = 180°$ and $\alpha_n = 0°$), this becomes

$$\nu = \arccos(\cos\alpha \cdot \cos\psi) \tag{2.20}$$

For eastward- or westward-oriented vertical surfaces ($\psi_n = 90°$ or $\psi_n = -90°$ and $\alpha_n = 0°$), Eq. (2.19) can be written as follows:

$$\nu = \arccos(\cos\alpha \cdot \sin(\pm\psi)) \tag{2.21}$$

where "+" indicates the eastward-oriented vertical surfaces and "−" the westward-oriented ones.

Further on, considering the direct solar irradiance (B) as a vector, its projection on the normal line (n) to the receiving surface; thus, the received direct solar irradiance (B_n) can be calculated using

$$B_n = B \cdot \cos \nu \tag{2.22}$$

The incidence angle should thus be as low as possible to increase the received direct solar irradiance. For usual (one-face) solar energy convertors, incidence angles higher than 90° will reduce to zero the received direct solar irradiance. However, there are on the market bifacial photovoltaic modules [61] that also exploit the direct solar radiation received from the backside.

The received diffuse solar irradiance (D_n) depends only on the altitude angle (α_n) and on the tilt angle (χ) of the solar energy convertor. According to Rakovec and Zaksek [126], the isotropic model of the sky is further considered, with the sky vault as a uniform source of diffuse solar radiation and the maximum diffuse solar irradiance received in the horizontal plane (D_h). Thus, for a tilted surface, the received diffuse solar irradiance can be calculated using

$$D_n = \frac{D_\mathrm{h} \cdot (1 + \sin \alpha_n)}{2} = \frac{D_\mathrm{h} \cdot (1 + \cos \chi)}{2} \tag{2.23}$$

Finally, the received global solar irradiance (G_n) represents the sum of the direct and diffuse components

$$G_n = B_n + D_n \tag{2.24}$$

To calculate the received global solar energy (E_{Gn}) over a time interval (e.g. for one day, month, season, year), Eq. (2.24) has to be integrated over the time interval (t_1, t_2) of interest

$$E_{\mathrm{G}_n} = \int_{t_1}^{t_2} G_n \mathrm{d}t \tag{2.25}$$

The yearly received global solar energy is plotted in Fig. 2.24 as a function of the surface's tilt and azimuth angles at the latitude $\varphi = 45.6$°N. The horizontally positioned surfaces ($\chi = 0°$) are not affected by the azimuth angle; thus, the yearly received global solar energy is 1096 kWh/m^2. For any other surface orientations, the azimuth angle has a major influence; as for a South-oriented vertical surface, the yearly received global solar energy is 711 kWh/m^2, while for a North-oriented, one is 278 kWh/m^2, resulting a reduction factor of 2.56. This factor decreases with the tilt angle: for a tilt angle of 10°, the reduction factor is 1.15 and for horizontal surfaces is 1.

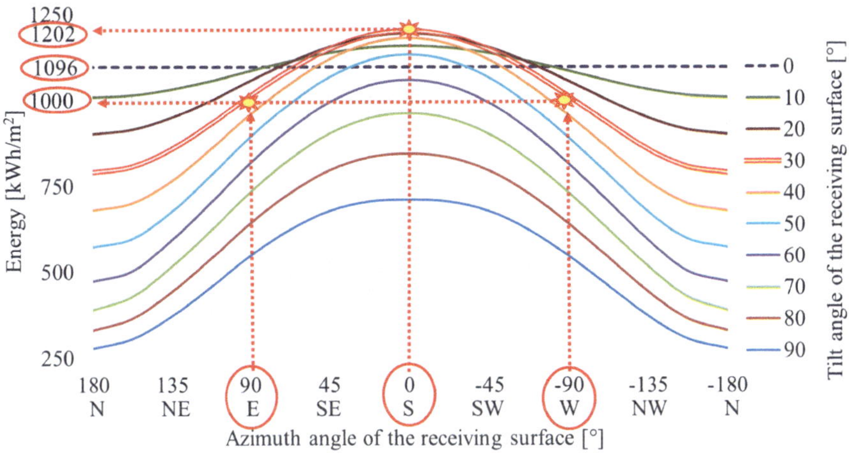

Fig. 2.24 Yearly received global solar energy in Brasov ($\varphi = 45.6°$N) on a surface as function of the surface orientation defined by its azimuth angle (ψ_n) and tilt angle (χ)

The tilt angle has a significant influence on the yearly received global solar energy as, in the considered location, the maximum yield (1202 kWh/m^2) is received by a surface tilted at 30° and an azimuth angle of 0° (South-oriented). If a similarly tilted surface is eastward- or westward-oriented ($\psi_n = 90°$ or $\psi_n = -90°$), the yield is reduced to 1000 kWh/m^2; for such an azimuth angle, to maximize the yearly received solar energy the tilt angle must be reduced to 0° (horizontal), when the yearly yield is 1096 kWh/m^2.

The optimal value of the surface azimuth angle is $\psi_n = 0°$ (South orientation) in the northern hemisphere or $\psi_n = 180°$ (North orientation) in the southern hemisphere; thus, the surface must be oriented towards the Earth's Equator.

The optimal tilt angle of the surface is site dependent and can be calculated as function of the latitude and meteorological conditions in the implementation location. Several studies were developed to find out the optimal tilt angle, and the results are synthesized in Table 2.3. Soulayman and Sabbagh [147] proposed an algorithm to calculate the optimum tilt angle for latitudes ranging between 0° and 60°, by simply searching for the values at which the solar radiation received on the solar energy convertor's surface is maximum in a particular day or during a specific period (month or year).

Stanciu and Stanciu [149] proposed a fourth-degree polynomial function to evaluate the optimum tilt angle for latitudes ranging between 0° and 80°. Siraki and Pillay [141] proposed a simple method based on a modified sky model to calculate the optimum tilt angle in urban applications at latitudes ranging between 15° and 55°; their report highlights the combined influence of the latitude in the implementation location and the weather conditions, on the yearly received solar energy.

These algorithms have a high generality degree and must be therefore validated by local experimental data. Research studies were developed all over the world (mainly

Table 2.3 Optimum tilt angle to install solar energy convertors as function of the site latitude

Site latitude (°)	Optimum tilt angle (°)		
	Soulayman and Sabbagh [147]	Stanciu and Stanciu [149]	Siraki and Pillay [141]
0	0.7	11	
5	5.4	10	
10	10.2	9	
15	14.9	16	19.5
20	19.7	21	
25	24.4	26	24
30	29.0	34	
35	33.7	40	32.5
40	38.3	45	
45	42.8	45	39
50	47.2	55	
55	51.3	57	45
60	55.1	60	
70		70	
80		80	

in the northern hemisphere) to evaluate the optimum tilt angle (Table 2.4). The differences between the optimum values observed for similar latitudes demonstrate the influence of the specific local weather conditions on the yearly received solar energy.

2.1.2.2 The Shading

The discussions so far formulated are valid for receiving surfaces that are not shadowed by the neighbouring obstacles. The geometries of these obstacles can be irregular (in the case of buildings, pieces of equipment, trees or various landforms) or regular (other solar thermal collectors or photovoltaic modules). In the first case, it may be difficult to calculate the shape of the shadow generated by each obstacle on the receiving surface.

A method to evaluate the time intervals when the receiving surface is shadowed consists of plotting the Sun position on the sky in the implementation location (solar altitude versus solar azimuth angles) and superposes on it the shading profile of the obstacle. As an example, Fig. 2.25 shows a ground-mounted photovoltaic (PV) platform (1) with a height of 4 m and a surface of 4 m × 4 m. This platform is installed 30 m West to a building (2) having a uniform height of 10 m, a length of 30 m of the South and North sides and a width of 15 m of the East and West sides. The northern side of the PV platform is aligned with the northern side of the building. The length

Table 2.4 Experimentally validated optimum tilt angles

References	Location	Latitude (°)	Optimum tilt angle (°)
Yan et al. [159]	Brisbane, Australia	27.47S	26
dos Santos and Ruther [32]	Brasilia, Brazil	16S	20
dos Santos and Ruther [32]	Boa Vista, Brazil	0	3
Elhab et al. [36]	Kuala Lumpur, Malaysia	3.14N	10
Yakup and Malik [158]	Bandar Seri Begawan, Brunei	4.90N	3.3
Li and Lam [90]	Hong Kong, China	22.31N	20
Shukla et al. [139]	Bhopal, India	23.26N	23.26
Benghanem [17]	Madinah, Saudi Arabia	24.5N	23.5
Chang [22]	Taipei, Taiwan	25.05N	23.25
Khahro et al. [84]	Sindh, Pakistan	25.89N	23
Moghadam et al. [99]	Zahedan, Iran	29.49N	27.9
Ibrahim [69]	Guzelyurt, Cyprus	35.21N	31
Gong and Kulkarni [56]	Carbondale, Illinois USA	37.73N	30
Bakirci [13]	Istanbul, Turkey	41.01N	32.6
Rowlands et al. [132]	Toronto, Canada	43.78N	32–35
Despotovic and Nedic [28]	Belgrade, Serbia	44.79N	40.6
Colli and Zaaiman [25]	Bolzano, Italy	46.47N	30
Hartner et al. [64]	Vienna, Austria	48.30N	36
Mondol et al. [101]	Aldergrove, UK	54.64N	45
Heller [67]	Marstal, Denmark	54.85N	40

of the building's shadow (L_S) can be calculated based on the solar altitude (α) and azimuth (ψ) angles and on the building's height (H), using

$$L_S = H \frac{\sin \psi}{\tan \alpha} \tag{2.26}$$

For the implementation location of the building and of the PV platform (Brasov, Romania, $\varphi = 45.6°N$), the solar altitude angles are plotted against the solar azimuth angles in Fig. 2.26, for the middle day in each month, resulting a set of twelve lines, marked with (1)–(12) for January to December. The marker dots on each line represent the solar time in the day. The shading profile (13) of the building relative to the location of the PV platform is also plotted in Fig. 2.26.

According to the results in Fig. 2.26, the photovoltaic platform is shadowed by the building only during morning time (approximatively between 6:00 and 7:30) in five months: February, March, April, September and October.

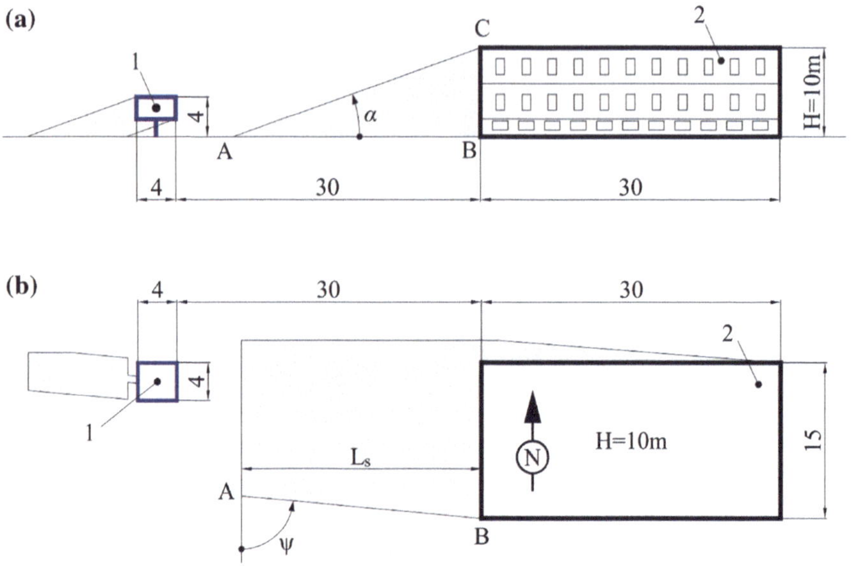

Fig. 2.25 Position of the ground-mounted PV platform (1) relative to the neighbouring building (2): **a** front view; **b** top view

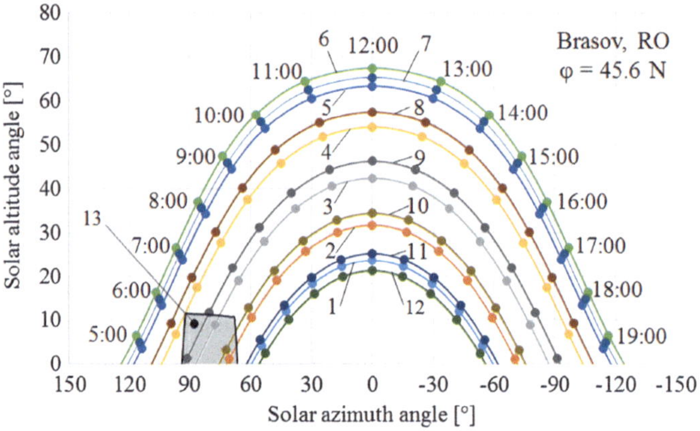

Fig. 2.26 Monthly correlations of the solar angles starting from January (1) to December (12) and the shading profile (13) of the building relative to the PV platform location

To fully avoid the PV platform shadowing all over the year, several measures can be applied: increasing the height of the PV platform or increasing the distance between the building and the PV platform or increasing both, height and distance. When increasing the height of the PV platform, the cost of the platform's structure will increase too; therefore, if possible, it would be more feasible to increase the distance between the PV platform and the building. However, this is not necessary

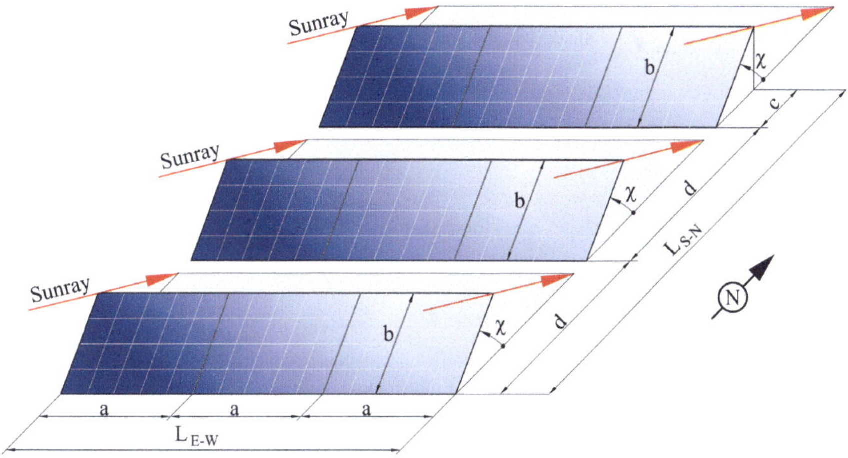

Fig. 2.27 Mutual shading between consecutive rows in an array of solar energy convertors

for a fixed photovoltaic platform since the solar azimuth angles are close to 90°; thus, the incidence angles will also have large values. Moreover, when using solar tracking systems, the increase of these parameters is subject of optimization considering the low solar irradiance during the early morning hours.

Mutual shading between consecutive rows of larger arrays of solar energy convertors (Fig. 2.27) can be easily assessed due to their regular geometry. Solar energy convertors are usually mounted in South-oriented rows, tilted at an optimal angle (χ) to maximize the received solar energy, equally distributed on the North–South direction. The distance (d) between two consecutive rows has to be calculated to avoid shadowing among the rows. In the first approach, this distance is considered equal with the longest shadow length on the North–South direction. This occurs at the winter solstice, at the lowest solar altitude at noon. When this prerequisite is respected, shading will never occur, but this solution implies large available surfaces that might be not feasible when in the implementation location the cost of the ground is high (e.g. in cities). This solution is therefore recommended when the ground availability and the ground cost do not rise any problems and the climate is steady. In temperate climates, the solar altitude angles and the irradiance are significantly lower during winter as compared with the summer. Thus, the PV rows can be implemented at a distance optimized considering that some shading can be accepted during the winter season when the solar energy is not so high and the losses in the received solar energy are compensated by the higher amounts received during the other periods in the year. Special attention should be paid to the type of solar energy convertors as partial shading of one PV module will significantly affect the power output of the entire PV row (it can reduce the output power close to zero) while in the case of solar thermal collectors, only the thermal output of the partially or totally shaded collector will be reduced.

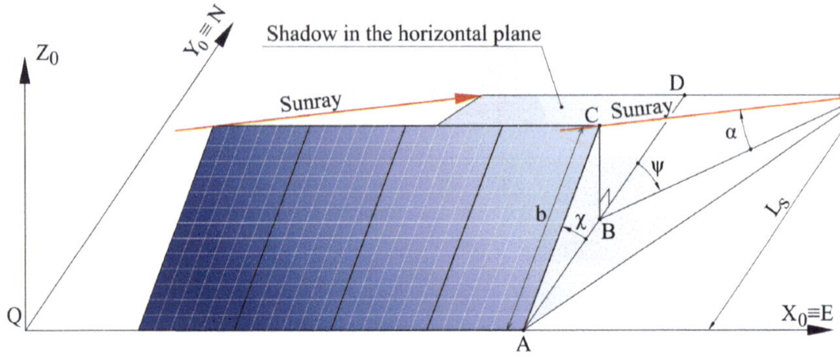

Fig. 2.28 Geometrical model for the length of the shadow of a row in the horizontal plane

The geometry of the shadow casted by a row is outlined in Fig. 2.28, and the length of the shadow can be calculated using

$$L_S = b \cdot \left(\cos \chi + \frac{\sin \chi \cdot \cos \psi}{\tan \alpha} \right) \tag{2.27}$$

where b is the width of the PV row.

The daily variation during the equinoxes and solstices of the shadow length in the horizontal plane of a row with a width $b = 1$ m tilted at $\chi = 40°$ is plotted in Fig. 2.29. The longest shadow occurs during the winter solstice (1) while the shortest during the summer solstice (3) as a result of the solar altitude angles (α) that are lower in winter and higher in summer.

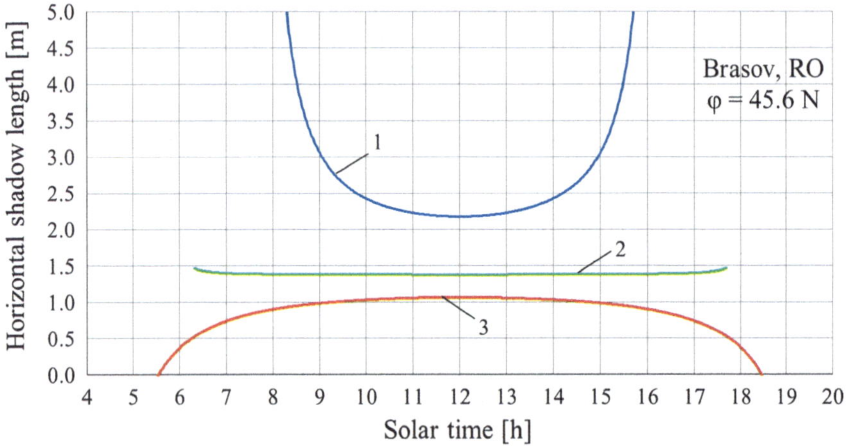

Fig. 2.29 Variation of the shadow length of a row in the horizontal plane during the winter solstice (1), the equinoxes (2) and the summer solstice (3)

The influence of the distance (d) between two consecutive PV rows is outlined in Fig. 2.30 where the yearly solar energy received by a row is plotted as function of the row's tilt angle (χ) for the number of rows (n_R) and different ratios between the distance (d) and the width (b) of each row. As presented in Fig. 2.30, to avoid reciprocal shading between the rows and thus the decrease of the solar energy received, the ratio *d/b* must be increased at values higher than 3, so that each row in the array receives the same amount of solar energy as the first one ($n_R = 1$) that is never shaded.

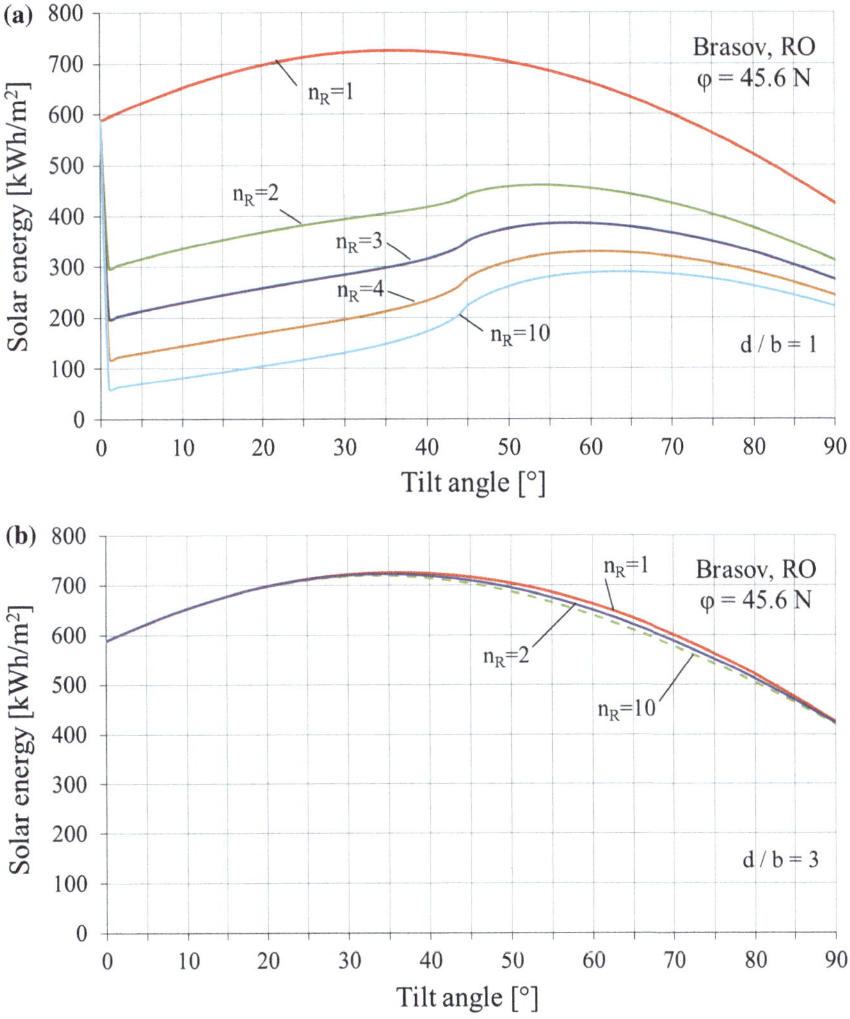

Fig. 2.30 The variation of the yearly received specific solar energy with the tilt angle (χ) for the number of PV rows (n_R) in an array with: **a** ratio $d/b = 1$; **b** ratio $d/b = 3$

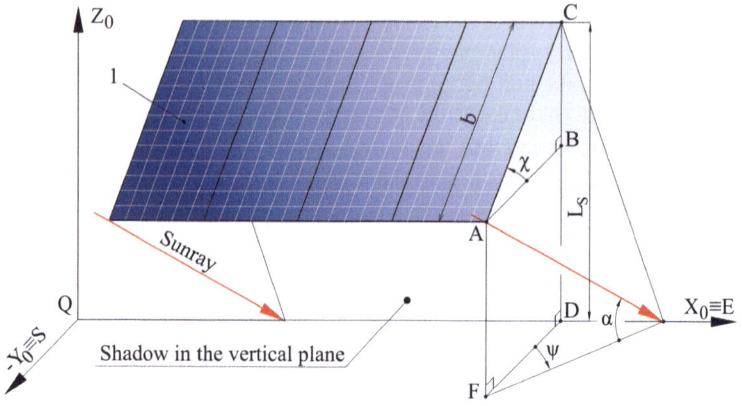

Fig. 2.31 Geometrical model for the length of the shadow of a row of solar energy convertors installed in the vertical plane

For rows of solar energy convertors installed tilted on a vertical plane, e.g. on the buildings' facades, the geometry is detailed in Fig. 2.31. Based on this sketch, the length of the shadow can be calculated using

$$L_S = b\left(\sin\chi + \frac{\cos\chi\tan\alpha}{\cos\psi}\right) \tag{2.28}$$

where b is the width of the row.

The daily variation during the equinoxes and solstices of the vertical shadow length of a row with a width $b = 1$ m tilted at $\chi = 40°$ is plotted in Fig. 2.32.

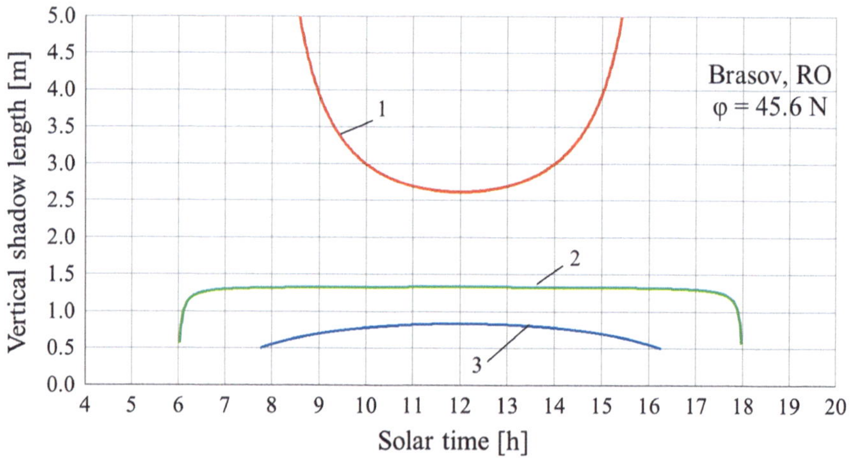

Fig. 2.32 Variation of the shadow length of a row of solar energy convertors installed in the vertical plane during the summer solstice (1), the equinoxes (2) and the winter solstice (3)

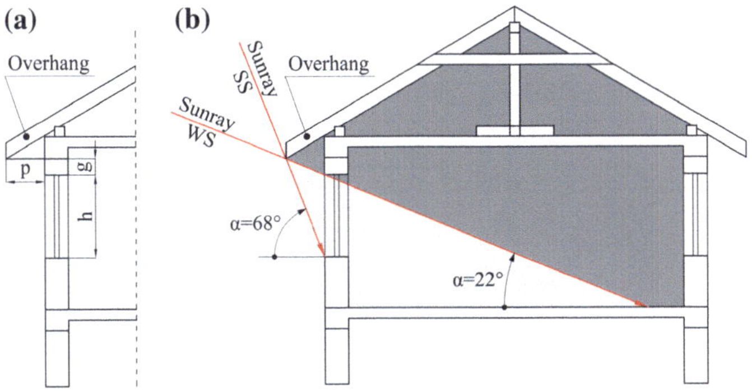

Fig. 2.33 **a** Overhang geometrical elements; **b** shadowing provided at winter solstice (WS) and summer solstice (SS), respectively

The shadow casted by the solar energy convertors can be beneficial when this effect is exploited to reduce the solar gain in the building, mainly during summer, thus decreasing the cooling energy demand. Following the same goal, there are used horizontal shading elements (overhangs) or vertical ones (wing walls) as architectural pieces applied to the buildings facades to shade the windows against the solar radiation during the warm season. A cross section through an exterior wall with a window and an overhang is detailed in Fig. 2.33a. The overhang is defined by the horizontal distance between its outer edge and the plane of the window, also called projection (p) and by the vertical distance to the upper edge of the window, also called gap (g). These elements are used in the passive building's design, where their dimensions and tilt angle are calculated to allow the sunrays with low elevation angles (e.g. 22° at noon at winter solstice—WS, at 45°N latitude) to reach the windows during wintertime and contribute to the passive heating of the building. The sunrays with higher solar altitude angles (e.g. 68° at noon during the summer solstice—SS) are blocked, and thus, the building interior is protected against overheating (Fig. 2.33b).

Different other methods can be used to provide shading during summertime allowing the passive use of solar energy during the cold season. A common and well-accepted solution consists of planting deciduous trees in front of the buildings, providing also wind and sound protection. Another option is represented by the use of climbing plants in the ground at the base of the walls, forming a green facade. A green wall is created when the wall includes a growing medium, such as soil or a substrate, and an integrated water delivery system supports the plants growing.

2.1.2.3 The Albedo in the Built Environment

The Albedo is the fraction of solar radiation reflected by the surrounding objects or materials. It is defined as the ratio between the global up-welling radiation and the global horizontal down-welling radiation, both measured with pyranometers over the entire wavelength range of the solar spectrum. Thus, high Albedo means a large

Table 2.5 Albedo of materials commonly found in the built environment [88]

Material	Albedo	Material	Albedo
Black body	0.00	Brick	0.30
Asphalt (new)	0.04	Glass	0.305
Black asphalt shingles	0.05	Red clay tiles	0.33
Asphalt (worn)	0.12	Sand	0.40
Wood	0.15	Water	0.50
Soil	0.17	Concrete (new)	0.55
Red concrete tiles	0.18	Ice	0.5–0.7
White asphalt shingles	0.21	Aluminium	0.61
Unpainted concrete tile	0.25	Galvanized steel	0.61
Green grass	0.25	White concrete tiles	0.73
Stone	0.275	Snow (fresh)	0.8

amount of light reflected and a brighter object and here comes the origin of the word: *albus* in Latin means white. For natural surfaces, the Albedo ranges between approximatively 0.04 for calm, deep water and overhead Sun, to higher than 0.8 values for fresh snow [4]. The Albedo of materials that are commonly used in the built environment is given in Table 2.5.

Most of the solar energy convertors are installed at an optimal tilt angle to increase their output. Therefore, in addition to the direct and diffuse components of the solar radiation, the reflected component needs to be evaluated as a third part of the global solar radiation received by the tilted surface of the solar energy convertor [33]

$$G_n = B_n + D_n + R_n \tag{2.29}$$

The reflected component (R_n) is assessed using

$$R_n = \frac{\sum_i (G_{ni} \cdot \rho_i \cdot A_i \cdot F_i)}{A} \tag{2.30}$$

where

- i refers to each reflecting surface contributing to the Albedo effect.
- G_{ni} is the global solar irradiance received on the ith surface.
- ρ_i is the Albedo of the ith surface.
- A_i is the area of the ith surface.
- F_i is the view factor from the ith surface to the receiver surface.
- A is the area of the solar energy convertor.

For a general assessment, it is recommended to use a fixed Albedo of 0.2 in the absence of snow [33]. For highly reflective foregrounds (e.g. snow, white exterior walls and rooftop coatings, glazed facades at the opposite buildings), the energy gains related to the Albedo effect can be substantial [62]. These must be carefully assessed and special attention must be paid to evaluate the conditions allowing the solar energy convertors to be covered with more than a very thin layer of ice blocking the solar irradiance [6].

The broadband solar spectrum (300–4000 nm) is beneficial for solar thermal collectors to produce thermal energy, while for photovoltaic modules the Albedo evaluation over a narrowband section must be considered, according to the spectral response of the photovoltaic module.

Recently, bifacial photovoltaic modules have been introduced to increase the specific output of the occupied surface; their backside is solely exposed to the solar radiation reflected by the surfaces behind them (e.g. snow-covered surfaces, white-painted rooftop or vertical walls, other photovoltaic modules). A particular case is represented by highly reflective coatings implemented on the buildings' envelope (facades and rooftops) aiming at decreasing the buildings' cooling demand and at adjusting the level of indoor natural light. However, this artificially generated Albedo, negatively affects the surrounding buildings, both visually and in terms of increased cooling energy demand.

In a recent literature review, Gueymard et al. [63] showed that a sizeable number of Albedo data sources exist, widely differing in terms of spatio-temporal and spectral resolution, method of derivation, geographical coverage and completeness. A 5 km resolution database was found ideal to evaluate the Albedo effect impacting the global horizontal solar irradiance; in the assessment of the global solar radiation received on tilted surfaces, a much higher (approx. 10 m) spatial resolution is required. Such detailed data can be obtained from space-borne measurements, at resolution ranging between 10 and 60 m.

2.1.2.4 The Heat Island Effect

The heat island effect represents a climatic change phenomenon that occurs in the built environment. It consists of a higher growth in the outdoor air temperature in the built areas than in the unbuilt ones. Thus, this effect is defined as the air temperature difference measured by urban and rural meteorological stations closely positioned. In hot climates, this temperature difference can be as high as 12 °C [3], and it is influenced by the climate and the specific features of the built environment. Several studies were developed to assess these temperature differences all over the world as presented in Table 2.6. The heat island effect is the result of the anthropic activities that modify the ground surface and atmospheric properties in the built environment and is considered as a good example of unsustainable climate change [114].

For solar energy convertors, the heat island effect increases the efficiency of the solar thermal collectors (the temperature difference between the solar thermal collector and the outdoor air is reduced, and therefore, the convection losses are decreasing). However, this effect reduces the efficiency of the photovoltaic modules that are negatively influenced by the rise of the outdoor temperature (the cooling effect of the outdoor air on the photovoltaic modules is reduced due to the lower temperature difference).

Table 2.6 Air temperature differences between urban and rural weather stations

References	Location	Latitude (°)	Temperature difference (°C)
Soltani and Sharifi [146]	Adelaide, Australia	34.9S	5.9
Alves [3]	Ceres, Brazil	15.3S	12
Ahmed et al. [2]	Putrajaya, Malaysia	2.9N	2
Wang et al. [156]	Shenzhen, China	22.5N	4
Lazzarini et al. [89]	Abu Dhabi, UAE	24.5N	6
Borbora and Das [18]	Guwahati, India	26.1N	2
Livada et al. [91]	Athens, Greece	38.0N	10
Ge et al. [51]	Beijing, China	39.9N	9.27
Marando et al. [95]	Rome, Italy	41.9N	3.17
Busato et al. [21]	Padua, Italy	45.4N	0.5
Lac et al. [87]	Paris, France	48.9N	7
Klok et al. [86]	Rotterdam, Netherlands	51.9N	1.3
Azevedeo et al. [11]	Birmingham, UK	52.5N	5
Lokoshchenko [92]	Moscow, Russia	55.7N	1.8

For buildings, this phenomenon has a positive effect during the cold season by decreasing the heating energy demand and a negative effect during summertime when the cooling energy demand is increased [65, 161].

The positive effect that occurs due to the increased outdoor air temperature is supported by the reciprocal shelter effect of neighbouring buildings, both decreasing the heating degree days thus the thermal energy demand during the cold season.

To mitigate the negative heat island effects, many studies focused on defining correlations between the increase in the outdoor air temperature and the temperatures of different built environment elements exposed to the solar radiation. Temperatures exceeding 50 °C for wall surfaces and between 20 and 55 °C for street surfaces were reported by Andreou and Axarli [7] in Tinos, Greece.

However, for an optimized built environment design, Ahmed et al. [2] reported lower temperatures of only 38 °C for the wall surface of buildings and 40 °C for street surfaces in Putrajaya, Malaysia. The heat migrates from the surface of the building envelope towards the colder interior and accumulates into the thermal mass of the building. Due to the thermal inertia of the building materials, this heat is felt indoors during the evening and the night. Thus, higher loads (for cooling) occur during these periods when electricity cannot be supplied by solar energy conversion systems. This increased summertime peak energy demand generates, besides higher air conditioning costs, air pollution and greenhouse gas emissions.

Many studies were reviewed by Santamouris et al. [134] focusing on heat island effect mitigation strategies, showing that highly reflective/emissive, light/cool coloured surfaces, and phase-change materials as well as the green coverage used for buildings roofs and/or facades can significantly contribute to the improvement of the built environment quality.

Another solution to reduce the temperature of the building elements consists of integrating solar thermal collectors in the facades or roofs, to remove the thermal energy and to use it e.g. for domestic hot water preparation. With this aim, Fekete and Farkas [46] proposed solar tile collectors, while O'Hegarty et al. [111] developed concrete solar collectors for façade integration and, further on, road pavement solar collector systems were assessed by Nasir et al. [102]. The use of photovoltaic modules can also reduce the heat island effect, but in a less extent compared to the solar thermal collectors, because a reduced part of the solar radiation is converted into electrical power.

2.1.2.5 The Street Canyon Effect

A street canyon is created when tall buildings are built on both sides of the street, resulting a canyon-like appearance [109]. To classify the street canyons, two aspect ratios are used [154] as further detailed.

Depending on the ratio between the length (L) and the width (W) of the street, the street canyons are grouped in three categories: long, medium and short canyons for length-to-height ratio higher than 7, between 7 and 3, and lower than 3, respectively (Fig. 2.34a).

Depending on the ratio between the height of the building (H) and the width (W) of the street, an avenue canyon is defined by an aspect ratio less than 0.5, a regular street canyon has an aspect ratio of 0.5–2, and a deep street canyon has an aspect ratio higher than 2 (Fig. 2.34b).

The street canyon effect has a major influence on the solar energy availability in the built environment, either for passive or for active use [57]. The shadow casted by the buildings (on the street and on the opposite buildings' façades and rooftops) and thus the available solar energy on these elements are highly influenced by four factors: the orientation of the street canyon, the height of the buildings, the day in the year and the hour in the day [100]. For East–West-oriented street canyons, at mid-latitude (45.6°N) and with buildings of equal heights (Fig. 2.35) the surface of the southern façade of the northern building is entirely exposed to solar irradiance during the summer solstice (SS), at noon, for an aspect ratio $H/W < 2.5$. During the winter solstice (WS), at noon, the surface of the southern façade of the northern building is partly exposed to the solar irradiance and the shadowed surface increases with the increase in the buildings' height.

To further evaluate the daily influence of the street canyon effect on the available solar energy on the southern facades of the northern buildings, the shading profile of the southern building must be evaluated and plotted as presented in Fig. 2.36. Thus, for an avenue canyon (Fig. 2.36a) the lower part of the facade is almost fully shadowed only between 15 November and 15 January, for a regular street canyon (Fig. 2.36b) between the 15 October and 15 March and for a deep street canyon

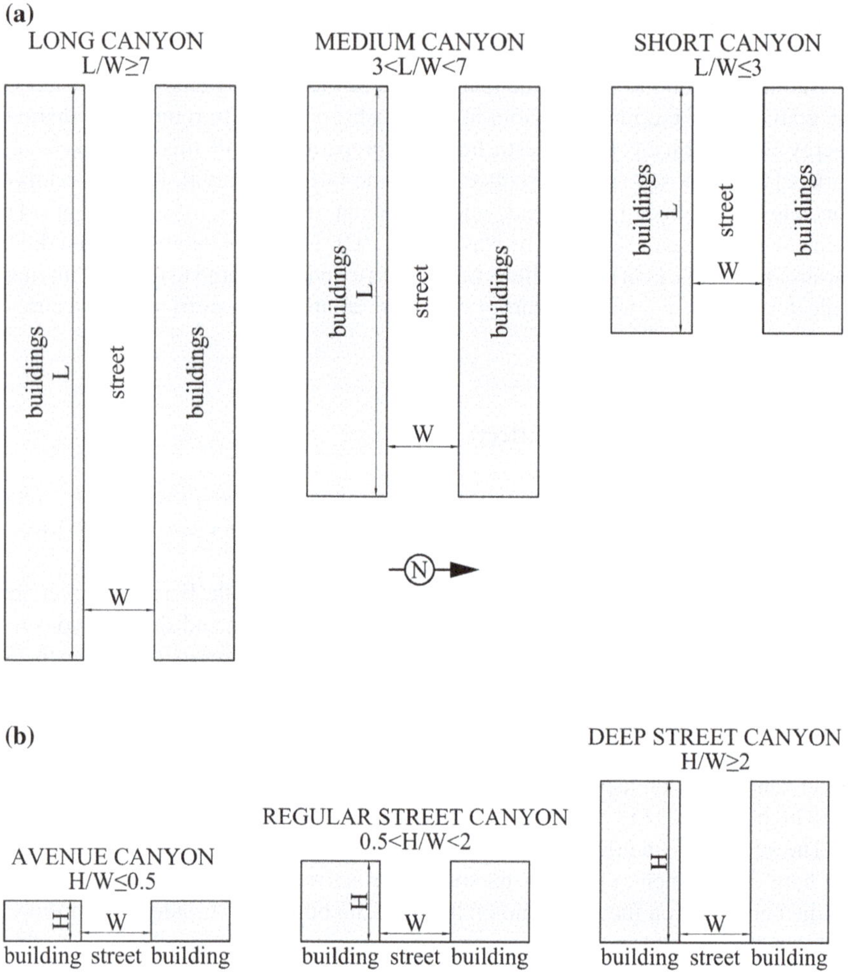

Fig. 2.34 Street canyon geometry with different: **a** *L*/*W* ratios; **b** *H*/*W* ratios

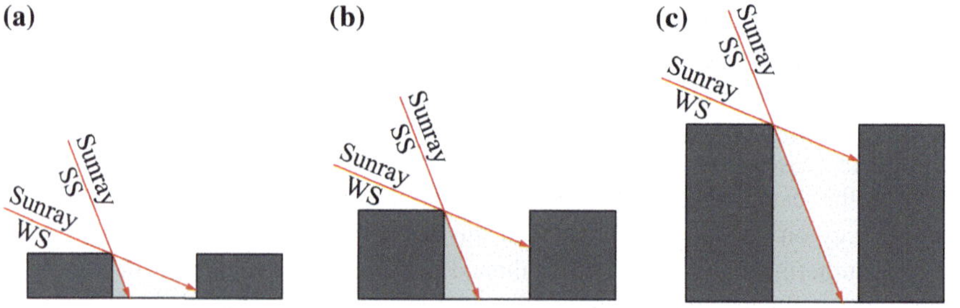

Fig. 2.35 Shadow casted by the southern building, in street canyons with buildings of equal height, at summer (SS) and winter (WS) solstice, at noon for: **a** an avenue canyon; **b** a regular street canyon; **c** a deep street canyon

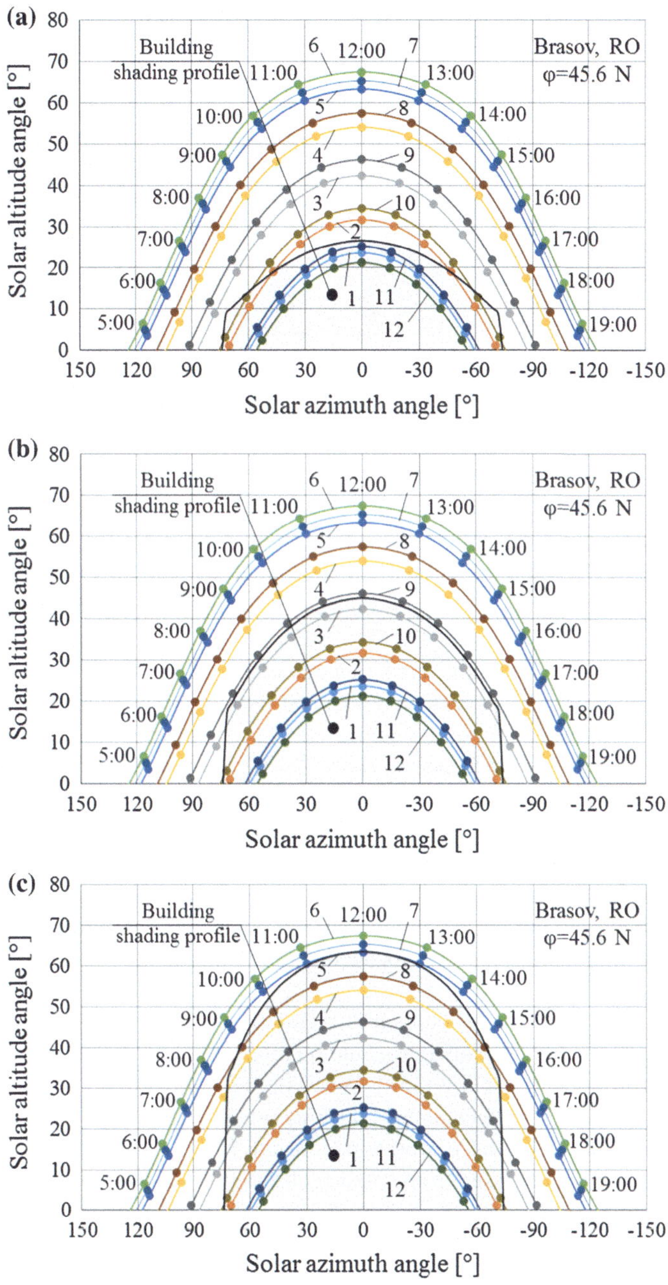

Fig. 2.36 Southern buildings monthly shading profile from January (1) to December (12) for: **a** an avenue canyon, $W/L = 0.5$; **b** a regular street canyon, $W/L = 1$; **c** a deep street canyon, $W/L = 2$

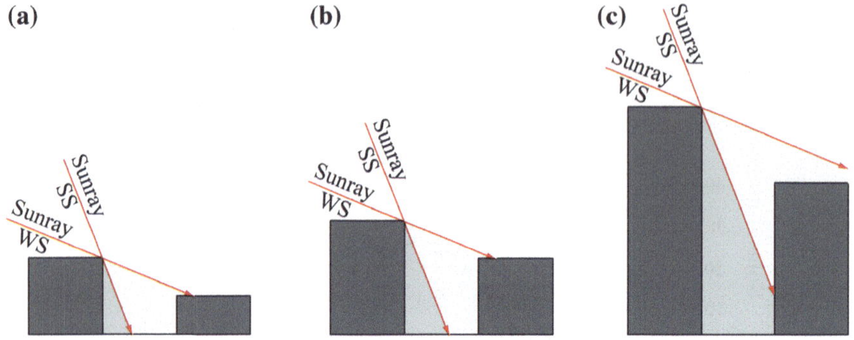

Fig. 2.37 Shadow casted by the southern buildings, in street canyons with buildings of uneven heights, at summer (SS) and winter (WS) solstice noon, for: **a** an avenue canyon; **b** a regular street canyon; **c** a deep street canyon

(Fig. 2.36c) between the 15 August and 15 April. The procedure must be repeated for the entire surface of the facades to fully assess the available solar energy on the buildings' vertical surfaces. Combining this analysis with the climatic profile in the implementation location, the optimum height/width ratio can be calculated to maximize the area of the building facades exposed to the solar radiation and thus the energy output of facade integrated solar energy convertors.

For East–West-oriented street canyons, at mid-latitude (45.6°N) with higher buildings located on the southern side, shading is more extensive as Fig. 2.37 shows. In this case, the entire southern façade and partially the rooftop of the northern building on the avenue and regular street canyons are shaded (Fig. 2.37a, b) at the winter solstice.

When the difference between the building heights increases, the southern façade is (partially) shadowed even during the summer solstice, at noon while the rooftop can be entirely shadowed during the winter solstice at noon, as the sketch in Fig. 2.37c outlines. It is thus recommended to locate the highest buildings on the northern side of the canyon.

Another effect of the street canyon is related to the air movement through the canyon that affects the air temperature and quality [151] and the buildings' energy demand for heating and cooling [153]. When the wind blows parallel to the street canyon, a channelization effect appears flushing out pollutants (e.g. exhausted gases from vehicles) but causing discomfort to pedestrians [113]. If the air pollution source is above (e.g. exhausted gases from fossil fuel-based boilers) or if the wind blows perpendicularly to the canyon, the pollutants are confined into the canyon worsening the air quality [160].

The wind speed is increased by the street canyon effect and will increase the thermal losses of the solar energy convertors integrated with the adjacent facades. This is beneficial for photovoltaic modules (the temperature decrease increases the electrical energy output of the photovoltaic modules), but for the solar thermal collectors this has a negative effect (the temperature difference between the solar thermal collector, and the outdoor air is higher, and thus, the conversion efficiency decreases due to these thermal losses).

2.2 Other Renewable Energy Sources in the Built Environment

Relying only on solar radiation to cover the energy demand in the built environment does not represent a feasible alternative in most of the locations on the Earth due to the variability of this energy source. Thus, other renewable energy sources have to be additionally considered, such as the geothermal and the biomass energy to meet the thermal energy demand or the wind and/or the water (hydro) energy to cope with the electrical energy demand.

2.2.1 Geothermal Energy

The geothermal energy represents the energy contained in the Earth's crust. Its main source is the huge temperature of the Earth's core estimated at 6500 K. The Sun is the other source that heats the Earth's crust, but its effect is significant only at a shallow depth [54].

The geothermal potential is influenced by the thickness of the crust and by its exposure to the solar radiation. The Earth's crust accounts for less than 1% of the Earth's volume; thus, it can be considered as a thin shell floating on the Earth's mantle due to its lower density than that of the mantle (Fig. 2.38). The crust is of two types: oceanic, that is denser and has a variable thickness between 5 and 10 km, and continental, less dense and 30–50 km thick. The crust is broken in tectonic plates,

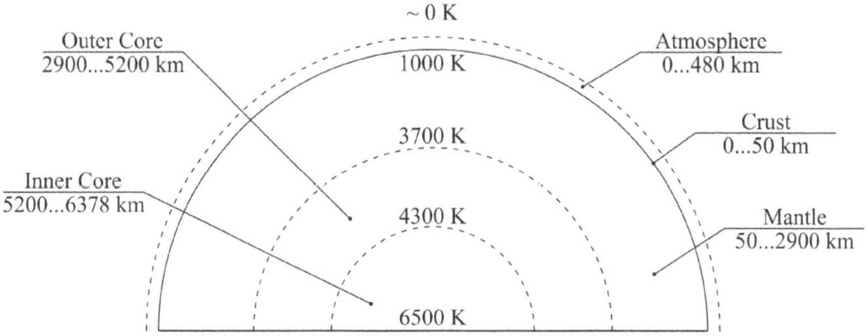

Fig. 2.38 Temperature distribution in the Earth's layers

their boundaries allowing the heat in the mantle to easily reach the Earth surface. In these areas, the geothermal gradient is maximum (up to 100 °C/km) and decreases to an average value of 30 °C/km in the thicker part of the plates [49].

The exposure to solar radiation is site dependent, with the maximum amount of solar energy received between the Earth's Tropics (yearly amount higher than 2500 kWh/m^2) and minimum at the Earth's Poles (yearly amount less than 700 kWh/m^2).

The geothermal potential can be assessed using the ground temperature. At the surface, the ground temperature is strongly influenced by the outdoor parameters: during summer, the ground surface temperature increases because of the received solar energy, acting as a heat storage medium. The high thermal inertia of the ground keeps the ground temperature at values higher than those corresponding to the air during winter; starting with a depth of 15 m below the surface, the ground temperature stabilizes at about 10 °C regardless of the season (Fig. 2.39a).

Once stabilized and not further influenced by the weather conditions, the ground temperature slowly increases with a gradient of 30 °C/km in the Earth's crust having on average a thickness of 30 km. Further on, the temperature gradient reduces to 1 °C/km in the Earth's mantle and to 0.26 °C/km in the outer core and slightly increases to 2 °C/km in the inner core (Fig. 2.39b). Thus, 99% of the Earth's volume has a temperature higher than 1000 °C being an immense source of thermal energy.

The geothermal energy has obvious advantages:

- It is a renewable energy as the heat is continuously recharged from the thermal energy of the Earth's core and seasonally from the received solar energy;
- It supports very low greenhouse gas emissions, the ones extracted from the ground along with the geothermal water that, in the case of closed-loop geothermal power plants, are reintroduced in the geothermal reservoirs;
- It is not significantly influenced by the day/night or seasonal climatic variations and can be considered as a base source of energy, available 24/7;
- It can be used for both electrical and thermal energy supplies;
- For electrical energy production, there is no need for the additional use of fossil fuels;
- It uses less land (404 m^2/GWh) than other renewable-based energy systems, e.g. 3237 m^2/GWh for a photovoltaic park, or fossil fuel-based power plants, e.g. 3642 m^2/GWh for a coal plant [52].

However, the geothermal energy also has several disadvantages:

- Deep geothermal energy can only be feasibly used in regions with hot rocks below the surface that produce steam over a long period of time;
- Initial geological studies and high associated costs are required;
- High initial equipment costs are involved;
- Drilling the boreholes can generate ground instability;
- The extracted thermal fluid must be re-injected in the ground causing operation and maintenance-related costs;

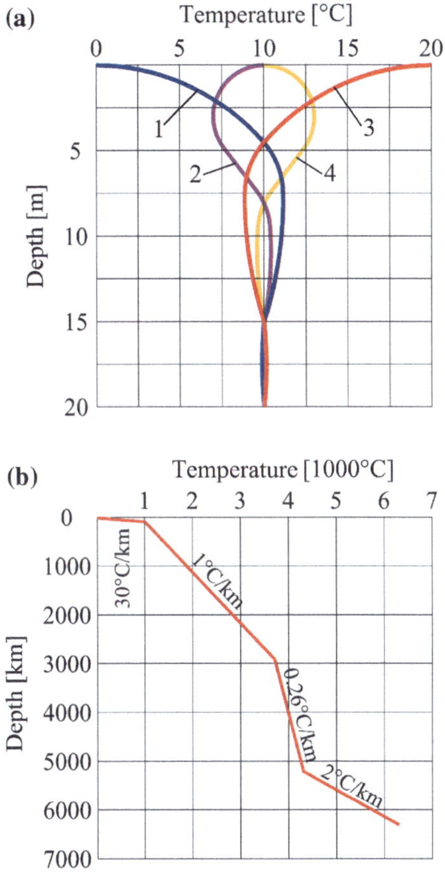

Fig. 2.39 Ground temperature variation: **a** over the year in the upper part of the Earth's crust in February (1), May (2), August (3) and November (4); **b** in the depth of the Earth structure

- Harmful gases and vapours (CO_2, CH_4, NH_3, H_2S, C_6H_6) and toxic elements (mercury, arsenic, boron and antimony) can affect the environment when geothermal water is extracted from the ground;
- Low extraction rates of the shallow geothermal energy sources can be considered, between 10 and 35 W/m^2 for horizontal geothermal heat exchangers, and from 30 to 100 W/m for vertical geothermal heat exchangers;
- The heat pumps used in the exploitation of the shallow geothermal sources need electrical energy to be driven;
- Undersized ground heat exchangers accelerate ground freezing reducing the heat pump efficiency, thus using more electricity;
- The difficulties in the thermal energy transport have to be considered.

Two categories of geothermal energy sources can be used for meeting the energy demand in the built environment:

- *High enthalpy sources* (temperature higher than 150 °C);
- *Low enthalpy sources* (temperature lower than 150 °C).

The high enthalpy geothermal energy sources provide a mixture of steam and hot water and are used for power generation through a turbine/generator coupled system while the thermal energy left in the condensate exiting the turbine can be further used for space heating and domestic hot water preparation in a district heating system. According to the International Energy Agency [74], the geothermal electrical energy production in 2017 was of 51.8 TWh that, compared to 28.6 TWh produced in 1990, represents an average annual growth rate of 2.2%. The first producer are the USA (34.9%) followed by New Zealand (14.4%), Italy (11.9%), Turkey (11.8%) and Mexico (11.4%). A total amount of 14,060 MW installed geothermal power capacity was reported as for January 2018, with four countries having an installed capacity over 1 GW and other three approaching this threshold value, as presented in Fig. 2.40.

Low enthalpy geothermal energy sources are used to cover the thermal energy demand in the built environment. According to Lund and Boyd [93], a total amount of 163,071 GWh of thermal energy was worldwide produced in 2014 using geothermal sources, based on a 70,329 MW total installed capacity, 38.7% higher than in 2010, resulting in an average annual growth of 7.7%. The frontrunner is China with an installed capacity of 17,870 MW (25.4% of the total installed capacity), closely followed by the USA with 17,416 MW (24.76%) and at a considerable distance by Sweden with 5600 MW (7.96%). Thirteen countries overpassed the 1 GW threshold at the end of 2014, and another five were close to it, as the data in Fig. 2.41 show.

Among the systems worldwide installed, almost 20,000 MW are directly using the geothermal energy sources with temperatures ranging between 30 and 100 °C (e.g. for bathing and swimming, space heating, greenhouse heating, aquaculture,

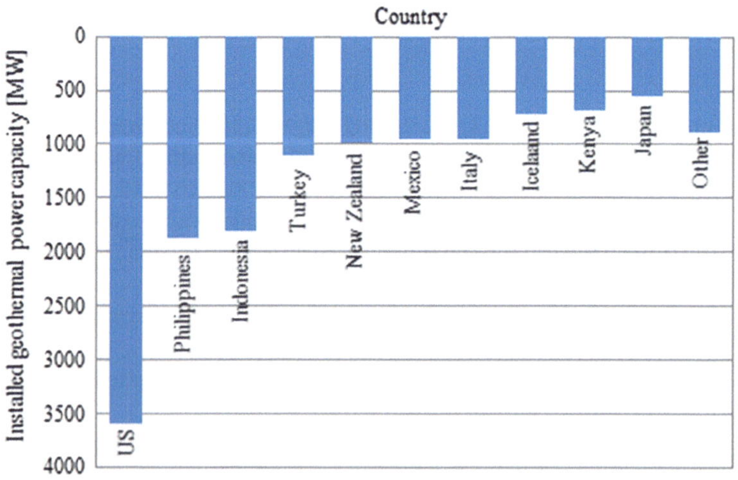

Fig. 2.40 Worldwide geothermal installed power capacity

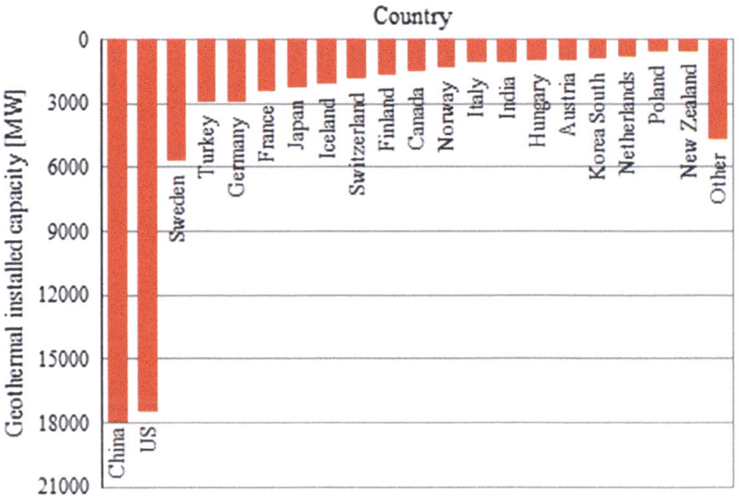

Fig. 2.41 Worldwide installed geothermal systems for thermal energy production

industrial use, agricultural drying processes). Moreover, systems amounting almost 50,000 MW indirectly use the low-temperature geothermal energy sources (T < 10 °C) for space heating and domestic hot water preparation based on geothermal heat pumps.

Ground-coupled heat pump systems are good candidates for the built environment as new or existing buildings can be equipped with such systems relying on the low enthalpy geothermal energy as a base source.

2.2.2 *Bioenergy*

Bioenergy is a form of renewable energy derived from biologically produced matter that originates from living or recently living organisms (plants or animals) that can be considered a form of solar energy storage. Currently, bioenergy is the mostly used renewable energy in the world [76] being employed for space heating and domestic hot water preparation, combined heat and power generation, vehicle's engines, etc. Its main advantage is that it is a base-load resource that is not (directly) influenced by the meteorological variability. Once stored as solid, liquid or gaseous fuels, biomass can be converted into thermal and/or electrical energy whenever needed, regardless of the time interval or season. Bioenergy plants are similarly operated to the fossil fuels ones and can be considered a solution in balancing the grid, thus allowing a more extensive solar and wind power integration [72].

The bioenergy use can be divided into two main categories: traditional and modern bioenergy. Traditional bioenergy refers to the combustions of biological materials such as wood, wood waste, straw and other crop residues, to cover the thermal energy demand for space heating and domestic hot water preparation. The main properties of the traditional biomass fuels are presented in Table 2.7 using coal as a reference.

Biomass is currently the primary energy source for about 2.5 billion people in developing countries, estimated at 28 EJ in 2015 [73]. The systems used in this case consist of basic stoves, usually of small capacity for residential houses, with rather reduced combustion efficiency (5–15%), harmful emissions of high particulate matter and issues related to the periodical feeding and maintenance.

Modern bioenergy conversion systems provided 12.9 EJ of the thermal energy in 2015 for industrial processes (63%), buildings (34%) and agriculture (3%) [73]. Modern biomass boilers, with highly efficient combustion burners fitted with control equipment, are currently used. Modern bioenergy systems include the use of *liquid biofuels*, *biogas* or *pellets*.

Liquid biofuels can be divided in three main categories related to the replaced fossil liquid fuels: *biodiesel*, *bioethanol* and *biokerosene*.

The biodiesel fuel is mainly obtained through transesterification from oil crops (e.g. rapeseed, soybeans, palm), animal fat or waste cooking oil. Its pure form, denoted B100, may only be used in modified diesel engines, while blends of 20% or lower concentration of biodiesel with petroleum diesel (B20) can be used in diesel engines with no or only minor modifications. Biodiesel is also used in heating systems (in domestic and industrial boilers) and in power generators as base-load or as backup source.

The bioethanol fuel is mainly produced using starch/sugar/lignocellulosic crops (e.g. corn, maize, wheat, barley, rye, sugar bet, sugar cane, willow, poplar, miscanthus, sorghum). The process is based on the sugar fermentation followed by distillation. Blends of 10% bioethanol and 90% gasoline (E10G) can be used in spark engines without any modifications and without affecting their warranties, while higher bioethanol concentration up to 85% (E85G) can be used only in specially designed engines. Bioethanol can be also used as fuel in cogeneration systems.

The biokerosene fuel is mainly obtained from oil-bearing crops (camelina, jatropha, etc.) but also from waste cooking oil or by chemically combining carbon extracted from CO_2 (captured from the atmosphere) and hydrogen extracted from water (by electrolysis with green electricity). In the European Union, the emissions

Table 2.7 Main properties of traditional biomass sources (dry state) [38]

Traditional biomass sources	Density (kg/m^3)	Higher heating value (MJ/kg)	Higher heating value (MJ/m^3)
Spruce	468	15.03	7035
Douglas fir	510	22.85	11,654
Birch	695	14.99	10,418
Beech	708	14.97	10,595
Oak	708	14.97	10,595
Locust	730	16.51	12,052
Coal	900	27.60	24,840

from civil aviation account for about 3% of the total greenhouse gas emissions. Since 2011, the European Commission in partnership with the European Airlines and biofuel producers (Choren Industries, Neste Oils, Biomass Technology Group and UOP) launched the *European Advanced Biofuels Flightpath*. Several companies in the aviation sector already use biokerosene blends (KLM, Lufthansa, SAS, Continental Airlines, etc.).

The heating values of these liquid biofuels are included in Table 2.8 and are comparatively presented with the values corresponding to the fossil liquid fuels and also inserted in Table 2.8.

The biogas consists of a mixture of gases (mainly methane and carbon dioxide with small amounts of hydrogen sulphide) and is used in the built environment as fuel in boilers for heating or in gas engines for combined heat and power generation. Biogas can also be compressed and used in the transportation sector. The main properties, at 0 °C and 1 atm, of the biogas, natural gas and hydrogen are comparatively presented in Table 2.9.

The biogas is produced using raw organic materials such as agricultural, forest, food or municipal wastes and manure, sewage sludge or wastewater cofermented in anaerobic digesters, biodigesters or bioreactors. The biogas production must be a mandatory process in activities generating large amounts of organic waste, since methane is a greenhouse gas with a global warming potential that is 86 times higher than that of carbon dioxide over a period of 20 years [71].

Pellets are obtained from compressed organic matter: wood waste, agricultural waste, industrial waste, specially grown plants such as willow and switchgrass. The

Table 2.8 Properties of liquid bio- and fossil fuels

Liquid fuel	Density (kg/m^3)	Higher heating value (MJ/kg)	Higher heating value (MJ/m^3)
Biodiesel B100[a]	880	40.24	35,409
Diesel[a]	850	45.13	38,363
Bioethanol[b]	789	29.7	23,433
Gasoline[b]	737	46.4	34,197
Biokerosene[b]	888	40.2	35,698
Kerosene[b]	821	46.2	37,930

[a]National Renewable Energy Laboratory [105]
[b]Engineering ToolBox [39]

Table 2.9 Properties of the gaseous bio- and fossil fuels [40]

Gaseous fuel	Density (kg/m^3)	Higher heating value (MJ/kg)	Higher heating value (MJ/m^3)
Biogas	1.02	26.32	25.74
Natural gas	0.85	45.35	35.44
Hydrogen	0.09	142.06	12.12

Table 2.10 Properties of the main types of pellets

Pellets types	Density (kg/m^3)	Heating value (MJ/kg)	Heating value (MJ/m^3)
Canola meal[a]	1079	23.80	25,680
Canola meal and hulls[a]	986	23.70	23,368
Sunflower seeds[b]	1105	19.54	21,592
Wood sawdust[b]	1120	18.43	20,642
Alfalfa[c]	1198	18.41	22,055
Miscanthus[d]	578	17.5	10,115
Bioethanol production residue[d]	714	17.2	12,281
Wheat straw[c]	1015	16.70	16,951
Coffee husk[e]	245	15.00	3675
Sugarcane bagasse[e]	90	11.81	1063
Rice husk[e]	160	10.72	1715

[a] Azargohar et al. [10]
[b] Simone et al. [140]
[c] Sarker et al. [135]
[d] Kallis et al. [81]
[e] Marugo et al. [96]

wood pellets generally obtained using compacted sawdust and forestry by-product (branches, leaves, cones, etc.) represent the most common type of pelletized biomass. Their main advantage resides in the low moisture content and high heating value, thus allowing a high combustion efficiency. Another advantage is represented by the pellets' small size and regular geometry allowing the automatic feed of the burners using pneumatic or mechanic conveyors, allowing implementing these burners in district heating systems. The high-density pellets require less storage volume and increase the transport efficiency. The properties of the main types of pellets are included in Table 2.10.

All these bioenergy sources are considered renewable because the carbon dioxide released in the atmosphere when these sources are converted to other forms of energy is actually absorbed from the atmosphere in the growth of the primary bioresource. The production and use of cycles can be designed on a repeating basis relying on the existent infrastructure for planting, growing, harvesting and processing the biomass as well as for the thermal and electrical energy distribution and storage. The use of bioresources supports the agricultural land preservation and reduces the negative impact on soil, water and air, the acid rain, the smog, etc.

The bioenergy technologies require public support and adequate decarbonization policies to compete with the fossil fuel plants. The state-of-the-art bioenergy technologies are required for a future with higher (and variable) shares of solar and wind energy, to actively balance the grid. Moreover, bioenergy must be considered an important part of the circular economy, strongly interacting with various sectors of activity (agriculture, forestry, transport, built environment, food, industry, etc.) with a specific focus on wastes, by-products and energy demand. Finally, bioenergy can contribute to increasing the number of jobs and the income in rural economies, while providing clean and affordable energy in these areas.

2.2.3 *Wind Energy*

2.2.3.1 Wind Parameters

The wind is one of the mostly used renewable energy sources. To check if a specific site is suitable for installing wind turbines, two aspects have to be analysed: the wind speed and its predominant direction.

The most important parameter is the *wind speed* that depends on weather, landscape, height from the ground and time. Since wind depends on the repeatable pattern of the solar radiation, the wind speed has also a yearly predictable variation. The wind speed is usually reported as average values (v_m) over a certain time interval that can range from several seconds to one year. Obviously, the smaller the time interval is, the better the energy output of a wind turbine can be predicted. Further on, annual averages have to consider at least 10 years [118].

The wind speed variation can be graphically presented in diagrams similar to the one in Fig. 2.42 that indicates the hourly, monthly and annual average values of the wind speed in Brasov, Romania, during 2017.

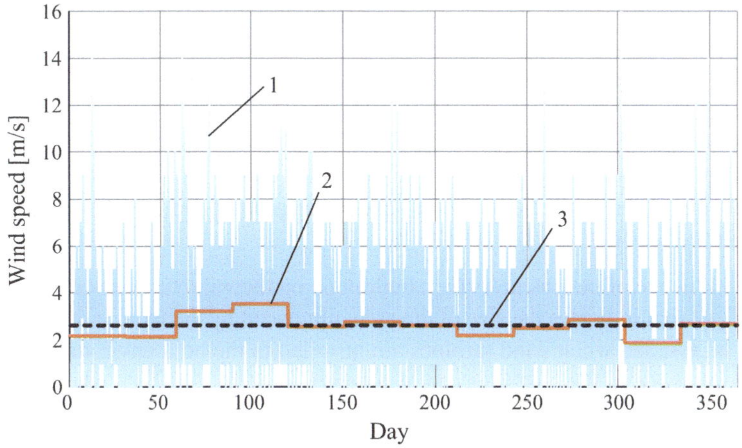

Fig. 2.42 Hourly (1), monthly (2) and yearly (3) average values of the wind speed during 2017, in Brasov, Romania

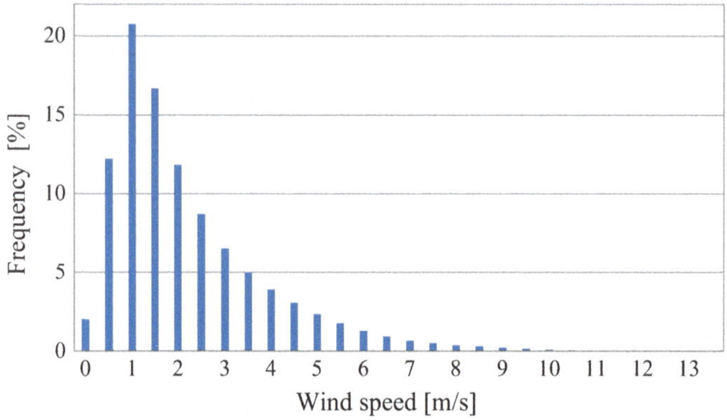

Fig. 2.43 Wind speed distribution during 2017, in Brasov, Romania

To estimate the wind potential in a location, the following parameters have to be considered:

- The *frequency distribution* of the wind speed measured in a given location. This is usually identified by separating the measured wind speeds into several smaller or larger classes/intervals and measuring their occurrence [66]. In this way, the professionals will have an overview of how often the wind is blowing with a specific speed and can decide whether a turbine type is suitable or not for the analysed location. The wind speed distribution can be represented in diagrams as the one in Fig. 2.43 that shows the results recorded in the outskirts of the Brasov city, Romania, using data from a local meteorological station installed in the R&D Institute of the Transilvania University of Brasov. The wind speed distribution can be approximated for high mean values using a Weibull or a Rayleigh curve that can be used to predict the available wind potential if only the average wind speed is known [53];
- *Wind turbulence* indicates instantaneous variations of the wind speed from the average value and depends on the topographic configuration of the terrain, the roughness of the surfaces, the height above the ground and the average available wind speed [20].

The influence of the obstacles (e.g. buildings) on the wind turbulences is outlined in Fig. 2.44 that indicates the size of the area influenced by the wind turbulences that should be avoided when wind turbines are installed in the built environment.

The influence of the obstacles can also be observed in Fig. 2.45, where the wind speed variation is presented considering two measurement points in the location of the R&D Institute from the Transilvania University of Brasov, Romania. The diagram indicates higher wind speeds near the institute's buildings (average 3.06 m/s) compared to the top of the buildings (average 2.24 m/s).

- *The wind shear* indicates how the wind speed increases with the distance from the ground (on height) and is useful for predicting high altitudes' wind speed values.

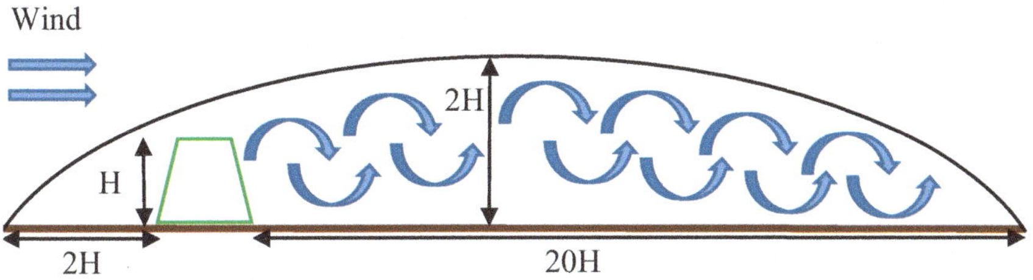

Fig. 2.44 Size of the area influenced by wind turbulences generated due to nearby obstacles

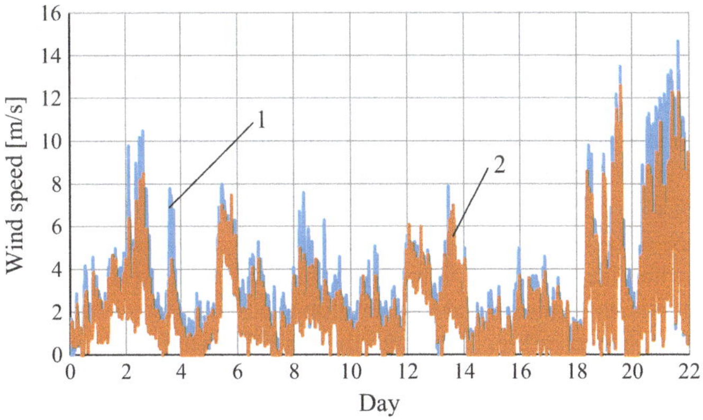

Fig. 2.45 Effect of wind turbulences during 13 January 2013–03 February 2013 in Brasov, Romania: in the field near the buildings (1) and on the rooftop of a building with $H = 13$ m (2)

This increase depends on the air temperature and humidity and can be estimated, according to Johnson [79] and Earth System Research Laboratory [35] using

$$v_H = v_{ref} \left(\frac{H}{H_{ref}} \right)^{\alpha} \tag{2.31}$$

where

- v_H is the estimated wind speed at the height H.
- v_{ref} is the measured wind speed at height H_{ref} (usually at 10 m).
- α is the Hellmann exponent (ground surface friction coefficient) with typical values given in Table 2.11;

- *The wind direction* is influenced by the general flow of the air currents; these currents may have a regular (daily or seasonal) variation according to the position of the analysed location.

Table 2.11 Usual values of the Hellman exponent, α [83]

Location	Hellman exponent
Unstable air above water	0.06
Neutral air above water	0.10
Stable air above open water surface	0.27
Unstable air above flat coast	0.11
Neutral air above flat coast	0.16
Stable air above flat coast	0.40
Unstable air above human areas	0.27
Neutral air above human areas	0.34
Stable air above human areas	0.60

Wind rose diagrams represent the usual instrument for indicating the prevailing wind direction in a location and can also include information about the wind speed. The diagram is divided into an even number of sectors each of them representing a different N–S–E–W direction [136] and indicates, for each direction, the time percentage when the wind has blown from that direction. At the city outskirts of Brasov, Romania, the typical wind rose developed based on the in-field data recorded in 2017 is represented in Fig. 2.46, outlining that the wind blows mostly from the NNW direction.

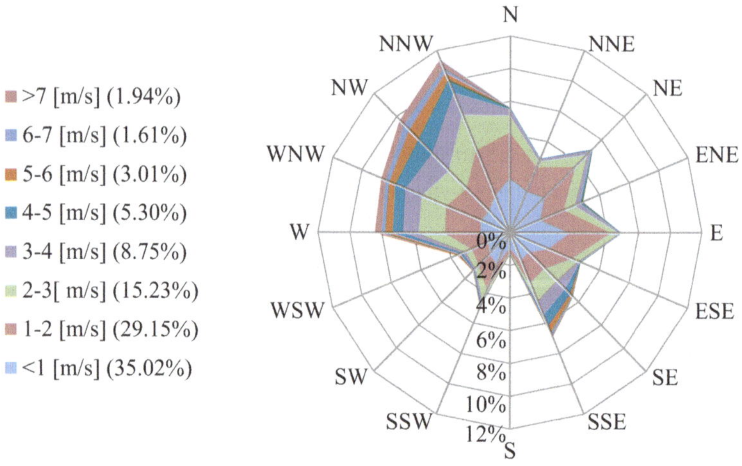

Fig. 2.46 Wind rose in the city outskirts of Brasov, Romania, during 2017

2.2.3.2 Wind Measurement Instruments

The instruments used for measuring the wind speed are called *anemometers* and can be divided, according to the *American Society of Mechanical Engineers* [5], based on their operation principle in three categories: *sensors with kinetic energy transfer* (cups, propellers, pressure plates, etc.), *sensors with heat transfer* (thin wires or films) or *sensors with Doppler effect* (acoustic or laser). The most commonly used sensors are with cups, propellers or ultrasonic with Doppler effect.

Cup and propeller anemometers (Fig. 2.47) use the rotation of a wind-driven axle to generate a signal proportional to the wind speed; the number of rotations can be measured using photoelectric switches, mechanical counters or the changes of an electrical parameter. The difference between the cup and the propeller anemometers is the angle between the wind direction and the anemometer axis. Cup-type anemometers are used for horizontal wind speeds while propeller-type anemometers are used for measuring the vertical wind speed component and for estimating wind turbulences.

Ultrasonic anemometers are based on the Doppler effect (Fig. 2.48) that involves the change in the frequency of an ultrasonic signal with the wind speed along the distance from the signal source to the receiver. This type of sensor has the ability to measure multidirectional wind speed with a frequency of up to 20 Hz, which makes it ideal for measuring the wind turbulences and for accurately estimating the wind direction.

If local measurements are not available in a specific location, the wind energy potential can be estimated using wind maps (Fig. 2.49) and various interactive online tools as those offered by Global Wind Atlas (https://globalwindatlas.info), Windy (https://www.windy.com) or https://earth.nullschool.net. Other solutions involve mathematical models or wind measurements from nearby airports, universities, etc., that need to be adjusted to the local topography, buildings and vegetation.

(a) (b)

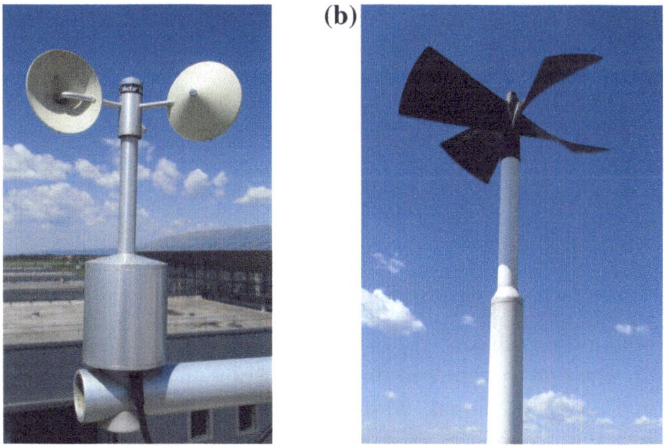

Fig. 2.47 Anemometers: **a** cup type; **b** propeller type

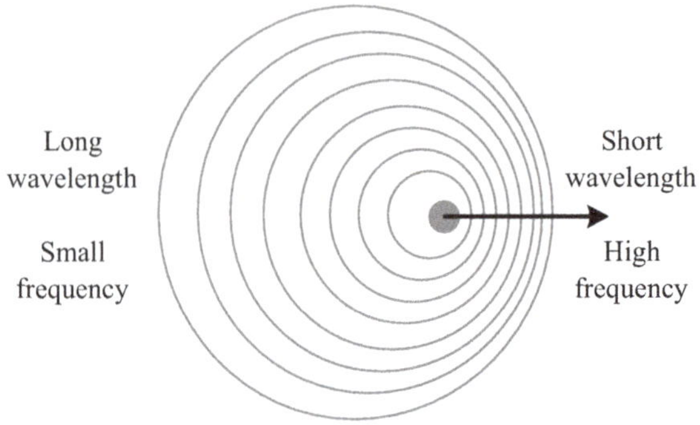

Fig. 2.48 Doppler effect principle that is exploited in ultrasonic anemometers

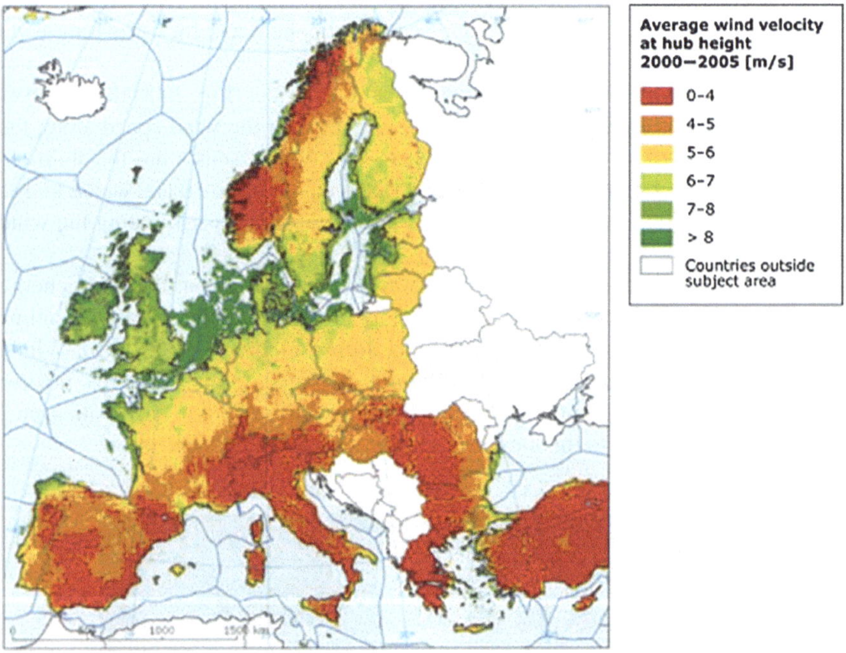

Source: EEA, 2008.

Fig. 2.49 Wind map of Europe [45]

The instruments used for measuring the wind direction are called *wind vanes*; these usually contain a rod with a vertical surface at one end and with a counterweight at the other end. The gravity centre of this rod is sitting on a vertical axle that generates a proportional electrical signal, while it rotates according to the wind direction. Another instrument for measuring the wind direction is the ultrasonic anemometer.

2.2.4 Hydro-energy

Water flows (e.g. rivers) represent the main renewable energy source globally used to produce electrical energy. In 2016, 71% of the renewable electricity was produced in hydroplants, with an installed hydroelectric capacity power of 1064 GW. This accounts for 16.4% of the worldwide electrical energy demand, generated from all power sources [157].

The energy potential of a hydrographic basin can be defined by the power or the energy that can be generated by a hydroelectric power plant using the water in that basin and the level difference that is specific for the local topography of the terrain.

Unlike solar and wind, the hydroelectric potential cannot be estimated based on maps; hydrologic maps may be used to identify the rivers with the required flows, suitable for certain hydroturbine types. However, to establish the hydroelectric potential in a given location, the following aspects need to be thoroughly investigated and the results can be used for a preliminary analysis of the opportunity of a large investment, usually required for the construction of the hydroelectric power plants [47]:

- The rainfall amount in the area where the installation of the hydroelectric power plant is targeted. The rainfall history for the last years can be usually obtained from local meteorological agencies or downloaded for a multitude of locations from http://rp5.md;
- The monthly and yearly values of the minimum and maximum flows of the river (or rivers) that will supply water to the hydroelectric power plant. These parameters can be obtained from local monitoring agencies of watercourses. It is to notice that the values of these parameters measured during one single year are not enough to justify the funds required for changing the local topography and the traditional use of large land areas. Also, the available rainfall required for the safe operation has to be predicted for the entire lifetime of the plant; this is usually done by flow prediction techniques based on ARIMA models and neural networks or genetic programming using the ECSID software [121];
- The topographical features of the hydrographic basin provide the water in the river(s) supplying the hydroelectric power plant, because its surface and slope can generate different friction levels in the water supply ducts;
- The geological structure of the soil and the existing vegetation is essential for the precipitations' retention and for the gradual release of water into the local hydrographic basin.

The rainfall amount is currently measured using pluviometers (Fig. 2.50) with two self-draining cups (2) connected to the data loggers of an automatic digital weather station. The collecting bowl (1) has usually a 203 mm diameter, and each tipping of the cups is equivalent to 0.2 mm of precipitation; thus, the rainfall amount can be easily calculated by counting the tipping with an electronic counter that sends the information through the cable (3) to a data logger.

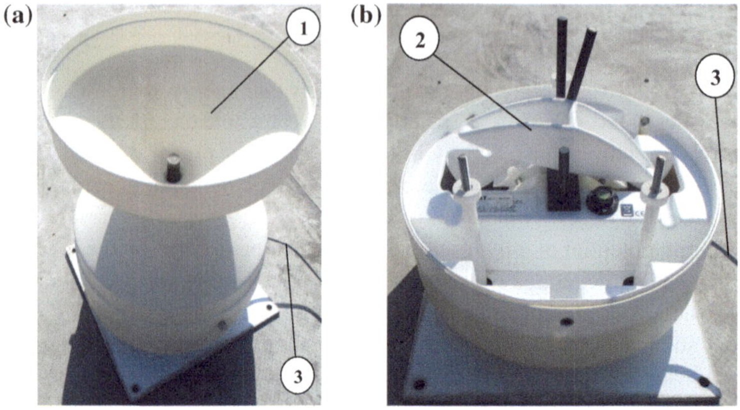

Fig. 2.50 **a** Delta-*T* self-draining pluviometer: collecting bowl (1), cable connection (3); **b** interior details of the pluviometer: self-draining cups (2), cable connection (3)

The variation in the rainfall amount in Brasov, Romania, during 2018 is presented in Fig. 2.51 and is typical for mountain locations in the temperate climate (http://rp5.md). The values in Fig. 2.51 during the winter months represent the total amount of rainfall and melted snow.

Two options are available for measuring the rivers' water flow rate (m^3/s): on small rivers, this can be directly measured by counting the time required to fill a vessel with a known volume; large rivers require an indirect method based on the product between the river cross section (Fig. 2.52a) and the average flow rate.

When using the indirect method for measuring the water flow rate of a river, the following aspects have to be considered [48]:

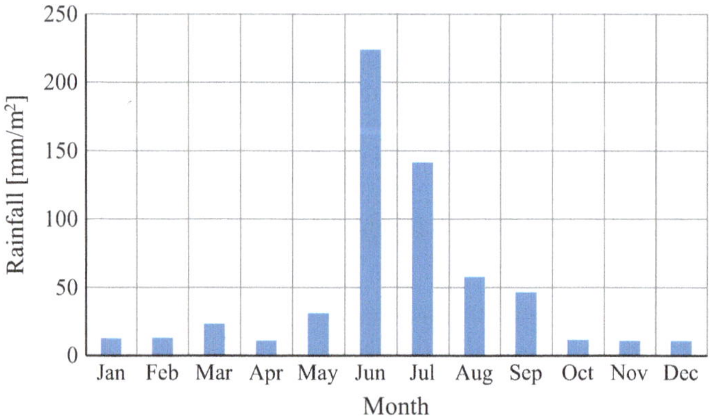

Fig. 2.51 Monthly rainfall in Brasov, Romania, during 2018 (total annual rainfall 597 mm/m^2)

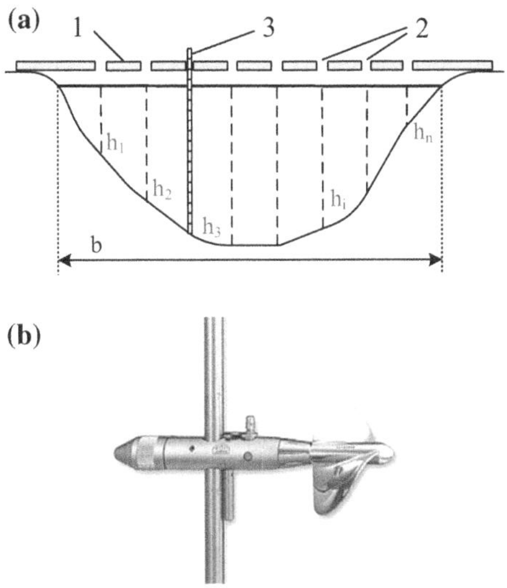

Fig. 2.52 **a** Measurement principle of the cross section of a river using a horizontal plate (1) with equidistant measuring points (2) and a measurement line (3); **b** OTT C31—Universal current meter for flow velocity measurement in rivers and open waterways [68]

- The measuring location has to be on a straight portion where the river is clean, has uniform width and smoothly flows. The cross-sectional area (S) of the river can be calculated as the sum of the trapeze surfaces that are part of the area, (Fig. 2.52a)

$$S = b \cdot \frac{\sum_{i=1}^{n} h_i}{n} \tag{2.32}$$

  where

  - b is the width of the river.
  - h_i is the ith depth measurement.
  - n is the number of measurements;

- The river level has to be daily measured, at the same hour, for at least one year including the drought intervals, to accurately calculate the average annual level;
- The flow rate varies with the position of the measurement device, as at the riverbanks, and at the bottom, it is smaller due to the larger roughness. Thus, the flow rate has to be measured in several points and the average value has to be calculated. There are four ways to measure the flow rate:

  - Measuring the time required by a floating submerged object, to cover a known distance in the river. To estimate the average flow rate, the measured speed has to be multiplied with a correction factor, usually around 0.7.

- Using a mechanical instrument with a horizontal shaft and a conical cups propeller, similar to an anemometer (Fig. 2.52b) able to generate a fixed number of electrical pulses during every rotation; these pulses can be converted into flow rate, recorded and stored in an automatic data logger.
- Using an electromagnetic instrument with no moving components but with a sealed induction sensor able to accurately measure even very low flow rates.
- Using dilution methods, especially for small and turbulent rivers that involve pouring a chemical agent and measuring its concentration at a certain downstream distance after it was completely mixed with the river water.

2.3 Solar Energy Conversion Systems

2.3.1 Solar Thermal Systems

A solar thermal system converts the solar energy into thermal energy that can be further used in the built environment to prepare domestic hot water, to heat the water in swimming pools, for space heating and cooling, etc.

Solar thermal systems can be divided according to the type of thermal fluid that transports the heat from the collector. In the case of solar thermal systems designed for the built environment pure water, a mixture of water and glycol, air or refrigerants can be used. The first two types are mainly used for domestic hot water preparation and for the systems that partly cover the heating energy demand of a building that uses a heating system based on water as thermal fluid. The use of water has main advantage its high specific heat, as compared to air, implying thus smaller flow rates and smaller diameter of the pipes in the system.

In locations where the outdoor air temperature decreases below the water freezing point, a mixture of water and propylene or ethylene glycol must be used, also known as *antifreeze solution*. The organic component in the antifreeze mixture degrades over time and has to be periodically changed (each 3–5 years). This degradation mainly occurs as a result of the high temperatures reached during the stagnation periods, usually in summer, when the storage capacity is met and the thermal energy is not removed from the solar thermal system.

The third type of thermal fluid, the air, is mainly used in air-based space heating systems. The use of air as thermal fluid has the main advantage that it can be employed in any location (as freezing does no longer represent an issue). Moreover, in the case of leakages, no damages occur and the system is automatically and easily refilled. The main disadvantage of the air, when used as thermal fluid in solar thermal systems, consists of the large duct dimensions required by the higher flow rates. This negatively affects the cost of the distribution system, which has certain implications on the available space for its installation and on the visual aspect of the entire system.

Refrigerants are used in hybrid systems consisting of solar thermal collectors and heat pumps (direct-expansion solar-assisted heat pumps). The main advantages of these hybrid systems are the higher efficiencies of the solar thermal collector working at lower temperatures and the higher coefficient of performance (COP) of the heat pumps when using higher temperatures provided by the solar thermal collector. Special attention must be paid to the negative impact of these refrigerants on the Earth's ozone layer when accidentally leaked in the atmosphere.

Other thermal fluids are used in solar thermal systems that are not particularly designed for the built environment. One example is represented by the solar thermal systems working at high temperatures that use *oil-based fluids* (synthetic hydrocarbons, paraffin hydrocarbons and/or refined mineral oils) or *molten salts*. These are recommended to be used because of their resistance at high temperature, up to 350 °C for oil-based fluids and up to 900 °C for molten salts. Such high temperatures occur in the solar thermal concentrated systems mainly used for power generation. Oils have a high boiling point, relatively low cost and low freezing point but are flammable and their viscosity increases at low temperatures. Molten salts have the advantages of a high thermal stability, resistance to flammability and low costs but rise corrosion issues. Moreover, this thermal fluid must be kept above its melting point when running through the pipes and must be drained back from the solar thermal system in a heated storage tank when there is not enough available solar energy.

New thermal fluids are permanently developed as the *nano-fluids* aiming at increasing the heat transfer efficiency. The influence of different types of nanoparticles on the thermal fluid properties was investigated, and it was found that pure metallic nanoparticles of silver [1] or copper [162] increase the performance with 10% and 1%, respectively. Ceramic nanoparticles such as titanium dioxide [152], aluminium oxide [162] or silicon dioxide [80] improved the performance with 12.47%, 8% and 8.1%, respectively, while carbon nanotubes increase the exergy efficiency with 17.48% [133] and the solar collector efficiency with 10% [94]. Tao et al. [150] reported nano-fluids based on dispersed silicon oil showing that the thermal properties were improved by adding Fe_3O_4@graphene hybrid nanoparticles. Silicone oil is attractive as heating thermal fluid, because it is non-corrosive and has a longer lifetime (compared with the antifreeze solutions or the hydrocarbon oils), but more pumping energy is required due to a higher viscosity [110].

The main physical properties of the common thermal fluids used in the solar thermal systems that have to be considered in the design step are synthetically presented in Table 2.12 and are explained as follows:

- *The boiling temperature* represents the temperature limit when the fluid changes from the liquid to the vapour phase. The boiling temperature is pressure dependent; therefore, one alternative to keep the thermal fluid in the liquid phase at temperatures above the normal boiling point (measured at $p = 1$ atm) is by increasing the pressure in the solar thermal system;
- *The freezing temperature* represents the temperature when the fluid changes into the solid phase, causing irreversible damage to the system components due to its volume expansion;

Table 2.12 Main properties of the thermal fluids used in the solar thermal systems [41]

Thermal fluid	Water	Water–Glycol 50%	Mineral oil
Boiling temperature (°C)	100	107.2	320
Freezing temperature (°C)	0	−36.8	−10
Density at 0 °C (kg/m^3)	1000	1090	873
Specific heat at 25 °C (J/(kg K))	4181.4	3412	1670
Enthalpy (kJ/kg)	2257	800	
Kinematic viscosity at 40 °C (mm^2/s)	0.6579	3	29.8
Kinematic viscosity at 100 °C (mm^2/s)	0.2938	0.8	4.5

- *The density* is temperature dependent, and this property is important during the thermal stratification that occurs in the natural circulation systems and in the storage tanks;
- *The specific heat* represents the energy required to raise the temperature of a unit mass of the heat transfer fluid by 1 K. This parameter is used in energy and mass balance calculations;
- *The enthalpy* represents the energy contained in the fluid, at constant pressure and under specific temperature conditions. Enthalpy is used to calculate the thermal power output of a component;
- *The fluid viscosity* influences the pumping energy required to circulate the fluid through the solar thermal system components, lower viscosities being preferred to decrease the pumping costs. The increase of temperature largely decreases the viscosity for most of the fluids.

Considering the thermal fluid circulation, there are two main types of solar thermal systems: *solar thermal systems with natural circulation* and *solar thermal systems with forced circulation.*

The first type of solar thermal systems (Fig. 2.53) only consists of the storage tank (1) and the solar thermal collector (2). The solar thermal collector can be either flat plate when pipes are required to connect the two components (Fig. 2.53a) or with evacuated tubes that can be directly inserted in the storage tank, thus simplifying the system (Fig. 2.53b). The thermal fluid circulates based on the density differences: the cold water (CW) with higher density enters the storage tank at the grid pressure when the users consume the domestic hot water (DHW) with lower density. In the initial stage, the thermal fluid in the entire system has the same (low) temperature. When the Sun heats the solar collector, the water temperature rises; thus, its density decreases. The cold water descends from the storage tank replacing the heated water that rises into the storage tank where, due to thermal stratification, the warm water remains in the upper part of the storage tank from where it is used when the warm water tap is open. The main advantage of this system relies on its simplicity, as no pump, temperature sensors, solar controller and expansion vessel are required. However, its use is limited to locations with temperatures above the freezing point of the thermal fluid. These systems are avoided in the built environment as they have

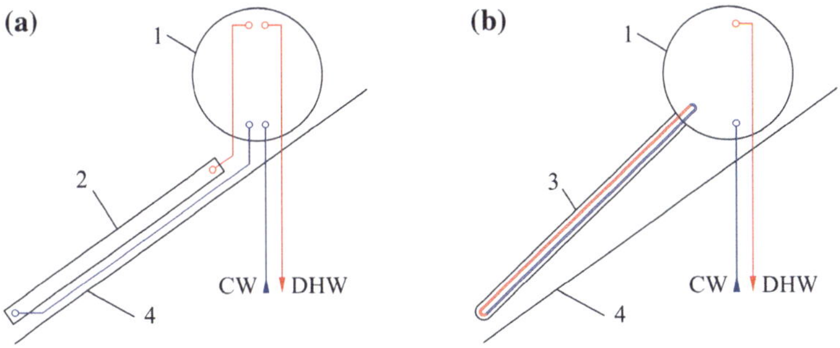

Fig. 2.53 Functional scheme of a solar thermal system with natural circulation: **a** with a flat plate solar thermal collector; **b** with an evacuated tube solar thermal collector: storage tank for domestic hot water (1), flat plate solar thermal collector (2), evacuated tube solar thermal collector (3), roof (4), cold water (CW); domestic hot water (DHW)

limited architectural acceptance, due to the aspect of the large storage tank installed on the roof. Special attention must be paid to the design of the pipes between the storage tank and the solar thermal collector to reduce the pressure losses as much as possible.

For solar thermal systems with forced circulation (Fig. 2.54), the storage tank (1) is installed at a lower level than the solar thermal collector (2) and a hydraulic pump (3) is required to circulate the thermal fluid between these. The pump is switched on/off by a solar controller (4) that compares the temperature of the water in the storage tank, measured by a temperature sensor (T_S), with the temperature in the solar thermal collector measured by another temperature sensor (T_c). When the solar energy does not cover the entire load, an auxiliary electric heater (5) immersed in the storage tank is activated. Usually, the components are connected through a closed loop (Fig. 2.54a) and are using water at the grid pressure as thermal fluid in the locations where the outdoor temperature does not decrease below 0 °C. A solution to avoid water freezing in the components installed outdoor is the use of an *open-loop system* (Fig. 2.54b). This system is designed to drain the water from the solar collector and outdoor pipes when freezing conditions occur. Therefore, two three-way valves (6) are installed on the flow and return pipes running/returning to/from the solar collector; these valves are controlled by the solar controller (4) that closes the way a-ab and opens the way b-ab when the temperature of the solar collector (T_c) gets close to 0 °C. In this case, the water from the storage tank cannot flow to the solar collector and the water from the solar collector is drained to sewage (Y). When the solar collector temperature (T_c) increases above the storage tank temperature (T_S), the solar controller switches the three-way valves (6) closing the b-ab way and opening the a-ab way to fill up the solar collector with water at grid pressure. The main advantage of this system is its simplicity; it does not need any antifreeze solution and heat exchanger; however, corrosion represents a risk as it may occur due to the access of the air in the system.

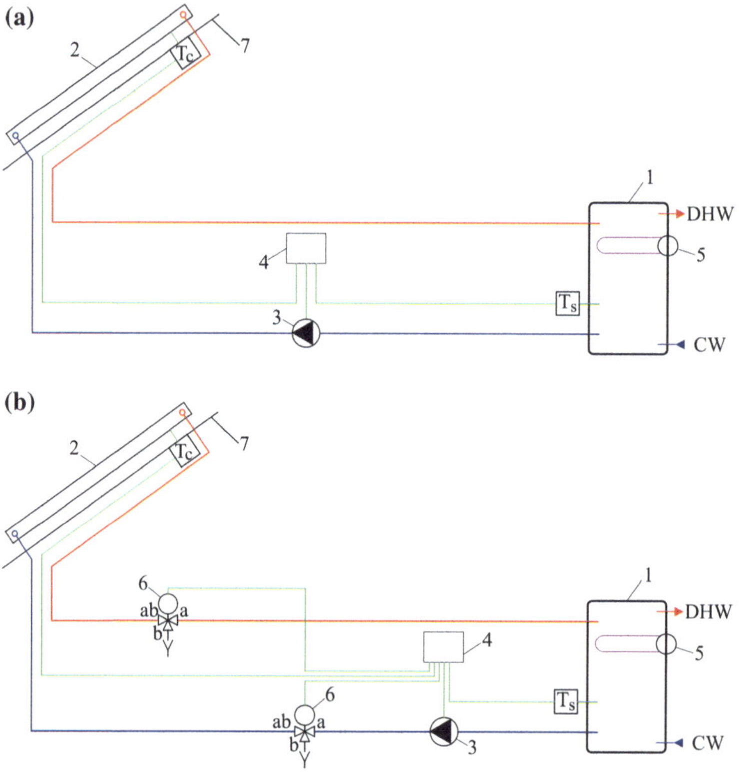

Fig. 2.54 Functional scheme of a solar thermal system with forced circulation of the water in: **a** a closed-loop system; **b** in an open-loop system: storage tank (1), solar thermal collector (2), hydraulic pump (3), solar controller (4), electric heater (5), three-way valve (6), roof (7), cold water (CW), domestic hot water (DHW), solar collector (T_c) and storage tank (T_s) temperature sensors

When the solar thermal system is installed in a location where the outdoor air temperature frequently decreases below 0 °C, an antifreeze solution is strongly recommended as thermal fluid in the outdoor components of the system (the solar thermal collector, the exterior pipes) and a heat exchanger (6) must be installed, inside (Fig. 2.55a) or outside (Fig. 2.55b) the storage tank, to separate the solar circuit from the domestic hot water circuit. An expansion vessel (7) is required to take over the volume variations of the thermal fluid with the temperature in the solar circuit. The first solution (Fig. 2.55a), consisting of an internal heat exchanger (6) installed in the storage tank, is simpler, but it cannot be applied for already existing storage tanks without internal heat exchanger. When an external heat exchanger is installed (Fig. 2.55b), an additional hydraulic pump (8) must be used in the closed circuit between the storage tank (1) and the heat exchanger (6).

The solar controller (4) that switches the circuit on/off along with the hydraulic pump (3) controls the hydraulic pump (8). For newly designed solar thermal systems, the choice between these two solutions depends on the costs of the storage

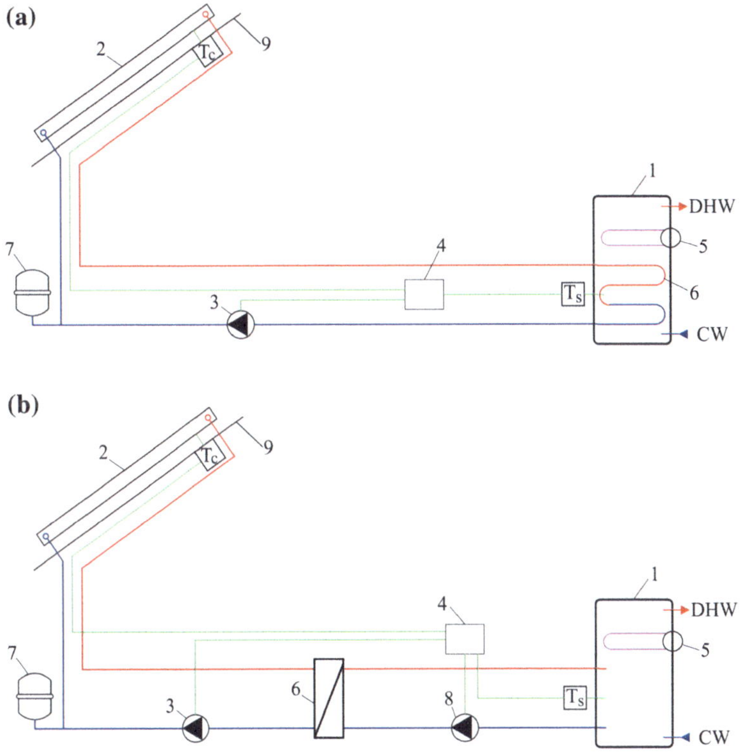

Fig. 2.55 Functional scheme of a solar thermal system with forced circulation of the antifreeze solution with: **a** an internal; **b** an external heat exchanger: storage tank (1), solar thermal collector (2), hydraulic pump (3), solar controller (4), electric heater (5), internal/external heat exchanger (6), expansion vessel (7), hydraulic pump (8), roof (9), cold water (CW), domestic hot water (DHW), solar collector (T_c) and storage tank (T_s) temperature sensors

tanks with/without internal heat exchanger, of the external heat exchanger, additional hydraulic pump and the more complex solar controller. An advantage of the external heat exchanger is its easier maintenance/replacement when leakages occur. Instead of the auxiliary electric heater (5), other thermal energy sources can be used (e.g. a natural gas/biogas/wood fired boiler, a heat pump) connected to the storage tank through an internal or an external heat exchanger.

The main component of a solar thermal system is the solar thermal collector that converts the received solar energy in thermal energy that is further transferred to a thermal fluid flowing through the collector. The solar thermal collectors are grouped into two categories: *concentrating* and *non-concentrating* solar thermal collectors, whether or not provided with solar radiation concentration devices.

Non-concentrating solar thermal collectors can be grouped in *flat plate solar thermal collectors*, *evacuated tube solar thermal collectors* and other types of solar thermal collectors.

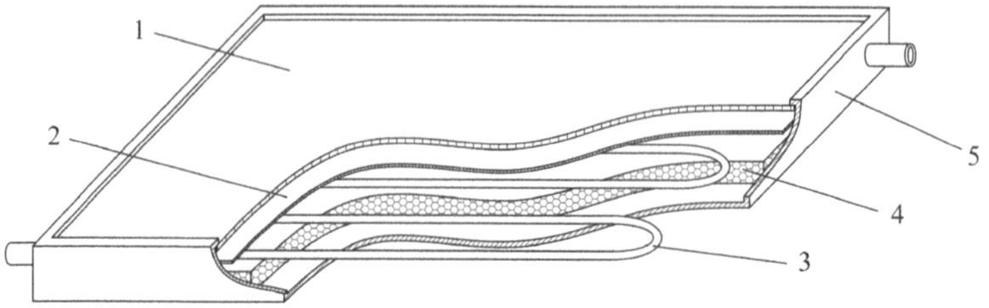

Fig. 2.56 Main components of a water heating flat plate solar thermal collector: transparent cover (1), absorber plate (2), pipes (3), thermal insulation (4) and casing (5)

A water heating *flat plate solar thermal collector* (Fig. 2.56) has the following main parts: a transparent cover (1), an absorber plate (2), a system of pipes (3), the thermal insulation (4) and a casing (5). The transparent cover consists of one or more sheets of regular (inorganic) glass, Plexiglas or other radiation-transmitting material, allowing the solar radiation to enter into the solar thermal collector and blocking the long-wave (IR) radiation to escape from it, reducing thus the thermal energy convection losses to the surrounding atmosphere. The material used for the transparent cover should thus have high transmittance over a broad part of the solar spectrum, low absorptance and low reflectance for the solar radiation wavelengths along with mechanical resistance to withstand hail, stones or loads from wind and snow. Antireflective and hydrophobic coatings are deposited on this glazing to increase the amount of solar radiation reaching the absorber plate.

For flat plate solar thermal collectors that work at low temperatures (e.g. plastic solar thermal collectors for swimming pools or those used as evaporators in the heat pump systems), the transparent cover is not required and these collectors are called *uncovered solar thermal collectors*.

The absorber plate usually consists of a thin sheet of metal (aluminium, copper, stainless steel, etc.) covered with a solar thermal coating or, in the case of reduced working temperatures, of plastic materials (polyethylene, polypropylene, ethylene-propylene-diene-polymer, etc.). The absorber plate is heated by the received solar radiation, and therefore, it must have a high solar radiation absorptance, a low reflectance and low emittance. To this end, spectrally selective coatings are deposited on the surface of the absorber plate enabling the conversion of a higher share of the solar radiation into heat. Once heated, the absorber plate has to transfer the thermal energy to the pipes where the thermal fluid flows; thus, the absorber plate must have a good thermal conductivity such as copper has, but the increasing price of this material makes solar thermal collectors too expensive and other materials are investigated to reach a cost-effective result. Moreover, the air gap thickness between the absorber plate and the transparent cover is subject of optimization aiming at reducing the top thermal losses in the solar thermal collector.

The pipes in the collector are made of an identical material as the absorber plate, for the same reasons: to increase the heat transfer to the thermal fluid flowing through them and to avoid corrosion. The connections between the absorber plate and the pipes are also important, because a poor connection decreases the heat transfer efficiency. Thus, bonding can be mechanically done or by embedding the pipes into the absorber plate, by laser or ultrasonic welding, etc. Solar thermal collectors without absorber plate (having only rectangular or flattened pipes with the exposed surface selectively coated) are also investigated to eliminate the thermal resistance of the bound between the absorber plate and the pipes. The pipes can be mounted in a *serpentine shape* or as a *register* aiming at reducing the dimensions, the pumping power and therefore the cost while maintaining a good heat transfer efficiency.

The thermal insulation usually consists of mineral wool, polyurethane foam or other materials with high thermal resistance, and it is located under the entire absorber plate/pipes' assembly and on the perimeter of the solar thermal collector. It has the role to reduce the bottom and the edge thermal losses, its thickness being subject of optimization.

The collector's casing is usually made of aluminium or plastic reinforced with carbon fibres to ensure the mechanical stability of the solar thermal collector and to protect the components from rain, snow, moisture, dust, etc.

The *evacuated tubes' solar thermal collector* (Fig. 2.57) consists of glass tubes (1), absorbers (2), pipes (3) or heat pipes (4) and insulated casing (5). The glass tubes play the role of the transparent cover in the flat plate solar thermal collectors, their cylindrical shapes being imposed to withstand the inside vacuum that ensures the reduction, down to zero, of the convection thermal losses from the absorber plate to the glass tube. Two types of evacuated glass tubes are used, with single (Fig. 2.57a) or double walls (Fig. 2.57b); the first type is cheaper, has a lower weight but, when broken, the entire assembly of the glass tube, absorber and pipe must be replaced, increasing thus the maintenance costs, while in the second case only the glass tube must be replaced.

The absorber plate is usually made of thin sheets of copper or aluminium that have a reduced width, simplifying thus the manufacturing process. Two types of absorbers can be used: plate (Fig. 2.57a) or cylindrical (Fig. 2.57b). Similarly the

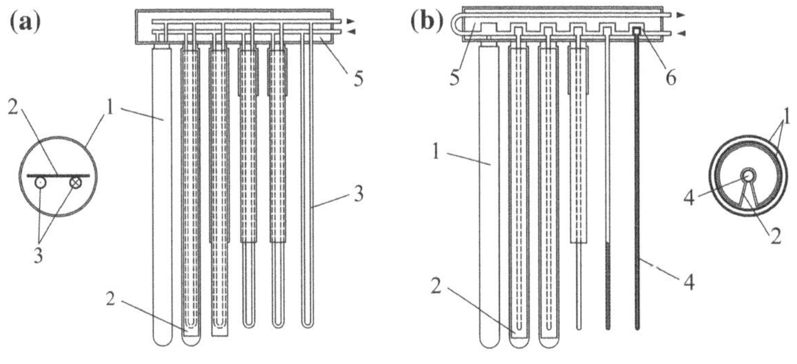

Fig. 2.57 Main components of an evacuated tube solar thermal collector with: **a** direct circulation; **b** heat pipes: glass tube (1), absorber (2), pipe (3), heat pipe (4), insulated casing (5) and heat exchanger (6)

flat plate solar thermal collectors, a spectral selective coating is used to increase the share of solar radiation transformed into heat. This selective coating is deposited either on the absorber plate or on the inner surface of the double-wall glass tube.

The heat transfer from the absorber coating to the thermal fluid can be directly done, by circulating the thermal fluid through the pipe in contact with the absorber (Fig. 2.57a) or indirectly using a heat pipe in contact with the absorber and having the evaporator in contact with a heat exchanger where the thermal fluid flows (Fig. 2.57b).

Thermal insulation is required only in the casing where the heat is transferred from the evacuated tubes to the thermal fluid. As requirements, the materials used for casings must have good mechanical and corrosion resistance and reduced weight.

2.3.2 Photovoltaic Systems

The photovoltaic (PV) conversion represents an advanced technology for converting the most abundant renewable energy at global scale, the solar energy, directly into electrical energy without using internal combustion systems or rotating equipment.

The photovoltaic effect was for the first time observed and analysed by Antoine Henri Becquerel in 1839 using selenium and Charles Fritts built the first photovoltaic cell in 1883. However, only in 1941 the first silicon photovoltaic cell was reported [112] with an efficiency of less than 1%. In 1954, the Bell Laboratories reported an efficiency of 4.5% [120] and of 6% [23], while a 10% efficiency was reported 18 months later.

Currently, there are various types of PV cells that differ considering the materials they are built of, as the data in Table 2.13 show.

According to the data in Table 2.13, the highest efficiency corresponds to the monocrystalline silicon solar cells, considering both infield values as those reported by Cotar and Flicik [26] and standard data (AM 1.5, solar irradiance of 1000 W/m^2 and module temperature of 25 °C) as those reported by Green et al. [58]. The high efficiency and the longer lifetime compensate, in some extent, the price of these cells (and modules). The CIGS modules are among the thin film PVs with the highest efficiency. As Green et al. [58] mentioned, there are more types of single-junction PV cells that qualifies as "notable exceptions", e.g. GaInP (efficiency in standard conditions: $21.4 \pm 0.3\%$) or GaInAsP/GaInAs (standard efficiency: $32.6 \pm 1.4\%$). Moreover, multijunction terrestrial cells were reported to have the highest efficiencies measured in standard conditions as the values in Table 2.14 show.

The terrestrial concentrator cells may reach even higher efficiencies, over 40%, as the GaInP/GaAs/GaInAs/GaInAs multijunction cell developed by the National Renewable Energy Laboratory, NREL [58] with an efficiency of $45.7 \pm 2.3\%$. However, these types of cells are mostly developed at laboratory scale and are not market available yet.

According to the International Energy Agency [75], the main PV cells and modules' manufacturers were located in 2017 in China (69% of the PV cells production and 72% of the PV modules), Taiwan (12% of the PV cells' production and 1% of

Table 2.13 Types of PV cells

Type of cell	Efficiency (%)	Expected lifetime (years)	References
Monocrystalline and polycrystalline silicon solar cells			
Monocrystalline silicon solar cells	17–22.7	25–30	International Energy Agency [75]
	26.7 ± 0.5		Green et al. [58]
Polycrystalline silicon solar cells	10–14	20–25	Cotar and Flicik [26]
	22.3 ± 0.4		Green et al. [58]
Thin films' solar cells			
Amorphous silicon solar cells	6	15–20	Cotar and Flicik [26]
	10.5 ± 0.3		Green et al. [58]
CdTe solar cells	18	15–20	Cotar and Flicik [26]
	21 ± 0.4		Green et al. [58]
$CuInS_2$ (CIS) solar cells	20	15–20	Cotar and Flicik [26]
$CuIn_xGa_{(1-x)}Se_2$ (CIGS) solar cells	21.7	15–20	Zentrum für Sonnenenergie- und Wasserstoff-Forschung Baden-Württemberg [163]
	22.9 ± 0.5		Green et al. [58]

Table 2.14 Efficiency of multijunction PV cells measured in standard conditions [58]

Type of cell	InGaP/GaAs/InGaAs	GaInP/GaAs/Si (wafer bonded)	Perovskite/Si (monolithic)	a-Si/nc-Si/nc-Si (thin film)
Efficiency (%)	37.9 ± 1.2	33.3 ± 1.2	25.2 ± 0.8	14 ± 0.4

the modules), Malaysia (7%—PV cells; 6%—PV modules), South Korea (5%—PV cells and 7%—PV modules), Japan (2% both in PV cells and in PV modules' production), Europe (1% PV cells and 2% PV modules) and the USA with no PV cells' production and 1% of the PV modules' production.

The classical, silicon-based photovoltaic cell consists of two semiconductor thin layers, differently doped (Fig. 2.58), similar to the well-known passive electrical component called *diode*. The upper layer (2), on which solar radiation is falling, is negatively doped using phosphorous, while the lower layer (3) is positively doped using boron. Thus, in the small area between the two layers, called *p-n* junction (1), several electrical charges combine and create an opposite electric field that does not allow all charges of the two layers to combine.

If the solar radiation is not reflected by the upper electrode and is received on/in the PV cell, the energy of the received photons is given to the electrons in the cell, which are initially in an energy state called *valence band*. If the electrons' energy gets higher

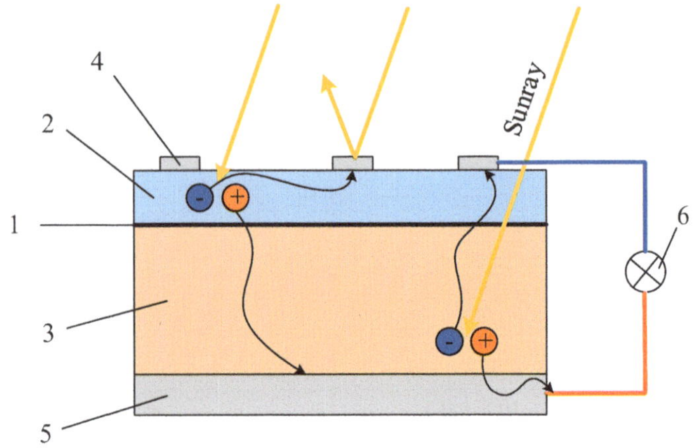

Fig. 2.58 Components and operational principle of a PV cell: *p-n* junction (1); negatively doped layer (2); positively doped layer (3); negative electrode (4); positive electrode (5); energy consumer (6)

than the attraction force of the nucleus, the electrons jump in a new energy state, called *conduction band,* and can travel free in the semiconductor material leaving a free space behind, called hole.

The electrical field at the *p-n* junction supports the electrons moving to the *n*-type layer/negative electrode (4) and the holes' movement to the *p*-type layer/positive electrode (5). If an external electrical consumer (6) is placed between the two layers, the electrons will pass through the exterior circuit to the *p*-type layer and recombine with the holes.

Metallic contacts are required to collect the electric charges on both parts of the cell: the entire lower surface of the cell can be covered with aluminium paste while the front contact is usually a thin metallic grid or an arborescent structure.

The photovoltaic cells have a relatively low conversion efficiency [106] that is the result of physical and functional limitations.

The physical limitations include losses due to that part of the solar radiation spectrum that cannot be used in the photovoltaic conversion (approx. 56%), mainly consisting of long-wave (IR) solar radiation that is only heating the PV cell and consequently decreases its conversion efficiency. Moreover, the electron–hole recombination (approx. 10%) and the losses generated by the equivalent serial and parallel resistances of the cell (approx. 4%) represent common physical limitations. Considering all these types of losses, Shockley and Queisser [138] stated that the maximal conversion efficiency in a single-junction (two-layer) photovoltaic cell is close to 30%.

The functional limitations of the photovoltaic cells' efficiency are the result of their construction and include losses due to the upper metallic grid contact that may reflect the solar radiation and shade the solar cell (approx. 3%), losses due to the electron–hole recombination inside the cell and ohmic losses on the equivalent series resistances (approx. 0.5%).

The upper part of the photovoltaic cells is covered with a thin, anti-reflecting coating based on silicon nitride or titanium dioxide [29] to reduce the solar radiation reflection.

To increase the photovoltaic conversion efficiency, different types of multijunction cells, with more than two semiconductor layers, were developed and their efficiency reached 46% in 2018 [59]. These new types of cells are mainly part of the systems that concentrate the direct solar radiation several hundred times.

2.3.2.1 The Equivalent Circuit of a Photovoltaic Cell

The photovoltaic processes can be modelled using equivalent electrical circuits. The most simple one contains a current source powered by solar radiation with a real diode parallelly connected (Fig. 2.59). Real diodes, in contrast to ideal ones, allow a small voltage drop across them if powered in forward direction and a small current if powered in reverse direction while ideal diodes have both voltage drop and reverse current equal to 0 [97].

In the equivalent circuit in Fig. 2.59, I_{d} represents the current that flows through the diode that, according to Masters [97], can be calculated using

$$I_{\mathrm{d}} = I_0 \cdot \left(\mathrm{e}^{\frac{q \cdot V_{\mathrm{d}}}{k \cdot T}} - 1 \right) \tag{2.33}$$

where

- I_0 is the reverse saturation current of the diode.
- q is the electrical charge of an electron (1.602×10^{-19} C).
- V_{d} is the voltage between the *p*- and *n*-terminals of the diode.
- k is the Boltzmann constant ($k = 1.381 \times 10^{-23}$ J/K).
- T is the absolute temperature of the *p-n* junction.

The connectors of the equivalent photovoltaic cell circuit can be in one of the following two particular situations:

- The connectors are shortened together; this situation corresponds to the highest current a cell can generate (I_{SC}) called *short-circuit current* (Fig. 2.60a);

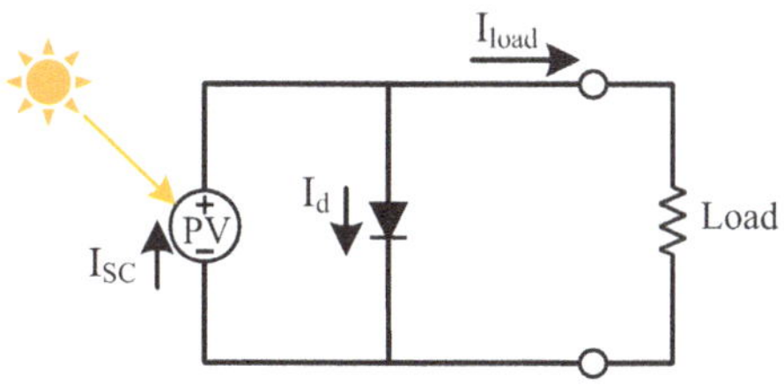

Fig. 2.59 Simple equivalent circuit of a photovoltaic cell

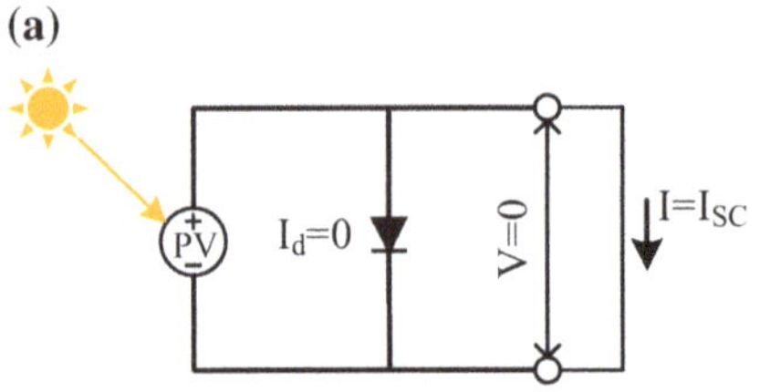

Fig. 2.60 Equivalent circuit of a photovoltaic cell under operation conditions that define: **a** the short-circuit current, I_{SC}; **b** open-circuit voltage, V_{OC}

- The connectors are not tied to any consumer and are left open; this situation corresponds to the highest voltage the cell can generate (V_{OC}) called *open-circuit voltage* (Fig. 2.60b).

This simple equivalent circuit described in Fig. 2.59 has several inconsistencies. For example, it is not able to describe the current that flows through a shaded cell (that, using this circuit, is limited to 0), and it is not able to describe the losses due to the internal resistance in the semiconductor material or at the bond between the metallic conductor and the semiconductor. The first issue can be addressed using a parallel resistance (R_P) while the second one, using a series resistance (R_S), as presented in Fig. 2.61 for a more accurate equivalent circuit.

Both resistances are important for an accurate modelling of the photovoltaic cell. The correlation between the current (I) and the voltage (V) that can be generated by the cell is described by Masters [97] based on this equivalent circuit

$$I = I_{SC} - I_d - I_P = I_{SC} - I_0 \cdot \left(e^{\frac{q \cdot (V + I \cdot R_S)}{k \cdot T}} - 1 \right) - \frac{V + I \cdot R_S}{R_P} \tag{2.34}$$

where

- I_P is the current that flows through the equivalent parallel resistance of the cell, R_P.
- R_S is the equivalent serial resistance of the cell.
- R_P is the equivalent parallel resistance of the cell.

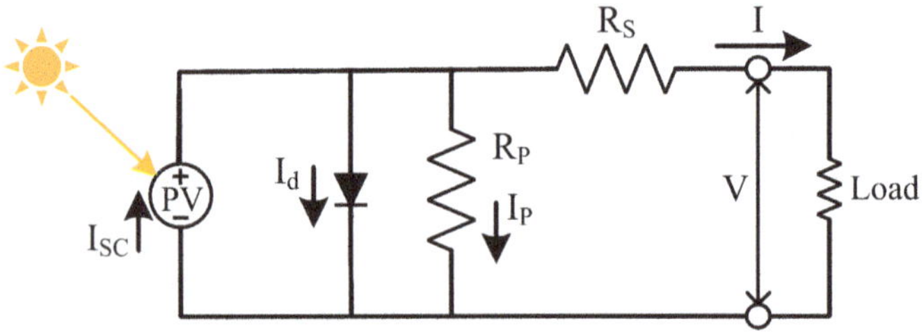

Fig. 2.61 Equivalent circuit of a photovoltaic cell, with serial and parallel resistances [97]

– V is the voltage of the photovoltaic cell.

The values of the two equivalent resistances in a photovoltaic cell should satisfy the following conditions to get losses lower than 1%

$$R_S < \frac{0.01 \cdot V_{OC}}{I_{SC}}; \quad R_P > \frac{100 \cdot V_{OC}}{I_{SC}} \tag{2.35}$$

2.3.2.2 Photovoltaic Cell, Module, Array

A classical crystalline silicon photovoltaic *cell* is approx. 6 inch wide and 6 inch tall (156 mm × 156 mm). Its electrical power is relatively small, approx. 3 W, while its voltage is only about 0.6 V. This is why the use of a single cell is feasible only in few applications as in small portable devices.

To use the photovoltaic power systems on a larger scale, several cells are serially interconnected by manufacturers forming strings (to increase the voltage to 16 V, usually required to charge 12 V batteries) that can be further parallelly connected (to increase the current) developing a *photovoltaic module* with electrical parameters usable in common applications. Currently, high-power modules are developed reaching 500 W [60].

The main two parameters of a PV module can be calculated using the corresponding parameters of the PV cell

$$V_{OC}^{M} = N_{SC} \cdot V_{OC}^{C} \tag{2.36}$$

$$I_{SC}^{M} = N_{PC} \cdot I_{SC}^{C} \tag{2.37}$$

where

– V_{OC}^{M} is the open-circuit voltage of the module.
– V_{OC}^{C} is the open-circuit voltage of the cell.
– I_{SC}^{M} is the short-circuit current of the module.
– I_{SC}^{C} is the short-circuit current of the cell.
– N_{SC} is the number of cells serially connected.
– N_{PC} is the number of parallel strings of cells.

The PV module encapsulates the fragile silicon cells and the electrical connections using weather-resistant materials to protect the cells from fingerprints, mechanical shocks or corrosion as a result of the environmental humidity. The upper material is usually a transparent layer of glass allowing the incident sunlight to reach the silicon cell with insignificant losses.

The PV module represents the main component of a photovoltaic system and contains all the elements required for its mounting on the support structure of the

PV system (metallic frame, spaces for fixture, etc.) and for wiring to other modules or adjacent electrical equipment (junction box, cables, etc.).

The number of modules in a PV system depends on the required power of the installed system. The modules of a PV system are usually connected in *arrays* that represent a group of modules connected in series and/or parallel according to the electrical limitations of the adjacent electrical components in the system. The modules are serially connected when high voltage is required and are connected in parallel when high current is targeted or when the series string voltage exceeds the accepted voltage thresholds of the electrical equipment used in the PV system.

A PV array that contains N_{PS} parallel strings, with N_{SM} modules serially connected in each string, can be described, according to Kalogirou [82] by

$$V_{OC}^{A} = \max_{j=1}^{N_{PS}} \left(\sum_{i=1}^{N_{SM}} V_{OCi}^{M} \right)_j \tag{2.38}$$

$$I_{SC}^{A} = \sum_{j=1}^{N_{PS}} I_{SCj}^{M} \tag{2.39}$$

where

- V_{OC}^{A} is open-circuit voltage of the array.
- V_{OCi}^{M} is the open-circuit voltage of the ith PV module in the string.
- I_{SC}^{A} is the short-circuit current of the array.
- I_{SCj}^{M} is the short-circuit current of the jth string of modules.
- N_{SM} is the number of modules serially connected.
- N_{PS} is the number of parallel strings of modules.

When the array contains identical modules receiving the same radiation on each module, the previous equations can be simplified as

$$V_{OC}^{A} = N_{SM} \cdot V_{OC}^{M} \tag{2.40}$$

$$I_{SC}^{A} = N_{PS} \cdot I_{SC}^{M} \tag{2.41}$$

This modular structure of the PV systems represents one of their advantages since new PV modules and electrical components can be relatively easy added to the system if the consumption requirements increase and extra power is required from the PVs.

2.3.2.3 Characteristic *I*-*V* Curves of the Photovoltaic Convertors

The parameters of a photovoltaic convertor (I_{SC}, V_{OC}, the nominal electrical power, etc.) are indicated by manufacturers as measured under either *Standard Test Conditions* (STC) or *Nominal Operating Cell Temperature* (NOCT) conditions. The nominal power is currently given in *Watt peak* (W_p) and indicates values obtained under STC.

STC represent an industry standard for testing PV modules that states during testing a 25 °C PV convertor temperature, 1000 W/m^2 (1 *Sun*) received irradiance and an air mass AM = 1.5. These conditions approximately correspond to the solar noon during equinoxes in the continental part of the USA.

However, photovoltaic systems seldom work in these conditions due to the variation of the available solar radiation and of the PV module temperature. This is why most manufacturers also test modules in more usual operating conditions (NOCT): 800 W/m^2 received irradiance, 1 m/s wind speed and 20 °C air temperature when the module temperature is about 45 °C.

During the day, the solar irradiance has a higher variation compared to the temperature variation and influences the PV cell output because the produced electrical current directly depends on the received (solar) radiation (Fig. 2.62). On the other hand, the module voltage at the maximum power point (V_{MPP}) remains roughly constant when the solar radiation varies above a specific threshold. As an example, for a 150 W polycrystalline silicon module the variation of the maximum power point voltage (ΔV_{MPP}) is of approx. 4 V; if the solar irradiance value falls at only several W/m^2, then the PV module voltage suddenly drops to nearly 0 V [29].

The voltage of a photovoltaic module mainly depends on the materials used to build it and is affected by the temperature variation. Using the same PV module (150 W polycrystalline silicon), the V_{MPP} value can be during hot summer days with approx. 10 V smaller and during cold winter days with approx. 10 V larger compared to the STC value, while the PV current only slightly increases with temperature [29], as presented in Fig. 2.63.

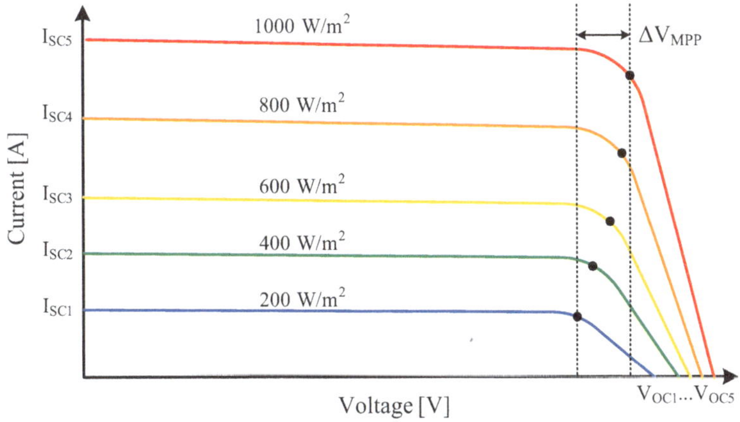

Fig. 2.62 I-V characteristic curve of a photovoltaic module at constant temperature of the PV module and variable received solar irradiance (adapted from [29])

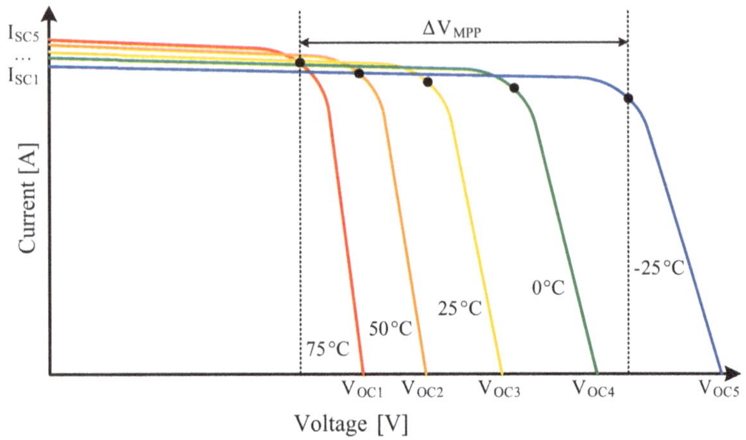

Fig. 2.63 I-V characteristic curves of a photovoltaic module, under constant received solar irradiance and variable temperature of the PV module (adapted from [29])

When several modules are interconnected in serial arrays, the voltages sum up and this seasonal difference (ΔV_{MPP}) should be carefully considered to not exceed the rated voltage of the electrical devices that receive the energy from the PV array.

2.3.2.4 Types of Photovoltaic Systems

Photovoltaic modules or arrays are seldom used without any additional components. This is why they are considered the main components of a system that allows power production, storage and/or distribution.

Three main configurations of PV power systems are currently developed: *stand-alone*, *grid connected* and *hybrid*.

The first applications of photovoltaics were mainly stand-alone (also known as *off-grid*), able to provide electrical power to various mobile devices such as computers and satellites, or remote residences with no connection to the local electricity grid.

The stand-alone PV systems can be divided into two subcategories. The simplest stand-alone system is *direct-coupled*, where the electrical load is directly connected to the photovoltaic module/array, usually without any power-conditioning devices or electricity storage facilities (Fig. 2.64). In this case, the load (usually a water pump motor) only works if solar radiation is available, and instead of storing the electrical

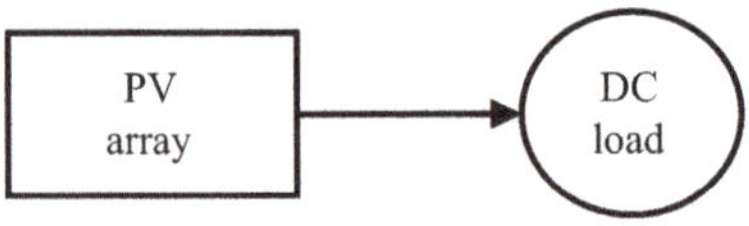

Fig. 2.64 Direct-coupled PV system

energy, the potential energy of water can be stored in a water tank for further use when needed.

The second type of stand-alone systems is able to provide cost-effective electrical energy to remote locations with no access to the electricity grid, due to its unavailability or due to the too long distances to this grid. Such a system contains at least the PV array, the batteries for storing the electrical energy to be used during time intervals when no solar radiation is available and the charge controller. If alternating current is required in the remote location, an inverter has to be also included in the system as presented in Fig. 2.65.

Most stand-alone systems do not operate at their maximum potential, because of various types of losses:

- Losses during the batteries charging and discharging;
- Losses because the PV array is usually not working at its maximum power point when charging the batteries;
- Losses due to the tilt angle of the PV array that is usually selected to generate similar amounts of energy during all seasons over the year, not to generate the maximum annual energy.

The maintenance of these systems has also to be considered as the batteries need to be replaced every 5–10 years during the life cycle of the PV system, and the consumption pattern of the inhabitants (users) needs to be adjusted according to the solar radiation available in the implementation location.

These limitations of the stand-alone PV systems are no longer valid for the second main type of PV configurations, the *grid-connected PV systems,* shortly known as *on-grid* PV systems. These systems allow the connection of the PV arrays to the local electricity grid using only a power-conditioning unit (an inverter) which mainly converts the DC power from the PV modules to AC power. The inverter also adapts the generated power to the grid parameters (Fig. 2.66).

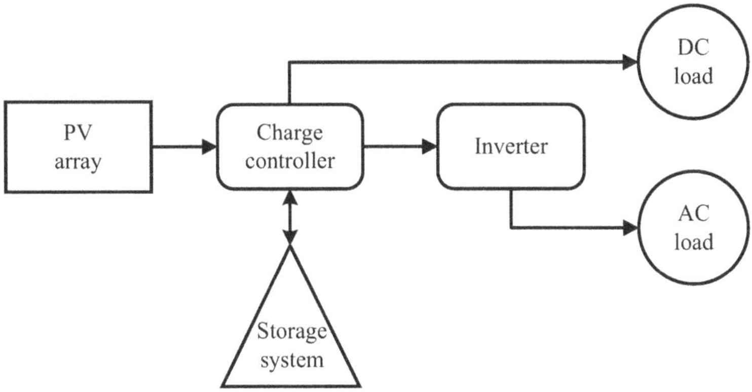

Fig. 2.65 Stand-alone PV system

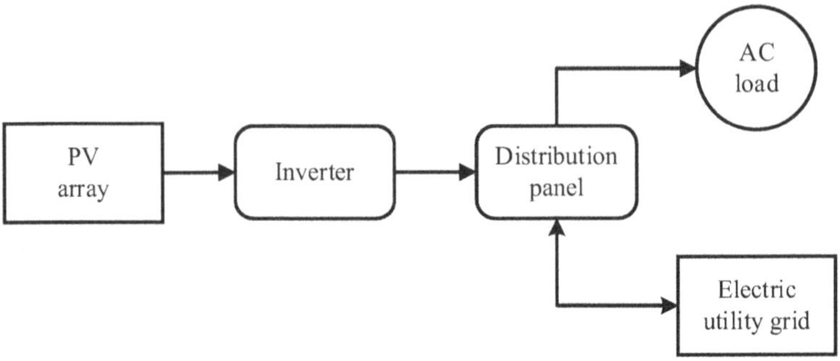

Fig. 2.66 Grid-connected PV system

The power generated by the PV system can be either locally used, if required by the local electricity consumers, or it can be injected into the grid when the production is higher than the consumption. If the local consumers require more power than that produced by the PV system, this can be taken (bought) from the grid, meaning that the grid can also act as a balancing device eliminating the high-cost batteries.

The main issue of the on-grid PV systems is that they are not producing electrical power when the electrical grid is disconnected. This happens because of safety reasons, to prevent injuring the personnel working on the grid.

To eliminate this issue, a third main type of PV configuration is commonly used, called *hybrid* system [37]. This PV system uses an inverter able to work both grid-connected when the grid is available and stand-alone when the grid is missing or fails to provide normal power parameters (Fig. 2.67).

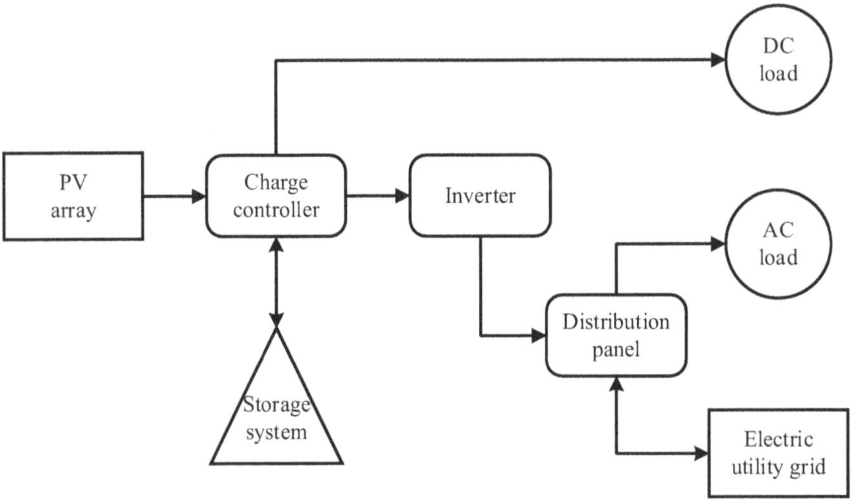

Fig. 2.67 Hybrid coupled PV system

As indicated in Figs. 2.65, 2.66 and 2.67, the photovoltaic systems also include additional equipment: on-grid systems require at least an inverter while stand-alone systems require at least rechargeable batteries, charge controller and a stand-alone inverter.

Moreover, as already presented, batteries and charge controllers are also part of the hybrid systems installed when the grid has time intervals either when it is missing or its voltage and frequency are unstable making the grid-connected system improper to function all the time.

Various overvoltage and overcurrent protection devices are also implemented on the DC and AC sides of the photovoltaic systems to eliminate any electrical hazards due to malfunctioning devices or atmospheric discharges.

2.3.2.5 Rechargeable Batteries

Although electricity is versatile and can be efficiently converted into other forms of energy; it has the disadvantage that it cannot be stored on a large scale; this is why conventional power plants have to adapt their energy production to the demand. This full adaptation is not possible in the case of the PV systems due to the intermittent nature of the solar radiation. This is why renewable energy systems usually also need an energy storage device which most of the time, due to its reduced price, is a rechargeable battery bank able to electrochemically store the energy.

The simplest equivalent circuit of a rechargeable battery consists of a voltage source *(E)* and a small serial resistance (R_S) as described in Fig. 2.68. During the battery discharge, the voltage (V) slightly decreases while the resistance increases [82]

$$V = E - I \cdot R_S \tag{2.42}$$

where I is current which charges or discharges the battery.

The nominal capacity of a battery (Q_{max}, usually measured in Ah) is one of the most important parameters and indicates the time interval when a predefined nominal current can be continuously extracted from a battery at 25 °C until it reaches the empty state. As an example, for a 12 V lead–acid battery, "empty" means 10.5 V between the battery electrodes [97]. The capacity of a battery depends on the current extracted

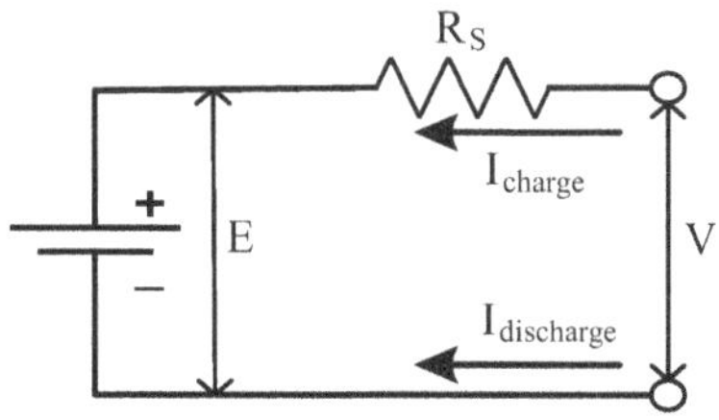

Fig. 2.68 Equivalent circuit of a rechargeable battery

from the battery: the larger the current is, the smaller the available capacity gets; thus, small discharge current is recommended [16].

A battery can be described using two additional main parameters: the state of charge (SOC) and the charging efficiency. According to Kalogirou [82], the SOC can be calculated as the ratio of the still available capacity (Q) and the nominal capacity ($Q_{\max}$) measured at 25 °C, considering the discharge time of 20 h [97]

$$\mathrm{SOC} = \frac{Q}{Q_{\max}} \tag{2.43}$$

The efficiency of a battery can be calculated as the ratio between the extracted energy and the energy required to bring the battery to its initial SOC; for a full charge–discharge cycle of lead–acid batteries, the efficiency is about 75%.

The major problem of rechargeable batteries is the relatively small number of charge–discharge cycles that they can support during their lifetime; commonly, lead–acid batteries with grid plates can be used up to 100–800 cycles according to the depth of discharge reached [29]. Except the normal degradation of the battery's electrodes, also random failures of individual serial cells inside the battery may occur that can be of three types [119]:

- Cell voltage drops below a certain value before the rated discharge capacity, and the battery may be functional since the other cells might compensate the voltage;
- Cell in short circuit, which might damage the charge controller and act as a resistive load for the other battery cells;
- Cell in open circuit that makes the entire battery not functional.

2.3.2.6 Charge Controller

If a PV system uses rechargeable batteries, then charge controllers are mandatory electrical components. Their main role is to adjust the PV power to the state of charge (SOC) of the batteries, more specifically to match the PV voltage to the voltage of the batteries. A second role is to protect the rechargeable batteries from deep discharging and from overcharging, which might increase their temperature and cause evaporation of the water solvent inside the cells. Both phenomena can shorten the battery lifetime or cause accidents (fire, explosion, etc.).

There are three types of charge controllers:

- *Series controllers* that switch off the PV modules by opening the circuit; their disadvantage is the constant switching when the battery is almost full and its voltage is near the cut-out voltage;
- *Parallel controllers* that connect the PV module to a shunt resistance when the batteries approach the full state of charge; in this way, the module is continuously producing power while only the current required to charge the batteries reaches them.

The series and the parallel controllers have the drawback that they are not able to extract the maximum power from the PV modules that depends on the received solar radiation and on the PV module temperature. These two charge controllers are only able to adjust the PV modules voltage to the voltage of the batteries, and in this way, part of the available PV power may be lost (between 10 and 40%).

- *Maximum power point tracking (MPPT) controllers* that identify on a predefined interval the I-V characteristic and the MPP power of the PV module and adjusts the internal resistance of the charge controller based on Ohm's law ($R_{MPP} = V_{MPP}/I_{MPP}$). The extracted maximal power is fitted to the battery voltage using a high-efficient (approx. 95%) DC/DC converter.

Considering how the batteries are charged, there are two types of algorithms implemented in controllers:

- Algorithms with multiple (three) charge rates: the *bulk (normal)* stage that recharges about 80% of the drained capacity with a constant current and increasing voltage; the *absorption* stage that fully recharges the battery with decreasing current and constant voltage and the *float (trickle)* stage that maintains a fully loaded battery using a lower fixed voltage and a small current, about 1% of the battery capacity compensating its self-discharge rate [119];
- Algorithms with single-charge rate current, while the charge regulator is either on or off depending on the value of the battery voltage, below or above a specific threshold. With this algorithm, a battery can be hardly fully loaded and maintained in this condition.

PV systems may also function without charge controllers but for a shorter lifetime. In this case, precisely choosing the PV modules is required so that their maximum voltage does not exceed the overcharge voltage of the batteries (as example 15 V in case of 12 V lead–acid batteries [119]) and at least, a transistor is required for transferring the PV power to the ground if the batteries are fully charged.

Most charge controllers available on the market use MPP tracking and multiple charge rate algorithms in order to increase the PV production and battery's life.

2.3.2.7 The Inverter

Photovoltaic systems produce *direct current* (DC), representing a flow of electrons in a single direction with an amplitude that can be either constant or variable (Fig. 2.69a). Electrical grids that provide electrical power to all consumer types and most of the electrical appliances used nowadays require alternating current (AC) to function. Thus, to be either locally used or transported to the consumers using the electrical grid, the energy provided by the PV systems has to be converted from DC to AC.

Alternating current is an electron flow that periodically reverses direction with a frequency and amplitude that can be either constant or variable (Fig. 2.69b). Alternating currents can have a periodic variation of their waveform, if repeated at a fixed

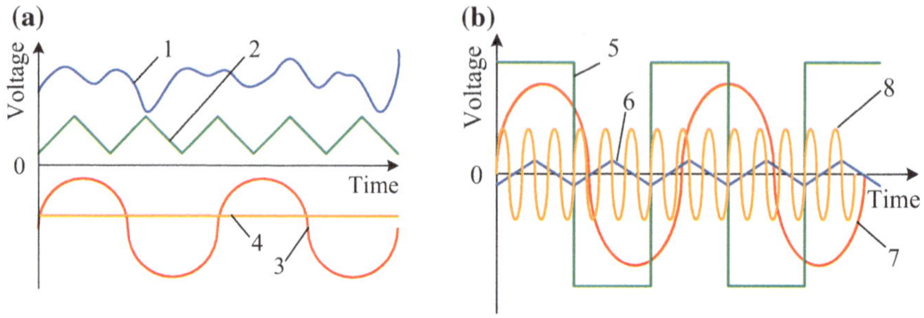

Fig. 2.69 Examples for the variation of the electrical parameters: **a** direct current (DC) signals with: non-uniform (1), uniform triangular (2), uniform sinusoidal (3) variations, constant (4); **b** alternating current (AC) signals with rectangular (5), triangular (6), sinusoidal with different amplitudes (7, 8) variation

time interval, or this variation can be a random one. The changing in the AC amplitude can be either continuously done or in discrete steps, while common used AC waveforms are sinusoidal, stepwise or modified.

The electrical converter of the DC to AC energy is commonly known in the photovoltaic industry as *inverter* and is represented on electrical diagrams with the symbol in Fig. 2.70.

The early inverters were based on a DC motor connected to an AC generator, while current inverters rely on static electronic power components like thyristors and transistors that are switched on/off either by the signal from the electricity grid or by microprocessors used in the inverter.

The invertor efficiency (η_{inv}) can be calculated using

$$\eta_{\text{inv}} = \frac{P_{\text{AC}}}{P_{\text{DC}}} = \frac{V_{\text{AC}} \cdot I_{\text{AC}} \cdot \cos\varphi}{V_{\text{DC}} \cdot I_{\text{DC}}} \tag{2.44}$$

where

- P_{AC} is the AC power delivered by the inverter.
- P_{DC} is the DC power produced by the PVs and delivered to the inverter.
- V_{AC} is the AC voltage of the electrical grid or of the stand-alone system.
- I_{AC} is the current injected to the electrical grid or powering AC loads.
- V_{DC} is the voltage of the photovoltaic array or of the storage system.

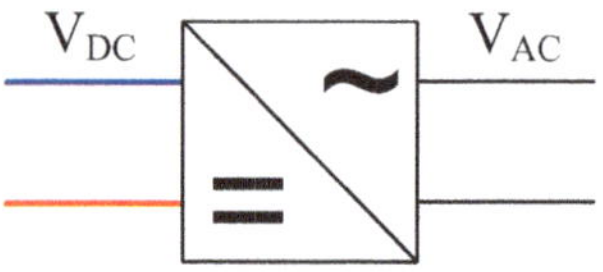

Fig. 2.70 Symbol used for representing an inverter in electrical diagrams

– I_{DC} is the direct current produced by the PV array or extracted from the battery storage system.
– cos φ is the AC power factor.

Except its main function of converting the input photovoltaic DC power (P_{DC}) to AC output power (P_{AC}) with a high efficiency (usually above 95%), the inverter also has to maintain a constant voltage and frequency on the AC output side.

According to the systems where they are used, inverters are of three types; due to the different objectives, they have to fulfil.

Grid-connected inverters inject the entire generated electrical energy in the local electricity grid. Therefore, on the AC output side, inverters need to continuously monitor the state of the grid and adjust the produced electrical energy to the voltage and frequency parameters of the grid. This is why grid-connected inverters are unable to operate when there is a problem on the AC connection (missing electrical grid or parameters out of the normal operation ranges).

The second objective of grid-connected inverters is the extraction of the maximum energy that the PV modules are able to generate for each solar irradiance—module temperature combination. This means that the inverter has to identify the maximum power point (MPP) of the PV system, similar to the charge controller with MPPT.

The inverters in the *stand-alone photovoltaic systems* have to comply with different restrictions: they have to be able to accept the DC voltage of the battery bank used for energy storage and to generate the standard AC voltage and frequency used in the implementation location. Moreover, the inverter has to support the output power for the local equipment both during continuous operation and during short intervals when start-up spikes occur.

Inverters used in stand-alone PV systems have a simpler construction and a lower price for the same rated power as compared to the ones for grid-connected PV systems, because:

- They do not need to monitor the grid parameters and synchronize to these;
- Their generated AC power does not always need to have a continuous sinusoidal waveform if no consumers with electrical motors are involved; in this case, low-cost inverters with simple or modified square signals can be used;
- There is no need for MPPT circuits since the batteries' charge regulator supports this task.

Hybrid inverters also represent a common alternative; they are a combination between a grid-connected system and a stand-alone system with batteries, with the advantage that the PV power can also be generated and consumers can be powered even if the electrical grid is missing, during power outages.

In the built environment, because of the limited space for installing photovoltaic systems and of the shadow that might be generated by nearby objects or buildings, small- and medium-sized inverters (Fig. 2.71a) are commonly used. A convenient alternative is represented by micro-inverters (Fig. 2.71b) for individual modules that have the possibility to mitigate the influence of the shadowed modules on the energy production of the entire PV system.

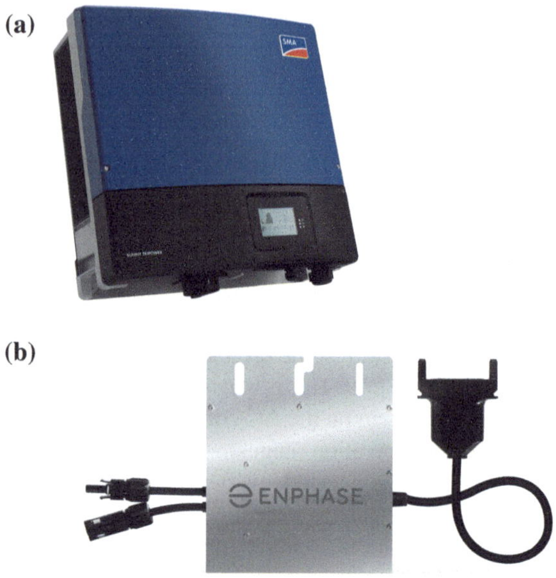

Fig. 2.71 Inverter for: **a** one or several PV arrays [142]; **b** individual PV modules [42]

2.4 Energy Mixes Based on Solar Energy Conversion Systems

The solar radiation variability represents a significant issue when the energy demand has to be fully or in a large extent covered using renewable energy sources. If the day/night variation can be balanced using feasible storage systems, with an acceptable capacity and cost, for the seasonal variability (summer/winter) huge storage capacity is required in the temperate and cold regions of the Earth.

The solutions to cover the energy demand during periods with low amount of solar energy consist of implementing energy mixes that firstly use the available solar energy and afterwards another renewable energy source such as geothermal, biomass, wind or hydro.

Depending on the type of the energy demand in the built environment (electrical or thermal energy), the possible mixes of renewable energy systems can be selected based on the data in Table 2.15. The energy mixes may consist of two or more renewable energy systems, according to the renewable energy potential in the implementation location of the building and to the available areas to install these systems.

These mixes usually rely on a renewable source of energy that is not strongly influenced by the meteorological conditions. An example of such a renewable energy source is the geothermal energy that is available all year round, day and night. Bioen-

Table 2.15 Renewable-based energy mixes in the built environment

Source	Mix		
	Thermal	Electrical	Thermal and electrical cogeneration
Solar energy	Solar thermal system	Photovoltaic system	PVT system
Geothermal energy	Heat pump system		Geothermal system
Bioenergy	Bioenergy system		Combined heat and power (CHP) plant
Wind energy		Small wind turbine	
Hydro-energy		Micro-hydroturbine	

ergy can be also considered a backup source of renewable energy. When considering the electrical energy demand, hydro-energy (if available) can be used both for energy production and for energy storage.

2.4.1 Solar Thermal–Heat Pump Systems

This type of energy mixes is used in temperate or cold climates, with rather cold winters, mainly for space heating and domestic hot water preparation.

There are three main types of solar thermal–heat pump systems: *parallel*, *serial* and *combined systems*.

The parallel systems (Fig. 2.72) consist of a classical solar thermal system where the solar thermal collector (1) is connected to the inferior coil (2) of a storage tank (3) and a heat pump (4) connected to the upper coil (5) of the storage tank (3). The heat pump extracts the geothermal energy through a geothermal heat exchanger (6). Hydraulic pumps (7) and expansion vessels (8) are used in each loop. A controller manages the heat pump and the three hydraulic pumps based on the measured temperatures in the solar thermal collector and of the water in the middle and upper part of the storage tank. An antifreeze mixture of water and ethylene glycol is used in the solar loop and in the primary loop of the heat pump, through the geothermal heat exchanger. In the secondary loop of the heat pump, deionized water is used to prevent clogging the heat exchangers. There is a broad variety of components for this type of system: the solar thermal collector (1) can be of flat plate type or with evacuated tubes; the storage tank (3) can be used for domestic hot water preparation or as buffer for space heating; the geothermal heat exchanger (6) can be either vertical or horizontal; the heat pump can extract the thermal energy from water or from air instead of the ground. In this connection type, the two systems independently heat the water in the storage tank.

The serial system (Fig. 2.73) consists of a solar thermal collector (1) directly connected in the primary circuit of the heat pump (2). An antifreeze solution circulated by a hydraulic pump (3) is used as thermal fluid in this circuit. The secondary circuit

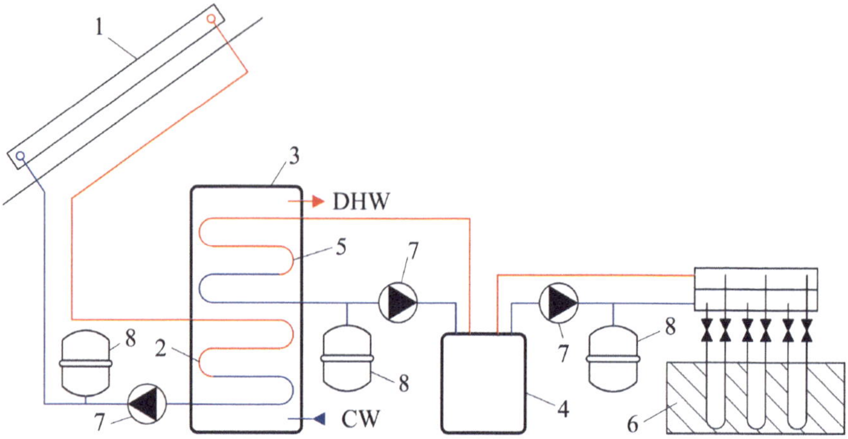

Fig. 2.72 Solar thermal–heat pump system of parallel type: solar thermal collector (1), inferior heat exchanger (2), storage tank (3), heat pump (4), superior heat exchanger (5), geothermal heat exchanger (6), hydraulic pump (7), expansion vessel (8), cold water (CW) and domestic hot water (DHW)

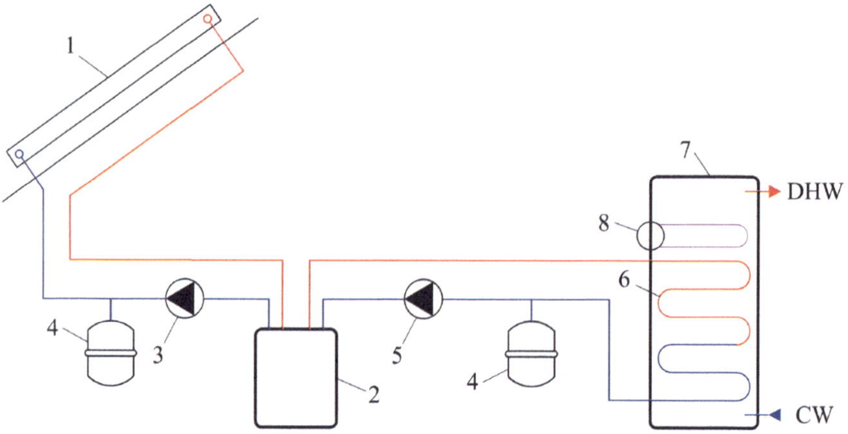

Fig. 2.73 Solar thermal–heat pump system of serial type: solar thermal collector (1), heat pump (2), hydraulic pump (3), expansion vessel (4), hydraulic pump (5), internal heat exchanger (6), storage tank (7), electric backup heater (8), cold water (CW) and domestic hot water (DHW)

of the heat pump (2) is connected to the coil (6) of the storage tank (7) that has an electrical heater (8) as auxiliary source. Deionized water circulated by the hydraulic pump (5) is used in this secondary circuit. The expansion vessels (4) compensate the volume variation of the thermal fluid during each heating/cooling phase. The solar thermal collector works at a higher efficiency due to the very low inlet temperature of the thermal fluid (usually negative). Unglazed solar thermal collectors are used in this type of system to fully exploit the thermal energy contained in the ambient air. In this

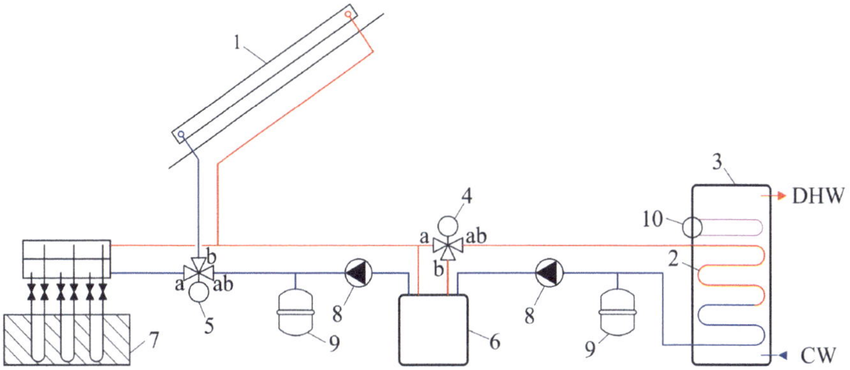

Fig. 2.74 Combined type of solar thermal–heat pump system: solar thermal collector (1), internal heat exchanger (2), storage tank (3), three-ways valves (4 and 5), heat pump (6), geothermal heat exchanger (7), hydraulic pump (8), expansion vessel (9), electric backup heater (10), cold water (CW) and domestic hot water (DHW)

type of systems, the heat pump has two sources of thermal energy: the solar radiation and the air. Another solution is used to directly connect the solar thermal collector to the heat pump's compressor when the refrigerant will be circulated through the solar thermal collector working as an evaporator. In this case, the direct expansion of the refrigerant occurring in the solar collector will increase the overall efficiency of the heat transfer while the system is simplified (the primary hydraulic pump (3) and the internal evaporator in the heat pump are no longer required). The disadvantages of larger amounts of refrigerant in the primary circuit and of possible leakage of the refrigerant into the atmosphere can be limited by the use of environmentally friendly refrigerants (e.g. CO_2 or non-halogenated hydrocarbons such as propane or isobutane when specific measures have to be considered to avoid the explosion risk in case of leakages).

The combined systems in Fig. 2.74 are more complex and can work either in parallel or in serial configuration depending on the level of the available solar energy and on the thermal energy demand. In the case of a parallel configuration, the solar thermal collector (1) is connected to the coil (2) of the storage tank (3) commuting the three-way valves (4 and 5) on the position a-ab and b-ab, respectively, and switching off the heat pump (6). When no solar energy is available, the heat pump (6) is switched on to extract the geothermal energy using the geothermal heat exchanger (7) to heat the water in the storage tank (3) through the coil (2) commuting the three-way valves (4 and 5) on the position b-ab and a-ab, respectively. In the serial configuration, at low solar irradiance levels, the primary circuit of the heat pump (6) is connected to the solar thermal collector (1) and the secondary circuit of the heat pump (6) is connected to the coil (2) of the storage tank (3) switching the three -way valves (4 and 5) in the position b-ab. The expansion vessels (9) compensate the volume variation of the thermal fluid during each heating/cooling phase, and the auxiliary heater (10) will cover the peaks of energy demand. An advantage of this system is the possibility to redirect into the ground the exceeding thermal energy produced by the solar thermal collector (1) during the summer period, after the energy demand is covered and the maximum temperature allowed in the storage tank is reached by switching the three-way valves (4 and 5) on the position b-ab and a-ab, respectively. Thus, the solar energy is stored into the ground, regenerating its heating capacity, which is mandatory especially in the case of deep vertical geothermal heat exchangers. For each functional regime, the hydraulic pumps (8) are switched on/off by a controller according to the temperatures measured in the solar thermal collector and in the storage tank.

2.4.2 Solar Thermal–Biomass Systems

The solar thermal–biomass systems are mainly used in rural areas as individual heating systems for residential houses but also for district heating in larger communities.

As high temperature is required in the biomass boilers, the two main components of a solar thermal–biomass system (Fig. 2.75), the solar thermal collector (1) and the biomass burner (2), are parallelly connected to the storage tank (3) through the

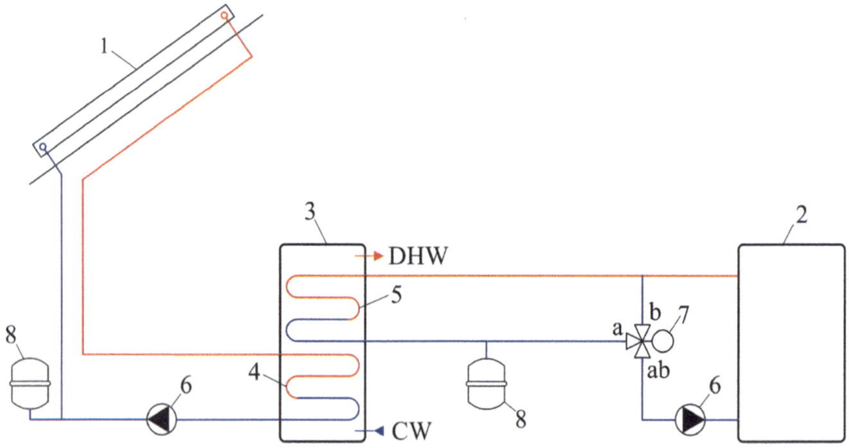

Fig. 2.75 Solar thermal–biomass system: solar thermal collector (1), biomass burner (2), storage tank (3), internal heat exchangers (4 and 5), hydraulic pump (6), three-way valve (7), expansion vessel (8), cold water (CW) and domestic hot water (DHW)

inferior (4) and superior (5) coils, respectively. Hydraulic pumps (6) circulate the thermal fluids in each circuit. A three-way valve (7) is used to keep the required high temperature level in the biomass burner circuit: after starting the burner, the three-way valve (7) is kept in the b-ab position until the temperature of the water increases enough (usually above 80 °C) when the three-way valve (7) is switched in a-ab position. The expansion vessels (8) compensate the volume variation of the thermal fluid during each heating/cooling phase. A controller manages the two hydraulic pumps (6) and the three-way valve (7) based on the temperatures measured in the solar thermal collector, in the water in the middle and upper part of the storage tank and at the return of the biomass boiler (2). This type of systems uses the solar energy to heat the water in the storage tank as long as there is enough solar irradiance; when it decreases, the water in the storage tank is only preheated and the biomass burner (2) is started. An advantage of such a system is that the thermal energy exceedingly produced during summer can be used to dry the biomass and to increase thus the solar fraction and the burning efficiency.

2.4.3 Solar PV–Wind Systems

Solar photovoltaic systems combined with wind turbine systems are common because solar radiation and wind speed resources are usually complementary. Such systems are able to ensure the required electrical power during a larger interval and may diminish the battery storage system.

The PV–wind hybrid systems can be grid connected or stand-alone. If the wind turbine produces alternative current (AC), this current has to be firstly converted to DC in the case of stand-alone systems, since the PV modules produce and the batteries store DC energy.

The energy transfer in a PV–wind system is done using *power buses*. One single bus bar is used for the systems with DC consumers only (Fig. 2.76), and two bus bars (Fig. 2.77) are used if either both types or at least AC consumers have to be powered [137]. The grey rectangles in Fig. 2.76 and in Fig. 2.77 represent components that are not mandatory in the design of a PV–wind system. For example if the charge controller and batteries are missing, Fig. 2.76, the system can be used only if the power produced using renewable energy is directly supplied, as in the case of a water pumping system. In Fig. 2.77, the DC consumers and the storage systems may not be included in the PV–wind system when only AC consumers need to be powered.

To meet the energy demand, the most important component is the inverter/rectifier able to convert one type of energy into another (from DC to AC or from AC to DC), each of them available on a different bus bar.

Another important component that supports the correct functioning of the wind turbine is the dump load. It has the role to consume the energy produced by the wind turbine when no consumers are connected to the DC bus of the system; this may happen when e.g. the DC load is disconnected and the rechargeable batteries are fully charged.

The hybrid systems described in Figs. 2.76 and 2.77, with single or double bus bars, are stand-alone systems. Grid-connected systems always require two bus bars (DC and AC), while the actual connection to the grid is done on the AC bus bar (Fig. 2.78), through a bidirectional energy metre.

The connection to a backup system is also possible in the hybrid systems with double bus bars at the AC bar; this is recommended when the renewable energy is not enough, the electricity grid is disconnected, and the batteries are discharged.

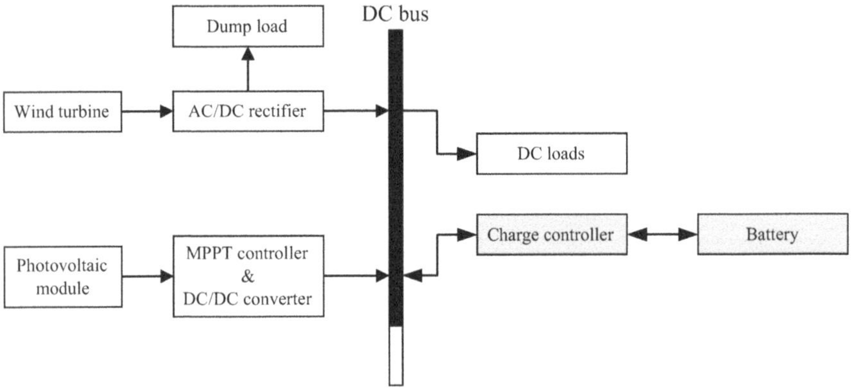

Fig. 2.76 Main components of a stand-alone PV–wind system with only DC loads

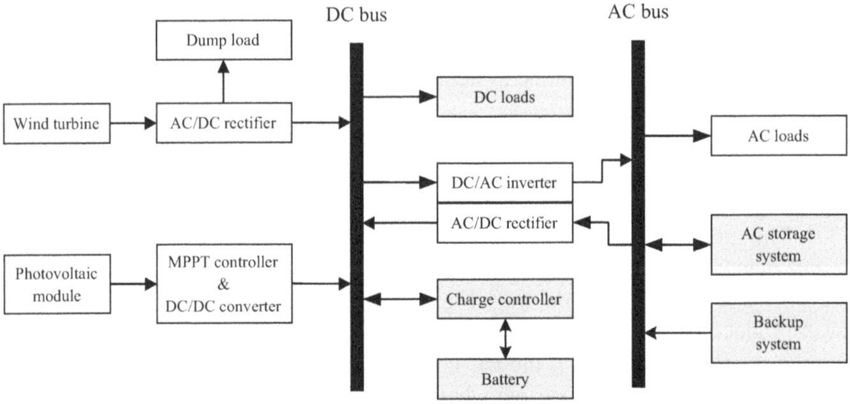

Fig. 2.77 Main components of a stand-alone PV–wind system with DC and AC loads backed up by another system

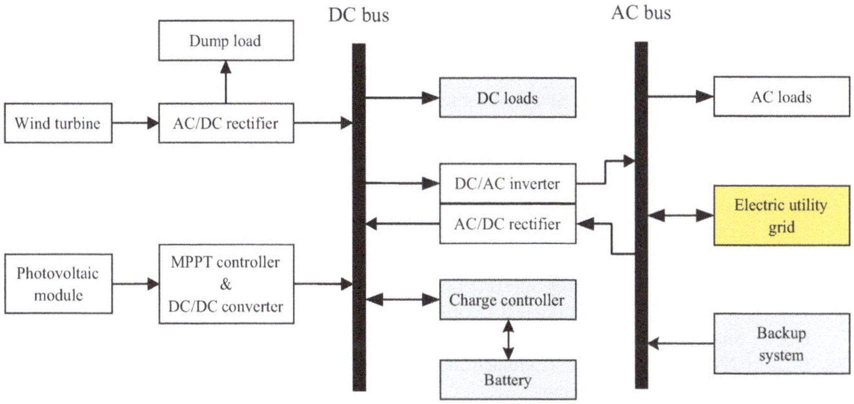

Fig. 2.78 Grid-connected hybrid system with photovoltaic modules and wind turbines

2.4.4 Solar PV–Micro-hydro Systems

Photovoltaic systems combined with micro-hydrosystems can also be used to meet the electrical energy demand. In this case, all aspects mentioned for the PV–wind systems are also valid for PV–micro-hydrosystems.

This type of hybrid systems is commonly employed to give full use to the water surface of the dam reservoir by installing PV floating arrays on it, as in Isawa, Japan [124]. Another application is when the load is increasing only during daytime and the micro-hydroturbine is not able to produce enough energy.

References

1. Aberoumand S, Ghamari S, Shabani B (2018) Energy and exergy analysis of a photovoltaic thermal (PV/T) system using nanofluids: an experimental study. Sol Energy 165:167–177
2. Ahmed A, Ossen D, Jamei E, Manaf N, Said I, Ahmad M (2014) Urban surface temperature behaviour and heat island effect in a tropical planned city. Theor Appl Climatol 115:493–514
3. Alves EDL (2016) Seasonal and spatial variation of surface urban heat island intensity in a small urban agglomerate in Brazil. Climate 4(61). https://doi.org/10.3390/cli4040061
4. American Meteorological Society (2019) American Meteorological Society Glossary of Meteorology. http://glossary.ametsoc.org/wiki/Main_Page. Accessed 20 July 2019
5. American Society of Mechanical Engineers (1989) Performance test code for wind turbines. ASME/ANSI PTC 42-1988, New York, USA
6. Andenæs E, Jelle BP, Ramlo K, Kolås T, Selj J, Foss SE (2018) The influence of snow and ice coverage on the energy generation from photovoltaic solar cells. Sol Energy 159:318–328
7. Andreou E, Axarli K (2012) Investigation of urban canyon microclimate in traditional and contemporary environment. Experimental investigation and parametric analysis. Renew Energy 43:354–363
8. Antonanzas-Torres F, Urraca R, Polo J, Perpiñán-Lamigueiro O, Escobar R (2019) Clear sky solar irradiance models: a review of seventy models. Renew Sust Energy Rev 107:374–387

9. Atwater MA, Ball JT (1981) A surface solar radiation model for cloudy atmospheres. Mon Weather Rev 109:878–888
10. Azargohar R, Sonil N, Dalai AK, Kozinski JA (2019) Physico-chemistry of biochars produced through steam gasification and hydro-thermal gasification of canola hull and canola meal pellets. Biomass Bioenerg 120:458–470
11. Azevedeo JA, Chapman L, Muller CL (2016) Quantifying the daytime and night-time urban heat island in Birmingham, UK: a comparison of satellite derived land surface temperature and high resolution air temperature observations. Remote Sens-Basel 8(153). https://doi.org/10.3390/rs8020153
12. Badescu V (2013) Assessing the performance of solar radiation computing models and model selection procedures. J Atmos Sol-Terr Phy 105–106:119–134
13. Bakirci K (2012) General models for optimum tilt angles of solar panels: Turkey case study. Renew Sustain Energy Rev 16(8):6149–6159
14. Bason F (2004) Diffuse solar irradiance and atmospheric turbidity. In: Proceedings of EuroSun Conference, Freiburg, Germany
15. Bason F (2007) Solar irradiance measurements from the Danish Galathea 3 expedition. In: Proceedings of ISES solar world conference, Beijing, China
16. Battery University (2019) How to calculate battery runtime. https://batteryuniversity.com/learn/article/bu_503_how_to_calculate_battery_runtime. Accessed 15 July 2019
17. Benghanem M (2011) Optimization of tilt angle for solar panel: case study for Madinah, Saudi Arabia. Appl Energ 88(4):1427–1433
18. Borbora J, Das AK (2014) Summer time urban heat island study for Guwahati city, India. Sustain Cities Soc 11:61–66
19. Brabec M, Paulescu M, Badescu V (2015) Tailored vs black-box models for forecasting hourly average solar irradiance. Sol Energy 111:320–331
20. Burton T, Sharpe D, Jenkins N, Bossanyi E (2001) Wind energy handbook. Wiley, West Sussex
21. Busato B, Lazzarin RM, Noro M (2014) Three years of study of the urban heat island in Padua: experimental results. Sustain Cities Soc 10:251–258
22. Chang YP (2009) Optimal design of discrete-value tilt angle of PV using sequential neural-network approximation and orthogonal array. Expert Syst Appl 36(3):6010–6018
23. Chapin DM, Fuller CS, Pearson GL (1954) A new silicon p-n junction photocell for converting solar radiation into electrical power. J Appl Physics 8:676–680
24. Charalambides AG, Burduhos BG (2014) Effect of clouds on solar irradiance. UEFISCDI, Module III—Bilateral Cooperation, contract no. 765/30.04.2014. http://old.unitbv.ro/ecsol-prognosis/ro-ro/results.aspx, Accessed June 2019
25. Colli A, Zaaiman WJ (2012) Maximum-power-based PV performance validation method: application to single-axis tracking and fixed-Tilt c-Si systems in the Italian Alpine region. IEEE J Photovolt 2(4):555–563
26. Cotar A, Flicik A (2012) Photovoltaic systems. REA Kvarner d.o.o., Rijeka, Croatia
27. Cotorcea A (2017) A review of methods for solar radiation estimation used in off-shore applications. "Mircea cel Batran" Naval Acad Sci Bull 20(2):8–11. https://www.anmb.ro/buletinstiintific/buletine/2017_Issue2/8-11.pdf. Accessed 15 July 2019
28. Despotovic M, Nedic V (2015) Comparison of optimum tilt angles of solar collectors determined at yearly, seasonal and monthly levels. Energy Convers Manag 97:121–131
29. Deutsche Gesellschaft für Sonnenenergie (2009) Planning & installing photovoltaic systems—a guide for installers, architects and engineers. Earthscan Ltd., London
30. Diaconescu DV, Visa I, Hermenean IS, Vatasescu MM (2009) Clouds influence on the solar radiation for a mountain location. Environ Eng Manag J 8:849–853
31. Donev JMKC, Yyelland B, Stenhouse K, Hanania J, Amin S (2018) Energy education—solar energy to the Earth. https://energyeducation.ca/encyclopedia/Solar_energy_to_the_Earth. Accessed 20 July 2019
32. dos Santos IP, Ruther R (2014) Limitations in solar module azimuth and tilt angles in building integrated photovoltaics at low latitude tropical sites in Brazil. Renew Energy 63:116–124

33. Duffie JA, Beckman A (2013) Solar engineering of solar thermal processes, 4th edn. Wiley, Hoboken
34. Dunlop JP (2010) Photovoltaic systems, 2nd edn. American Technical Publishers, Orland Park
35. Earth System Research Laboratory (2019) LLLJP wind shear formula (power law). https://www.esrl.noaa.gov/csd/projects/lamar/windshearformula.html. Accessed 15 July 2019
36. Elhab BR, Sopian K, Mat S, Lim C, Sulaiman MY, Ruslan MH, Saadatian O (2012) Optimizing tilt angles and orientations of solar panels for Kuala Lumpur, Malaysia. Sci Res Essays 7(42):3758–3765
37. Energy Informative (2019) Grid tied, off-grid and hybrid solar systems. https://energyinformative.org/grid-tied-off-grid-and-hybrid-solar-systems. Accessed 15 July 2019
38. Engineering ToolBox (2019a) Combustion of wood—heat values. https://www.engineeringtoolbox.com/wood-combustion-heat-d_372.html, Accessed 01 Feb 2019
39. Engineering ToolBox (2019b) Fuels—higher and lower calorific values. https://www.engineeringtoolbox.com/fuels-higher-calorific-values-d_169.html. Accessed 01 Feb 2019
40. Engineering ToolBox (2019c) Fuel gases heating values. https://www.engineeringtoolbox.com/heating-values-fuel-gases-d_823.html. Accessed 01 Feb 2019
41. Engineering ToolBox (2019d) Ethylene glycol heat-transfer fluid. https://www.engineeringtoolbox.com/ethylene-glycol-d_146.html, Accessed 01 Feb 2019
42. Enphase (2019) http://www.enphase.com. Accessed May 2019
43. Eppley (2019a) Automatic solar tracker, model SMT. http://www.eppleylab.com/instrument-list/automatic-solar-tracker. Accessed 15 July 2019
44. Eppley (2019b) Black & white pyranometer—the diffuse pyranometer model, pp 8–48. http://www.eppleylab.com/instrument-list/black-white-pyranometer. Accessed May 2019
45. European Environment Agency (2009) Europe's onshore and offshore wind energy potential: An assessment of environmental and economic constraints. EEA Technical report 6. https://www.energy.eu/publications/a07.pdf. Accessed 8 July 2019
46. Fekete I, Farkas I (2019) Numerical and experimental study of building integrated solar tile collectors. Renew Energy 137:45–55
47. Ferreira F, Camacho JR (2016) Small hydro plants—estimated hydropower potential. Lambert Academic Publishing
48. Fraenkel P (1997) Hydrosoft: a software tool for the evaluation of low-head hydropower sources. In: Proceedings of Hidroenergia97 Conference, p 380
49. Fridleifsson IB, Bertani R, Huenges E, Lund JW, Ragnarsson A, Rybach L (2008) The possible role and contribution of geothermal energy to the mitigation of climate change, Proceedings of IPCC scoping meeting on renewable energy sources, Luebeck, Germany, pp 59–80
50. Gautier C (1986) Evolution of net surface shortwave radiation over the Indian Ocean during MONEX 1979: a satellite description. Mon Weather Rev 114:525–533
51. Ge R, Wang J, Lixiao Z, Guangjin T (2016) Impacts of urbanization on the urban thermal environment in Beijing. Acta Ecol Sin 36. https://doi.org/10.5846/stxb201409301935
52. Geothermal Energy Administration (2007) A Guide to Geothermal and the Environment
53. Gipe P (2019) Fundamentals of wind energy. http://www.wind-works.org/cms/fileadmin/user_upload/Files/presentations/Gipe_ZNE-SCE-Wind_02_Sizing.pdf. Accessed 15 May 2019
54. Glassley WE (2014) Geothermal energy: renewable energy and environment, 2nd edn. CRC Press, Taylor & Francis Group, Boca Raton, London, New York
55. Global Solar Atlas (2019) https://globalsolaratlas.info. Accessed 01 May 2019
56. Gong X, Kulkarni M (2005) Design optimization of a large scale rooftop photovoltaic system. Sol Energy 78(3):362–374
57. Gong FY, Zeng ZC, Ng E, Norford LK (2019) Spatiotemporal patterns of street-level solar radiation estimated using Google street view in a high-density urban environment. Build Environ 148:547–566
58. Green MA, Hishikawa Y, Dunlop ED, Levi DH, Hohl-Ebinger J, Ho-Baillie AWY (2018) Solar cells efficiency tables (version 52). Prog Photovolt 26(7):427–436

59. Green MA, Hishikawa Y, Dunlop ED, Levi DH, Hohl-Ebinger J, Yoshita M, Ho-Baillie AWY (2019) Solar cell efficiency tables (version 53). Prog Photovolt 27(1):3–12
60. GreensunSolar (2019) https://www.greensunpv.com/greensun-mono-48v-solar-panel-480w-490w-500w_p26.html. Accessed 15 June 2019
61. Guerrero-Lemus R, Vega R, Kim T, Kimm A, Shepard LE (2016) Bifacial solar photovoltaics—a technology review. Renew Sust Energy Rev 60:1533–1549
62. Gueymard CA (2017) Cloud and Albedo enhancement impacts on solar irradiance using high-frequency measurements from thermopile and photodiode radiometers. Part 2: performance of separation and transposition models for global tilted irradiance. Sol Energy 153:766–779
63. Gueymard CA, Lara-Fanego V, Sengupta M, Xie Y (2019) Surface albedo and reflectance: review of definitions, angular and spectral effects, and intercomparison of major data sources in support of advanced solar irradiance modeling over the Americas. Sol Energy 182:194–212
64. Hartner M, Ortner A, Hiesl A, Haas R (2015) East to west—the optimal tilt angle and orientation of photovoltaic panels from an electricity system perspective. Appl Energy 160:94–107
65. Hassid S, Santamouris M, Papanikolaou N, Linardi A, Klitsikas N, Georgakis C (2000) The effect of the Athens heat island on air conditioning load. Energy Build 32:131–141
66. Hau E (2006) Wind turbines: fundamentals, technologies, application, economics. Springer, Germany
67. Heller A (2000) 15 years of R&D in central solar heating in Denmark. Sol Energy 69(6):437–447
68. OTT HydroMet GmbH (2019) OTT C31 universal current meter for discharge measurements. https://www.ott.com/products/water-flow-3/ott-c31-958. Accessed May 2019
69. Ibrahim D (1995) Optimum tilt angle for solar collectors used in Cyprus. Renew Energ 6(7):813–819
70. Ineichen P, Perez R (2002) A new air mass independent formulation for the Linke turbidity coefficient. Sol Energy 73:151–157
71. Intergovernmental Panel on Climate Change (2013) Climate Change 2013: the physical science basis. Cambridge University Press, Cambridge
72. International Energy Agency (2017a) Bioenergy's role in balancing the electricity grid and providing storage options—an EU perspective, IEA Bioenergy: task 41P6. https://www.ieabioenergy.com/wp-content/uploads/2017/02/IEA-Bioenergy-bio-in-balancing-grid_master-FINAL.pdf. Accessed 15 July 2019
73. International Energy Agency (2017b) IEA Bioenergy, global wood pellet industry and trade study 2017, IEA Bioenergy Task 40 Report. http://task40.ieabioenergy.com/wp-content/uploads/2013/09/IEA-Wood-Pellet-Study_final-2017-06.pdf. Accessed 15 July 2019
74. International Energy Agency (2018a) Renewables information: overview. https://webstore.iea.org/renewables-information-2018-overview. Accessed 15 July 2019
75. International Energy Agency (2018b) Trends 2018 in photovoltaic applications—survey report of selected IEA countries between 1992 and 2017. http://www.iea-pvps.org/fileadmin/dam/intranet/task1/IEA_PVPS_Trends_2018_in_Photovoltaic_Applications_03.pdf. Accessed 17 June 2019
76. International Energy Agency (2019) Annual report 2018 IEA bioenergy. Pearse Buckley Ed., Dublin, Ireland
77. Japan Meteorological Agency (2019) Advanced information on solar and infrared radiation. http://www.data.jma.go.jp/gmd/env/radiation/en/nmhs/nmhs_rad.html. Accessed May 2019
78. Jerlov NG (1968) Global radiation incident on the sea surface (Chap. 4). In: Optical oceanography. Elsevier Oceanography Series, vol 5, pp 65–68
79. Johnson G (2006) Wind energy systems. Manhattan, KS
80. Jouybari HJ, Saedodin S, Zamzamian A, Nimvari ME, Wongwises S (2017) Effects of porous material and nanoparticles on the thermal performance of a flat plate solar collector: an experimental study. Renew Energy 114:1407–1418
81. Kallis KX, Pellegrini Susini GA, Oakey JE (2013) A comparison between Miscanthus and bioethanol waste pellets and their performance in a downdraft gasifier. Appl Energy 101:333–340

82. Kalogirou SA (2009) Solar energy engineering—processes and systems, 2nd edn. Academic Press, London
83. Kaltschmitt M, Streicher W, Wiese A (2007) Renewable energy: technology, economics, and environment. Springer, Berlin, p 55
84. Khahro SF, Tabbassum K, Talpur S, Alvi MB, Liao X, Dong L (2015) Evaluation of solar energy resources by establishing empirical models for diffuse solar radiation on tilted surface and analysis for optimum tilt angle for a prospective location in southern region of Sindh, Pakistan. Int J Electr Power Energy Syst 64:1073–1080
85. Kipp & Zonen (2019) CM121B/C shadow ring. https://www.kippzonen.com/Product/42/CM121B-C-Shadow-Ring#.XRwvd_5S-M8. Accessed May 2019
86. Klok L, Zwart S, Verhagen H, Mauri E (2012) The surface heat island of Rotterdam and its relationship with urban surface characteristics. Resour Conserv Recycl 64:23–29
87. Lac C, Donnelly RP, Masson V, Pal S, Riette S, Donier S, Queguiner S, Tanguy G, Ammoura L, Xueref-Remy I (2013) CO_2 dispersion modelling over Paris region within the CO_2-megaparis project. Atmos Chem Phys 13:4941–4961
88. Lawrence Berkley National Laboratory (2019) Cool Roofing Materials Database, https://heatisland.lbl.gov/resources/cool-roofing-materials-database. Accessed February 2019
89. Lazzarini M, Marpu PR, Ghedira H (2013) Temperature-land cover interactions: the inversion of urban heat island phenomenon in desert city areas. Remote Sens Environ 130:136–152
90. Li DHW, Lam TNT (2007) Determining the optimum tilt angle and orientation for solar energy collection based on measured solar radiance data. Int J Photovolt 1–9
91. Livada I, Santamouris M, Niachou K, Papanikolaou N, Mihalakakou G (2002) Determination of places in the great Athens area where the heat island effect is observed. Theor Appl Climatol 71:219–230
92. Lokoshchenko MA (2014) Urban 'heat island' in Moscow. Urban Clim 10:550–562
93. Lund JW, Boyd TL (2015) Direct utilization of geothermal energy 2015 worldwide review. Proceedings of the world geothermal congress 2015
94. Mahbubul IM, Khan MMA, Ibrahim NI, Ali HM, Al-Sulaiman FA, Saidur R (2018) Carbon nanotube nanofluid in enhancing the efficiency of evacuated tube solar collector. Renew Energy 121:36–44
95. Marando F, Salvatori E, Sebastiani A, Fusaro L, Manes F (2019) Regulating ecosystem services and green infrastructure: assessment of urban heat island effect mitigation in the municipality of Rome, Italy. Ecol Model 392:92–102
96. Marugo G, Valdes CF, Gomez C, Chejne F (2019) Pelletizing of Colombian agro-industrial biomasses with crude glycerol. Renew Energy 134:558–568
97. Masters GM (2004) Renewable and efficient electric power systems. Wiley, Hoboken
98. Meliss M (1997) Regenerative energiequellen—praktikum. Springer, Heidelberg
99. Moghadam H, Tabrizi FF, Sharak AZ (2011) Optimization of solar flat collector inclination. Desalination 265(1–3):107–111
100. Mohajeri N, Gudmundsson A, Kunckler T, Upadhyay G, Assouline D, Kämpf JH, Scartezzini JL (2019) A solar-based sustainable urban design: the effects of city-scale street canyon geometry on solar access in Geneva, Switzerland. Appl Energy 240:173–190
101. Mondol JD, Yohanis YG, Norton B (2008) Solar radiation modelling for the simulation of photovoltaic systems. Renew Energy 33:1109–1120
102. Nasir DSN, Hughes BR, Calautit JK (2017) Influence of urban form on the performance of road pavement solar collector system: symmetrical and asymmetrical heights. Energ Convers Manag 149:904–917
103. National Aeronautics and Space Administration—Earth Observatory (2015) Climate and Earth's energy budget. http://earthobservatory.nasa.gov/Features/EnergyBalance. Accessed July 2019
104. National Renewable Energy Laboratory (1992) Shining on—a primer on solar radiation data. https://rredc.nrel.gov/solar/pubs/shining/chap4.html. Accessed May 2019
105. National Renewable Energy Laboratory (2009) Biodiesel handling and use guide, 4th edn. https://www.nrel.gov/docs/fy09osti/43672.pdf. Accessed 01 Feb 2019

106. National Renewable Energy Laboratory (2018) PV efficiency chart. https://www.nrel.gov/pv/assets/pdfs/pv-efficiency-chart.20181221.pdf. Accessed May 2019
107. National Renewable Energy Laboratory (2019) Solar spectra—air mass zero. http://rredc.nrel.gov/solar/spectra/am0. Accessed May 2019
108. Newport (2019) Calibrated reference cell, meter, quartz window model: 91150 V. https://www.newport.com/p/91150V. Accessed May 2019
109. Nicholson SE (1975) A pollution model for street-level air. Atmos Environ 9:19–31
110. Nkwetta DN, Smyth M, Zacharopoulos A, Hyde T (2013) Experimental performance analysis and optimisation of medium temperature solar thermal collectors with silicon oil as a heat transfer fluid. Int J Energy Res 37:570–581
111. O'Hegarty R, Kinnane O, McCormack SJ (2017) Concrete solar collectors for façade integration: an experimental and numerical investigation. Appl Energy 206:1040–1061
112. Ohl RS (1941) Light-sensitive electric device including silicon. US Patent 2443542, filed 27 May 1941
113. Oke TR (1988) Street design and urban canopy layer climate. Energy Build 11:103–113
114. Oke TR (1995) The heat island of the urban boundary layer: characteristics, causes and effects. Wind Clim Cities 277:81–107
115. Pandey CK, Katiyar AK (2009) A note on diffuse solar radiation on a tilted surface. Energy 34(11):1764–1769
116. Pandey CK, Katiyar AK (2009) A comparative study to estimate daily diffuse solar radiation over India. Energy 34(11):1792–1796
117. Pandey CK, Katiyar AK (2013) Solar radiation: models and measurement techniques. J Energy, Article ID 305207
118. Patel MR (1999) Wind and solar power system. CRC Press LLC, USA
119. Patel MR (2006) Wind and solar power systems—design, analysis, and operation. Taylor & Francis, New York
120. Pearson GL (1985) PV founders award luncheon. Conference record. In: 18th IEEE photovoltaic specialists conference, Las Vegas
121. Pena R, Medina A (2010) Capacity estimation methods applied to mini hydro plants. In: Gaonkar DN (ed) Distributed generation. InTech. http://www.intechopen.com/books/distributed-generation/capacity-estimation-methods-applied-to-mini-hydro-plants. Accessed May 2019
122. Perez R, Ineichen P, Moore K, Kmiecik M, Chain C, George R, Vignola F (2002) A new operational model for satellite-derived irradiances: description and validation. Sol Energy 73:307–317
123. Pinker RT, Frouin R, Li Z (1995) A review of satellite methods to derive surface shortwave irradiance. Remote Sens Environ 51:108–124
124. Prouvost B (2017) Creating the ultimate hybrid system by mixing solar energy and hydroelectricity. Renew Energy Focus. http://www.renewableenergyfocus.com/view/45793/creating-the-ultimate-hybrid-system-by-mixing-solar-energy-and-hydroelectricity. Accessed 25 July 2019
125. PVGIS (2017) http://re.jrc.ec.europa.eu/. Accessed May 2019
126. Rakovec J, Zaksek K (2012) On the proper analytical expression for the sky-view factor and the diffuse irradiation of a slope for an isotropic sky. Renew Energy 37:440–444
127. Reed RK (1977) On estimating insolation over the ocean. J Phys Oceanogr 7:482–485
128. Renne D, Perez R, Zelenka A, Whitlock C, DiPasquale R (1999) Use of weather and climate research satellites for estimating solar resources. Adv Sol Energy 13:171
129. Rigollier C, Bauer O, Wald L (2000) On the clear sky model of the ESRA—European solar radiation atlas with respect to the Heliosat method. Sol Energy 28(1):33–48
130. Rigollier C, Lefevre M, Wald L (2004) The method Heliosat-2 for deriving shortwave solar radiation from satellite images. Sol Energy 77:159–169
131. Rohde RA (2007) Global warming art. http://web.archive.org/web/20151013063240/http://www.globalwarmingart.com/wiki/File:Solar_Spectrum_png. Accessed 15 June 2019

132. Rowlands IH, Kemery BP, Beausoleil-Morrison I (2011) Optimal solar-PV tilt angle and azimuth: an Ontario (Canada) case-study. Energy Policy 39(3):1397–1409
133. Said Z, Saidur R, Sabiha MA, Rahim NA, Anisur MR (2015) Thermo-physical properties of Single Wall carbon nanotubes and its effect on exergy efficiency of a flat plate solar collector. Sol Energy 115:757–769
134. Santamouris M, Synnefa A, Karlessi T (2011) Using advanced cool materials in the urban built environment to mitigate heat islands and improve thermal comfort conditions. Sol Energy 85(12):3085–3102
135. Sarker S, Arauzo J, Nielsen HK (2015) Semi-continuous feeding and gasification of alfalfa and wheat straw pellets in a lab-scale fluidized bed reactor. Energy Convers Manag 99:50–61
136. Sathyajith M (2006) Wind energy—fundamentals, resource analysis and economics. Springer, Berlin
137. Sawle Y, Gupta SC, Bohre AK (2016) PV-wind hybrid system: a review with case study. Cogent Eng 3:1189305
138. Shockley W, Queisser HJ (1961) Detailed balance limit of efficiency of p-n junction solar cells. J Appl Phys 32:510–519
139. Shukla KN, Rangnekar S, Sudhakar K (2015) Comparative study of isotropic and anisotropic sky models to estimate solar radiation incident on tilted surface: a case study for Bhopal, India. Energy Rep 1:96–103
140. Simone M, Barontini F, Nicolella C, Tognotti L (2012) Gasification of pelletized biomass in a pilot scale downdraft gasifier. Bioresour Technol 116:403–412
141. Siraki AG, Pillay P (2012) Study of optimum tilt angles for solar panels in different latitudes for urban applications. Sol Energy 86:1920–1928
142. SMA (2019) http://www.sma.de. Accessed May 2019
143. Soares J, Oliveira AP, Božnar MZ, Mlakar P, Escobedo JF, Machado AJ (2004) Modeling hourly diffuse solar-radiation in the city of Sao Paulo using a neural-network technique. Appl Energy 79(2):201–214
144. SoDa (2019) http://www.soda-pro.com/maps. Accessed 01 May 2019
145. SolarGIS (2019) https://solargis.com. Accessed 01 May 2019
146. Soltani A, Sharifi E (2017) Daily variation of urban heat island effect and its correlations to urban greenery: a case study of Adelaide. Front Arch Res 6:529–538
147. Soulayman S, Sabbagh W (2014) An algorithm for determining optimum tilt angle and orientation for solar collectors. J Sol Energy Res Updat 1:19–30
148. Sözen A, Arcaklioglu E, Ozalp M (2004) Estimation of solar potential in Turkey by artificial neural networks using meteorological and geographical data. Energy Convers Manag 45(18–19):3033–3052
149. Stanciu C, Stanciu D (2014) Optimum tilt angle for plate collectors all over the World—a declination dependence formula and comparisons of three solar radiation models. Energy Convers Manage 81:133–143
150. Tao P, Shu L, Zhang J, Lee C, Deng T (2018) Silicone oil-based solar-thermal fluids dispersed with PDMS-modified Fe_3O_4@graphene hybrid nanoparticles. Prog Nat Sci-Mater 28(5):554–562
151. Thaker P, Gokhale S (2016) The impact of traffic-flow patterns on air quality in urban street canyons. Environ Pollut 208:161–169
152. Tiwari AK (2015) Application of nanoparticles in solar collectors: a review. Mater Today: Proc 2(4–5):3638–3647
153. Vallati A, Grignaffini A, Romagna M, Mauri L, Colucci C (2016) Influence of street canyon's microclimate on the energy demand for space cooling and heating of buildings. Energy Proc 101:941–947
154. Vardoulakis S, Fisher BEA, Pericleous K, Gonzalez-Flesca N (2003) Modelling air quality in street canyons: a review. Atmos Environ 37(2):155–182
155. Visa I, Jaliu C, Duta A, Neagoe M, Comsit M, Moldovan M, Ciobanu D, Burduhos B, Saulescu R (2015) The role of mechanisms in sustainable energy systems. Transilvania University of Brasov Publishing House, Brasov

156. Wang W, Liu K, Tang R, Wang S (2019) Remote sensing image-based analysis of the urban heat island effect in Shenzhen, China. Phys Chem Earth. https://doi.org/10.1016/j.pce.2019.01.00
157. World Energy (2019) Energy resources, hydropower. https://www.worldenergy.org/data/resources/resource/hydropower. Accessed May 2019
158. Yakup MAHM, Malik AQ (2001) Optimum tilt angle and orientation for solar collector in Brunei Darussalam. Renew Energy 24(2):223–234
159. Yan R, Saha TK, Meredith P, Goodwin S (2013) Analysis of yearlong performance of differently tilted photovoltaic systems in Brisbane, Australia. Energy Convers Manag 74:102–108
160. Yang F, Gao Y, Zhong K, Kang Y (2016) Impacts of cross-ventilation on the air quality in street canyons with different building arrangements. Build Environ 104:1–12
161. Yuan J, Emura K, Farnham C (2016) Highly reflective roofing sheets installed on a school building to mitigate urban heat island effect in Osaka. Sustainability 8:514
162. Yurddas A, Çerçi Y (2017) Numerical analysis of heat transfer in a flat-plate solar collector with nanofluids. Heat Transf Res 48(8):681–714
163. Zentrum für Sonnenenergie- und Wasserstoff-Forschung Baden-Württemberg (2014) ZSW brings world record back to Stuttgart: new best mark in thin-film solar performance with 21.7 percent efficiency. Stuttgart, Germany

Chapter 3
Increasing the Solar Share in Electricity Production in the Built Environment

3.1 Photovoltaic Systems at Building and Community Level

In the beginning, because of their high cost (300 USD/W in 1956, compared to 0.5 USD/W for a conventional power plant), the PV cells were used to power low energy consumers, as radios and toys. However, the silicon PV cells represented a key component for the Earth-orbiting satellites designed by the US Army. The PV cells could be used to power satellites starting with 1958, beginning with the Vanguard satellite. The use of PV cells on the Earth in specific applications as offshore warning lights for oil rigs represented a feasible alternative that was implemented based on PV cells manufactured using a poorer grade silicon and cheaper packaging materials, as proposed by Berman (Exxon Company). The next step in a broader use of the PV cells was related to powering the lighthouses of almost all Coast Guards in the world that started in the 1970s; further on, in 1974 the first solar power source was used to track railway safety devices in Georgia, USA. These types of applications, restricted to remote areas were then further extended all over the world, are the solar-powered repeaters that supported the telecommunications in Australia. The system worked very well thus, starting with 1985, remote communications are generally relying on PV modules. Another application was the use of solar power for water pumping and treatment in remote areas and this is currently applied in many African and other developing countries. During the 1980s, the electrical energy produced using solar energy became common in various parts of the world, e.g. in Kenya, Mexico or Tahiti where the power was used to cover the needs of individual houses. As stated by the World Energy Council [103], this was possible because during 1960s and 1970s a PV cell could not produce energy, during its operation period, at least equal with that used for manufacturing it but the energy payback time has been reduced to 2–4 years depending on the implementation location and on the use, while the lifetime of the cells and modules continuously increased (close to 30 years). During the 80s, the Swiss engineer Marcus Real demonstrated the economic advantages of installing PV arrays on the buildings by selling 333 rooftop systems to homeowners in Zurich,

I. Visa et al., *Solar Energy Conversion Systems in the Built Environment*, Green Energy and Technology, https://doi.org/10.1007/978-3-030-34829-8_3

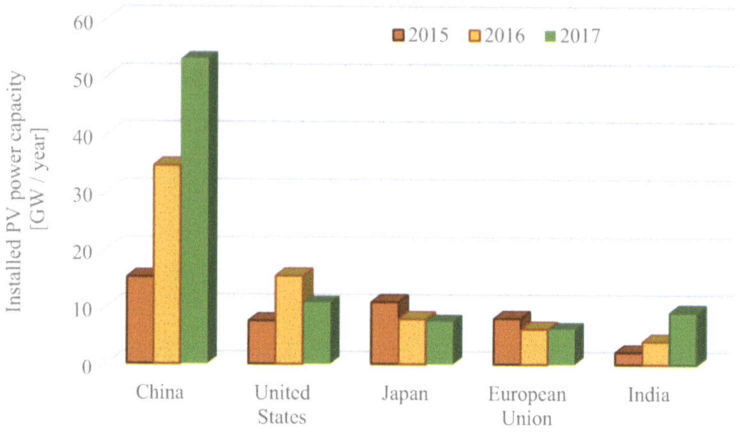

Fig. 3.1 Frontrunners in installed PV capacity in 2015, 2016 and 2017 [44]

Switzerland. After this success, the use of PV arrays in the built environment became more popular. Most of the issues that limit the PV systems implementation in the built environment are currently related to the cost of the silicon-based PV modules. However, the increase in the cost of the fossil fuel-based resources (e.g. oil and gas) and the increasing energy demand at world's population level are expected to support the solar energy use to produce electrical energy. Moreover, this path also well supports sustainability, as it is not producing any emissions of greenhouse gases during operation [95].

The implementation and use of photovoltaic systems recorded a dynamic that started about twenty years ago, reached worldwide 8 GW in 2007 and 403.3 GW in 2017 as presented in the Report of the International Energy Agency [44]. It is to mention that this value represents a 33% increase compared to 2016.

As the data in Fig. 3.1 show, China was the country with the highest installed PV capacity (53.1 GW in 2017), followed by the USA (10.7 GW), India (9.1 GW) and Japan (7.5 GW). In the entire European Union, the PV systems installed in 2017 represented 6.1 GW, the largest solar power systems being implemented in Germany (1.8 GW).

The 403.3 GW cumulative PV capacity at the end of 2017 is mainly split among 12 countries: China (32%), USA (13%), Japan (12%), Germany (11%), Italy (5%), India (5%), UK (3%), France (2%), Australia (2%) and Korea, Spain and Belgium with 1% each. The rest of 12% corresponds to other countries that each represents less than 1% of the total installed PV capacity [44]. Some countries have already installed large PV systems during the previous years and therefore are not well positioned in the 2017 ranking or are not included at all, e.g. Italy (19.7 GW), Belgium (3.8 GW), The Netherlands (2.9 GW), Thailand (2.7 GW), Romania (1.4 GW) and Brazil (1.1 GW). Moreover, there are several newcomers, countries that more recently decided to support the PV systems installation as Malaysia (60 MW installed in 2017), Taiwan (523 MW), Chile (892 MW), Mexico (285 MW), Jordan (117 MW) and Algeria (80 MW).

The grid-connected, centralized PV systems installed in 2017 represented over 60% of all the newly installed systems, proving that most of these systems were implemented as a decision towards sustainability. This percentage can be compared with that recorded in 2009 (about 22% in grid-connected, centralized systems and the rest of the systems installed as grid-connected decentralized systems). The grid-connected distributed systems can be applied on existing buildings (BAPV) or can be integrated in the buildings by replacing the conventional building materials (BIPV). The grid-connected centralized PV systems are not associated to a particular user (or group of users) and are usually ground-mounted, e.g. the PV parks or the PV arrays used for powering various agricultural devices, as the pumping systems [44].

Moreover, estimated data on the stand-alone newly installed PV systems show significant values, as 36 MW implemented in 2017 in Australia or 34 MW installed in 2016 in Japan.

The stand-alone PV systems may represent a feasible alternative in countries with enough solar energy resources, as in Africa or on remote islands, where mainly the energy storage issues have to be solved when designing the system, along with the local grid management.

Small PV systems with battery storage (solar home systems, SHS) reached six million units at world level, installed in 2017. This option was adopted, e.g. by Bangladesh that installed about four million SHS to deliver electrical energy to over thirty million people. Poverty alleviation programmes developed in several countries included installing PVs on the buildings in remote areas, using backup systems (based on diesel generators or chemical batteries), e.g. in China, while Colombia started to build in 2018 a 82 MW solar PV plant for rural electrification [44].

Installing PV systems to meet the electrical energy demand represented a continuous target during the past years. In Australia, the overall installed PV capacity is 7.49 GW (out of which about 1.3 GW installed in 2017) and most of these are implemented on the rooftop of over 1.8 million buildings, showing a PV penetration of over 20% in the residential sector. In Japan, the earthquake in 2011 outlined the need for decentralized electrical energy sources and the future 30 GW PV projects already approved are part of the development strategy, supported by the Feed-in Tariff schemes. In 2017, there were installed about 35 MW systems on or in the envelope of new buildings as BIPV systems. All over the world, several schemes support the PV systems implementation to cover the energy demand in buildings:

- In Korea, the Government supports up to 50% of the PV system cost for PV capacities of less than 50 kW installed on buildings, excluding homes, while in the new buildings with surfaces over 1000 m^2, at least 15% (in 2016) from the total energy demand has to be covered using renewable energy sources. It is planned to increase this percentage up to 30% by 2020;
- The PV rooftop program was launched in 2016 in Thailand and received over 350 applications (for 32.72 MW). The systems were installed and were connected to the grid in 2017 and will be monitored to register their output and to formulate future installation rules and prerequisites;

- The directive of the European Parliament and Council [29] entered in force in 2019 for public buildings and is expected to strengthen the cooperation between the buildings construction companies and the PV manufacturers, to comply with the Directive.

A synthesis of the key figures on PV installed all over the world is inserted in Table 3.1. The data show various trends that influence the PV systems installation. For countries that started years ago the quest for sustainability and have already installed PV systems that are close to meeting the demand, the installed PVs in 2017 represent a small share out of the total installed PV capacity, as, for example, in South Africa or in Italy. A rather low installed PV capacity in 2017 may also be the result of an inadequate stimulating system, as example in countries in Central and Eastern Europe.

There are countries that are using other renewable energy sources as the hydro-ones in Norway. This makes the use of PVs in these countries only of recent date and as result of different events that proved the advantages of the solar electricity.

A general comment is devoted to the legal frame that supports the PVs implementation: a steady and predictable legal frame (as the German system) always supports the large-scale implementation while sudden changes that are not justified by objective facts are decreasing the investors and the users trust.

The Feed-in Tariff is the mostly used supporting scheme representing 63.1% in 2017, followed by direct subsidies or tax breaks (12.6%), incentives for self-consumption or net-metering (8.5%), green certificates (2.1%), etc. Specific incentives for BIPV are part of the supporting schemes in Austria, China, France, Malaysia, Spain and Switzerland [44].

The PV market is dynamic and differs from one country (or region) to another. This is why the future development has to be well planned to largely support the electricity production using photovoltaic systems and to mitigate the potential restrictions that are currently foreseen when the electricity produced using PVs will represent a more significant share from the total electricity production (over 10–15%). These type of issues are already well considered when designing the plans for communities where PV systems are integrated as part of the electrical energy produced to meet the demand not only at buildings level but also for different other various uses as for traffic lights, for powering electric vehicles, etc. The declining cost of the PV electricity brings the PV systems in direct competition with the retail electricity providers that are using the grid and therefore schemes that regulate the use of locally produced electrical energy are required, as several countries already started to do. To avoid the grid unbalance, the excess electrical energy produced by a PV system is currently not paid when injected in the system, as in Spain or is paid at the market price as in Germany.

The near future is expected to rely on power sources based on renewables, and the PV systems represent a promising potential candidate for the electrical energy production. The smart use of the solar radiation to produce electrical energy represents the core topic of this chapter that focuses on the most important issues related

Table 3.1 Key figures on PV systems installed all over the world [44]

Country	PV installed capacity in 2017/cumulative installed capacity in 2017 (MW)	BAPV and BIPV systems installed in 2017 (MW)	Incentive schemes
Austria	173/1271	5	Feed-in tariff (FIT), investment grants, subsidy for BIPV
Belgium	892/3877		Green certificates, net-metering system, regional incentive programs, self-consumption schemes
Bulgaria	10/1040		FIT scheme
Denmark	61/910		Net-metering and self-consumption schemes
Finland	43/80.4	43	25% investment subsidy of the total costs of grid-connected PV projects; investment subsidy for renewable energy production in agriculture
France	875/8076	87	Self-consumption schemes, FIT and lump sum incentives
Germany	1776/42,491		FIT concentrated on financing smaller systems (<100 kW). Incentive program for storage units. The market integration model
Greece	2445/–		FIT scheme
Hungary	90/380		FIT for systems <500 kW and feed-in premium (FIP) for systems of 0.5–1 MW
Israel	103/978		Net-metering scheme
Italy	400/19,682	320	Tax credit (for systems <20 kW) and net-billing scheme

(continued)

Table 3.1 (continued)

Country	PV installed capacity in 2017/cumulative installed capacity in 2017 (MW)	BAPV and BIPV systems installed in 2017 (MW)	Incentive schemes
Morocco	4.7/14.6		Projects supported by the World Bank
Norway	17/45	17	Weak incentives for small systems (<15 kW); capital subsidy (offered by the Oslo municipality) to cover up to 40% of the PV system investment cost; renewable energy certificates (RECS)
Poland	100/199	About 50 for residential small systems	Auction scheme
Portugal	64/2581	24	Investment subsidy when integrating PV systems in buildings subject of refurbishment. The green tax reform; municipal real estate Tax reduced by 50% for RES power producing buildings
Romania	3/1385		Green certificates scheme
Slovakia	5/545		
South Africa	13/1482	217	Subsidy mechanism for large scale, grid-connected renewable energy systems. A rooftop solar PV wiring code is under development
Spain	148/5331		Self-consumption schemes

(continued)

Table 3.1 (continued)

Country	PV installed capacity in 2017/cumulative installed capacity in 2017 (MW)	BAPV and BIPV systems installed in 2017 (MW)	Incentive schemes
Sweden	118/322	108	Green certificates scheme. Subsidies covering 30% of the total installation costs (in 2018); tax deduction for the excess energy fed into the grid
Switzerland	242/1907	142	Self-consumption and direct subsidies schemes for systems <30 kW
The Czech Republic	20/2180		FIT scheme (that was no longer paid after 2015)
The Russian Federation	60/600		
The Netherlands	853/2938	Most PV systems are installed in the residential environment	Net-metering scheme for systems up to 15 kW and incentives for larger systems (SED)
Turkey	2588/3427		FIT schemes, net-metering draft regulation
Ukraine	220/750		FIT scheme and local content premium scheme
United Kingdom	1000/12,700		FIT scheme

to the photovoltaic conversion particularly in systems installed on buildings or in the building's neighborhood.

3.2 Design of the PV Systems Implemented in the Built Environment

Photovoltaics already have a long history considering their use as renewable energy systems. Their extended implementation started in the US, during the 80s, after President Jimmy Carter promoted, in 1977, the *National Energy Act* that supported the use of solar energy.

In the built environment, PV systems are commonly implemented as building applied PV systems (BAPV), building integrated PV systems (BIPV) and PV systems installed near the building.

The BAPV and BIPV consist of photovoltaic modules installed on/in the available surfaces that are suitable for this purpose. These surfaces are rather small and difficult to find in the build environment due to the position of the buildings, not compulsory oriented towards Equator, and the shadows generated by the nearby objects (other buildings, trees, etc.). This is why the PV systems implemented in the built environment are either small, of several kW installed power as the systems installed on residential buildings, or medium sized (several tens to hundreds of kW installed power) as for industrial/commercial buildings with larger roofs or more available space in their neighborhood. Further on, the PV modules can be integrated as part of the construction materials, with photoelectric proprieties, for the functional and design elements: bricks, façade's components, roof tiles, semi-transparent windows, shading elements, etc. These systems have a broader impact, covering part of the building's construction costs and decreasing the operation cost of the building by producing part of or the entire locally required electrical energy. An important aspect related to the BIPV is represented by the solutions identified for the periodical inspection and for the maintenance access to the PV system components that may need replacement or repairing during the expected lifetime.

Another possible option for installing photovoltaic modules is outside the built environment, at the outskirts of human settlements where more lower-cost implementation area is available. These systems may reach installed power of several MW. Moreover, these can be a financing and clean electrical energy source for the entire local community, if installed and managed by the local authorities or by associations of local community members, e.g. the Westmill Solar Coop. near Oxford, England, where 5000 inhabitants own a £16 million photovoltaic park [14, 37].

3.2.1 Special Requirements for Installing PV Systems

When installing photovoltaic systems in the built environment the access of the solar radiation to the PV surface is very important and depends on the horizon line and on the shadows of nearby objects. This can be evaluated [48] either visual, by the installer, based on Sun path diagrams (Fig. 3.2) or using electronic automatic skyline scanning devices able to generate a horizon file to be imported and further embedded in other evaluation software. Sun path diagrams allow to identify the altitude and the azimuth angles of the Sun during each moment of the year; in this way, it can be verified if the solar radiation is obstructed (e.g. by the relief).

A detailed shadow analysis supports a correct assessment of the power output of a PV system installed in the built environment, because even a small shadow on the PV module (of several cm^2) can generate an electrical power loss of more than half of the module nominal power [58].

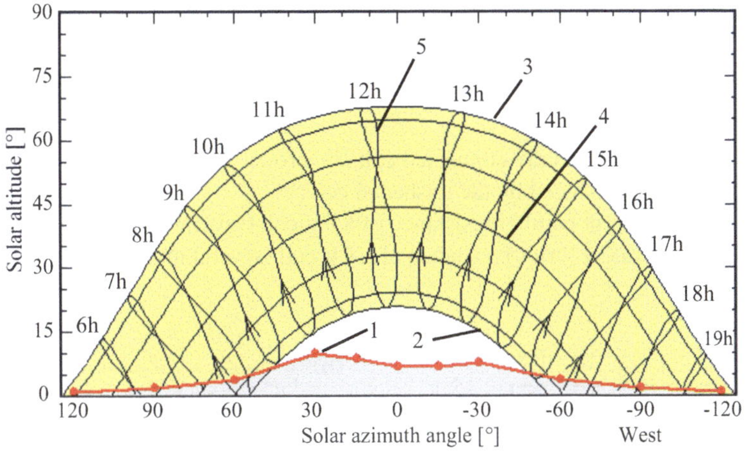

Fig. 3.2 Sun path diagram in Brasov, Romania (45.6°N, 25.5°E), obtained using the PV_{syst} software and indicating: the horizon line (Meteonorm) (1), the altitude variation during the winter solstice (2), the summer solstice (3) and equinoxes (4) at different local time (5)

To limit the shadowing effect on the power output of the PV systems, external diodes can be added to the module by the manufacturer or to parts of the PV system by the system designer, reducing thus the shading influence on the characteristic I–V curve of the PV.

Usually, a crystalline silicon photovoltaic module consists of 2–4 parallel rows with about 20 cells serially connected. A photovoltaic system (array) is similarly built and can contain several parallel strings with serially connected modules.

If all n serially connected cells in a PV module receive the same solar radiation (with the same irradiance) then the same current (I) passes through each cell (Fig. 3.3a). However, if one of the cells is shaded, this cell is not able to produce any electrical power and functions as an electrical consumer, decreasing the voltage already generated by the other $n - 1$ cells (Fig. 3.3b). Moreover, the power consumption generates heat in the shaded cell and this heat can irremediably deteriorate the module if hotspots occur [58]. The voltage drop appears because the cell current (I_C) of the shaded cell is 0, therefore the current I produced by the other cells has to pass through the equivalent parallel and series resistances of the cell (Fig. 3.3).

The voltage difference (ΔV) between a series with n unshaded cells and the same series with one shaded cell (Fig. 3.4) can be calculated using

$$\Delta V = V - V_{SH} = \frac{V}{n} + I \cdot (R_P + R_S) \cong \frac{V}{n} + I \cdot R_P \qquad (3.1)$$

where

- V is the voltage of the PV cell row with n unshaded cells;
- V_{SH} is the voltage of the same PV cell row containing $n-1$ unshaded cells and one shaded cell;
- n is the number of serially connected cells in a row;
- I is the current produced by the unshaded cells;
- R_P is the equivalent parallel resistance of a PV cell (several Ω);
- R_S is the equivalent series resistance of a PV cell (several mΩ).

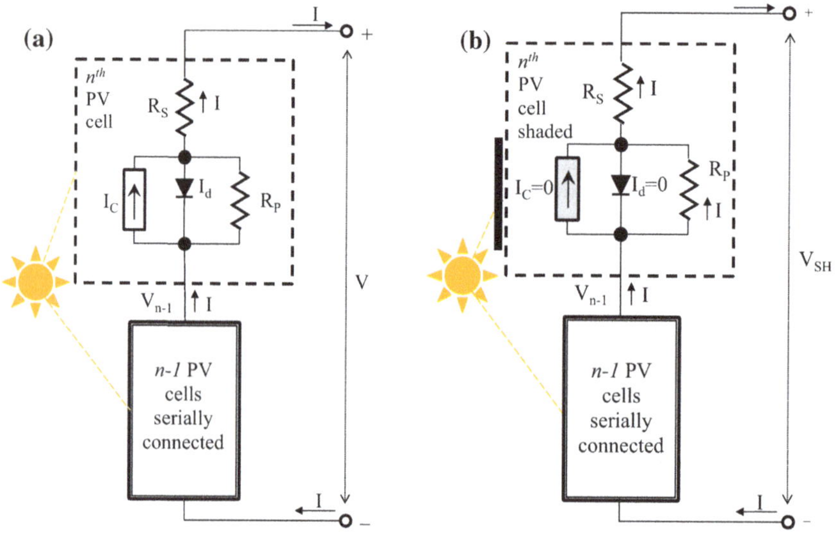

Fig. 3.3 Module with n cells serially connected: **a** no shadows on any cell; **b** shadow is casted on the nth cells

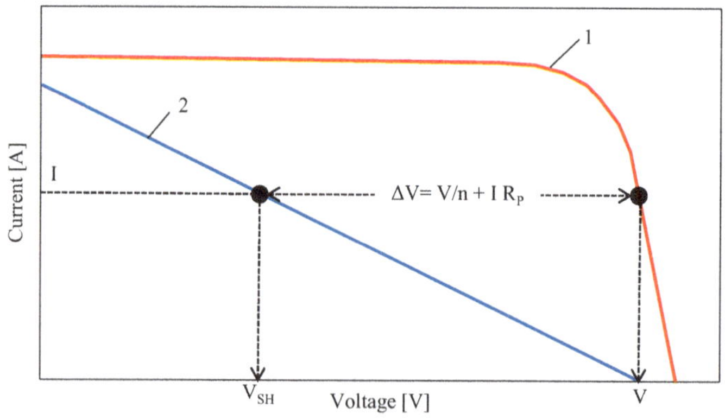

Fig. 3.4 Characteristic I–V curves of a module with n cells serially connected: all cells are non-shaded (1) and one PV cell is shaded (2) [58]

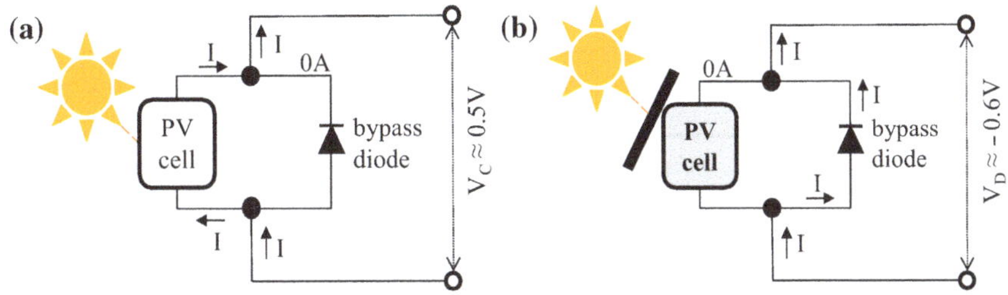

Fig. 3.5 Parallel bypass diode operation in: **a** sunny and un-shaded conditions; **b** shaded conditions

In a series of cells that is part of a crystalline silicon PV module, each individual cell contributes with about 0.5 V to the series voltage when operating under Standard Test Conditions (STC). When a cell is shaded it decreases the series voltage according to Eq. (3.1). To limit this negative effect *bypass diodes* can be mounted in parallel with each PV cell, when if shadowed, a PV cell will decreases the series voltage with only 0.6 V, that represents the voltage required for opening the diode (Fig. 3.5).

However, the installation of a bypass diode in parallel with each cell in a PV module is unpractical and increases the cost. Since the same benefit can be reached in the case of PV strings, module manufacturers usually provide at least one bypass diode for each module in order to protect the PV strings when shading occurs on the PV modules. Moreover, up to four diodes can be installed in PV modules to protect groups of cells.

Another type of diodes used in photovoltaic systems to mitigate the shadowing effects are the *blocking (insulating) diodes*. If a PV system contains parallel strings of PV modules, blocking diodes are mounted by the system installers on each parallel string to limit the current consumed by the shaded strings (Fig. 3.6).

An obvious solution in the built environment with shadowing conditions is to remove the shadow generating objects. If this is not possible, the shadowed areas should be avoided when installing PV modules. Another option that, however, increases the system's costs is the use of micro-inverters or DC-DC converters for each individual or pair of PV modules, so that each module will work independently, uninfluenced by the others.

There are additional parameters with influence on the photovoltaic response in the build environment that should be considered when choosing the optimum PV modules type in a photovoltaic system.

The type of the PV module glazing may have various wetting proprieties, being thus differently affected by surface soiling and with different residence time of water droplets or snow. Thin film modules are usually covered with a flat and smooth layer of glass that very well supports wetting. Silicon based modules, especially of monocrystalline type, usually have a textured surface that supports the accumulation of larger water droplets sliding easier when they reach the critical size. As Sarver et al. [78] mention, surface soiling can be limited by using hydrophobic and low energy surfaces, therefore textured surfaces are recommended.

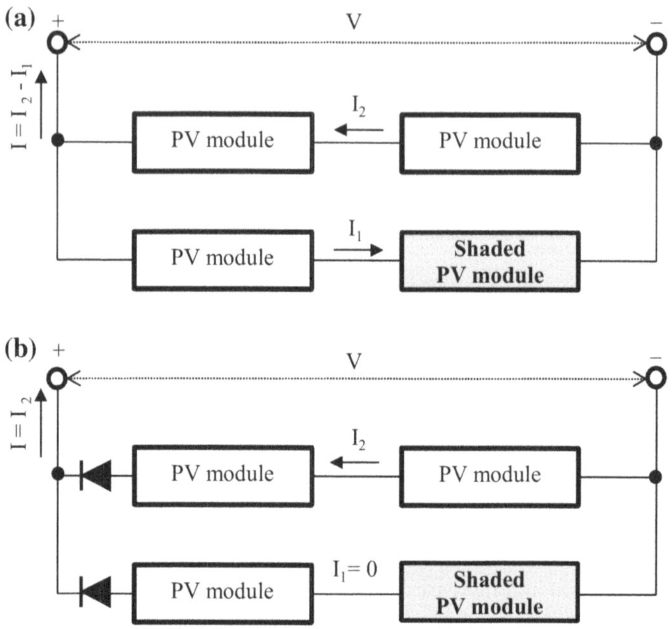

Fig. 3.6 Electrical current produced by a PV array with parallel strings and a shaded PV module: **a** without blocking diodes; **b** with blocking diodes installed on each parallel PV string

The relative air humidity is responsible for the broad spectral variation of the solar radiation received on the surface of the photovoltaic modules. The spectral shift mainly occurs during periods with low solar irradiance that differently influences the photovoltaic response, according to the module type. The CdTe modules have the lowest variation in the conversion efficiency because they use the narrowest part of the solar spectrum as presented by Mittal et al. [60]. This part of the spectrum is only little affected by the high relative air humidity, that mainly influences the solar radiation with longer wavelengths [51].

A similar experimental investigation is described by Schweiger and Herrmann [80], using 15 PV modules tested in 4 different locations (Germany, Arizona-USA, India and Italy). The results indicate an increase in the annual energy yield in locations with high relative humidity, of 5.3% for CdTe modules compared to 1.6% for crystalline silicon (c-Si) and 2.8% for the CIGS modules.

Burduhos et al. [10] also described the influence of the high relative humidity and of the solar radiation (Fig. 3.7) on the conversion efficiency of five types of PV modules (Fig. 3.8), during the morning hours in a mountain area, in the temperate climate. They outlined significant differences in the PV modules behavior although the climatic parameters during the selected summer intervals were similar.

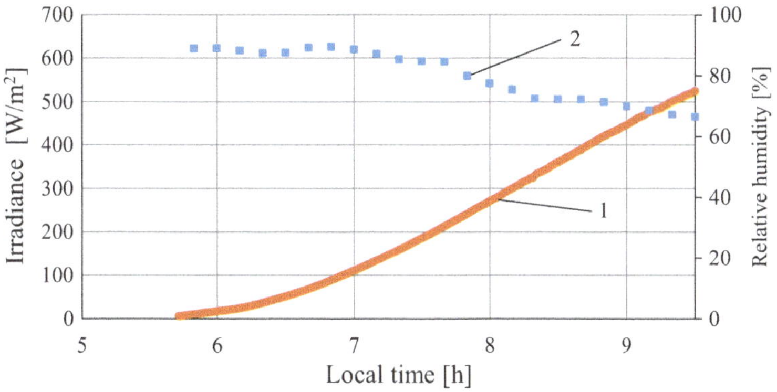

Fig. 3.7 Irradiance of the received global solar radiation (1) and relative humidity of the air (2) during the morning hours on August 28, 2015, in Brasov, Romania

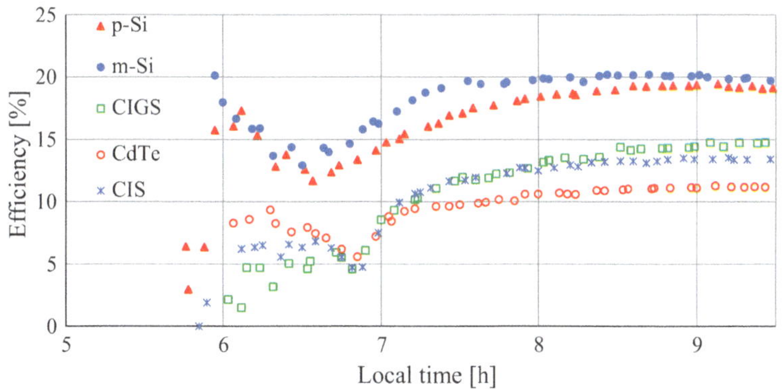

Fig. 3.8 The infield conversion efficiency of five PV module types during the morning hours on August 28, 2015

The relative efficiency loss ($\Delta\eta$) is calculated according to Visa et al. [101] using

$$\Delta\eta = \frac{\eta_0 - \eta}{\eta_0} \cdot 100 \quad (3.2)$$

where

- η is the infield conversion efficiency of the photovoltaic module;
- η_0 is the nominal efficiency in Standard Test Conditions (STC).

As Fig. 3.9 shows, the relative efficiency losses of five PV modules types reached high values during the early morning hours, when the relative air humidity was high and the solar irradiance was low. These conditions support the formation of dew/water droplets on the module's surface that partially reflect the small amount of incident solar radiation. PV modules based on crystalline silicon (Fig. 3.9a) have a smaller relative efficiency loss during these intervals than thin-film modules (Fig. 3.9b). Moreover, the negative relative efficiency losses of all PV module types indicate a

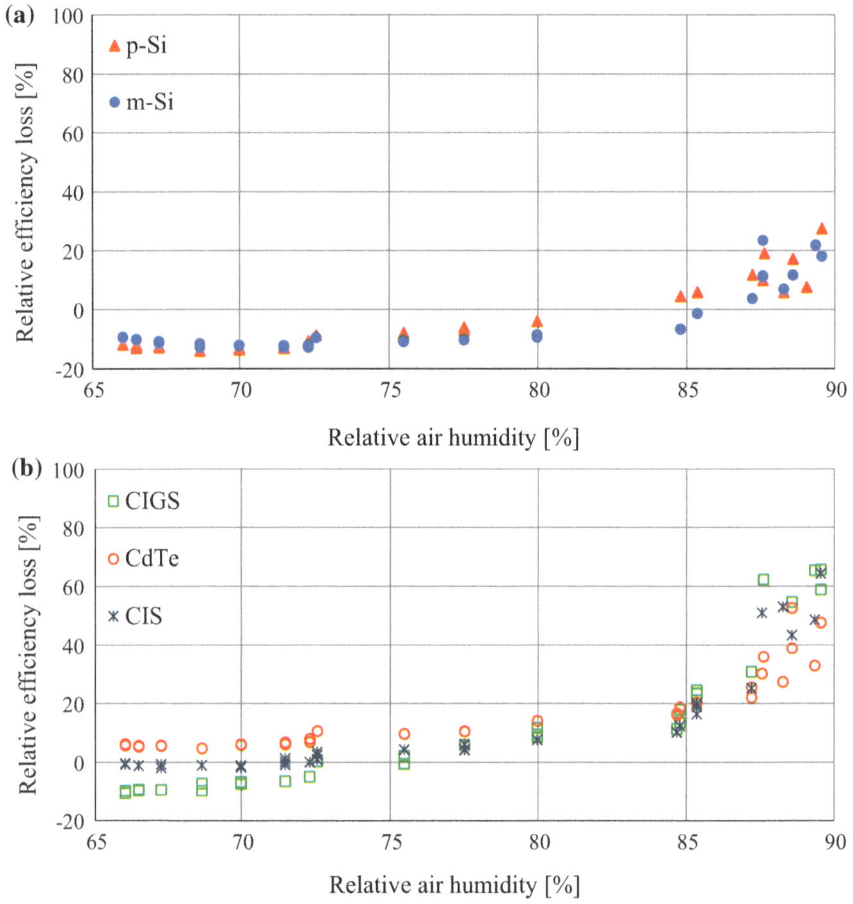

Fig. 3.9 Variation of the relative efficiency losses of five PV module types with the relative humidity in the air, during the morning hours of August 28, 2015: **a** p-Si and m-Si modules; **b** CIGS, CdTe, CIS modules

slight increase of the PV module conversion efficiency compared to the nominal efficiency at relative humidity values lower than 80% for the silicon modules. Moreover, the results show that this positive effect was not observed for the CdTe and for the CIS modules at relative humidity values higher than 65%.

3.2.2 *The Design Algorithm of PV Systems*

PV systems can be installed in the built environment consisting of new or already existing buildings for various reasons, the most likely being:

- To produce electrical energy on an available surface, within an available budget;
- To locally produce the energy required for the building's operation;
- To develop a sustainable building, able to produce energy using renewable energy systems, by including new active architectural components.

For the first reason the limitations are the available budget and the implementation area of the project. These will influence the selection of the PV modules type and number. Optimization in this case is mainly related to identifying the adequate tilt angles of the PV modules and to avoiding total or partial shadowing of the PV modules.

The second reason for installing PV systems in the built environment requires a more accurate analysis, as a certain amount of energy needs to be produced. Thus, except optimizing the tilt angle and avoid shadowing, the analysis of the energy demand, of the available implementation area and of the solar energy potential in the implementation location are part of the design process.

The last reason does not require any extensive optimization because installing the photovoltaic elements is limited by the building architecture, the Sun-facing, non-shaded parts of the building and the relatively fixed angles available for mounting the PVs. Nevertheless, an accurate estimation of the energy that can be produced by the building using solar radiation is useful to estimate the cost - benefit ratio of the system to be installed.

All these options also require a correct sizing of the PV system components for using, storing or feeding the produced electrical energy to the electrical grid. Therefore, a design algorithm for selecting and sizing the optimum components of the PV system is useful. The steps of such an algorithm are presented in Table 3.2.

Step 1. Estimation of the electrical energy demand

Identifying the electrical energy demand is required only when the system owner targets an electrical energy production that has to meet (in a certain extent) the local

Table 3.2 Algorithm for sizing the components of a PV system

Step number	
1. Estimation of the electrical energy demand 2. Estimation of the received solar energy	
A. Grid-connected PV system 3.a. Estimation of the required PV power 4.a. Estimation of the required PV surface 5.a. Selection of the grid-connected solar inverter 6.a. Selection of the interconnection type of the PV modules	B. Stand-alone PV system 3.b. Selection of the system voltage 4.b. Estimation of the required storage capacity and selection of the interconnection type of the batteries 5.b. Estimation of the required PV module number and selection of their interconnection type 6.b. Selection of the charge controller and inverter
7. Sizing the cables and the protection equipment	

demand. The average daily energy demand (E_{d}) estimated for a given period (month, year) can be calculated using

$$E_{\mathrm{d}} = \frac{E_{\mathrm{d,DC}} + E_{\mathrm{d,AC}}}{1 - z} \tag{3.3}$$

where

- $E_{\mathrm{d,DC}}$ is the average daily DC energy required to power the DC loads;
- $E_{\mathrm{d,AC}}$ is the average daily AC energy required to power the AC loads;
- z represents the share of electrical energy losses that occur in the photovoltaic system.

The most common electrical losses are caused by the deviation from the STC efficiency due to the solar irradiance values lower than 1000 W/m^2, due to soiling the module's glazing and because of the PV temperature higher than 25 °C. Moreover, losses can be the result of shading, maximum power point (MPP) mismatches of the modules, inverter losses and DC/AC losses on the electrical lines. For most PV systems, the share of these losses ranges between $z = 0.2$–0.3 [23].

There are at least two ways to identify the average daily AC and DC energy required by the local loads. The first one requires a monthly ($E_{\mathrm{m,DC}}$, $E_{\mathrm{m,AC}}$) or an annual ($E_{y,\mathrm{DC}}$, $E_{y,\mathrm{AC}}$) history of the electricity consumption. Based on it, the average daily energy demand can be estimated using

$$E_{\mathrm{d,DC}} = \frac{E_{\mathrm{m,DC}}}{N_{\mathrm{m}}}; \quad E_{\mathrm{d,AC}} = \frac{E_{\mathrm{m,AC}}}{N_{\mathrm{m}}} \tag{3.4}$$

$$E_{\mathrm{d,DC}} = \frac{E_{y,\mathrm{DC}}}{N_y}; \quad E_{\mathrm{d,AC}} = \frac{E_{y,\mathrm{AC}}}{N_y} \tag{3.5}$$

where

- N_{m} is the number of days in the month;
- N_y is the number of days in the year.

The second option implies an estimation based on the power of the electrical consumers and their average operation duration in the implementation location

$$E_{\mathrm{d,DC}} = \sum P_{ni} \cdot t_i; \quad E_{\mathrm{d,AC}} = \sum P_{ni} \cdot t_i \tag{3.6}$$

where

- P_{ni} is the nominal power of the ith consumer;
- t_i is the number of daily functioning hours of the ith consumer.

Step 2. Estimation of the received solar energy

The received global solar energy on the PV surface ($E_{\mathrm{G}n}$) in the implementation location depends on at least the following:

- The *location on Earth* that gives the general conditions for estimating the available solar energy. Following the information included in Sect. 2.1.1, the solar energy potential can be estimated using local measurements, mathematical estimation models or solar energy maps;
- The *tilt angle* at which the PV modules are installed; this angle may be constant over the entire year or it can be modified at specific time intervals (Sect. 2.1.2). Moreover, tracking systems for the PV modules can be considered to more accurately follow the Sun (Sect. 3.3);
- The distance between the rows of fixed PV modules or tracked PV platforms, as described in Sect. 2.1.2;
- The *horizon line* that may significantly influence the number of lighting hours over the day, thus the available solar energy in locations with irregular mountain- or hill-based topography. The horizon line and the nearby objects and buildings may generate shadows on the photovoltaic modules (Fig. 3.2).

The global solar energy that can be received by a PV surface in the implementation location has to be estimated as input data in the design of the PV systems. Section 2.1.2 (Eq. 2.25) contains a description on how this value can be estimated, for different time intervals, according to the PV system type and to the owner needs. For example, in the case of grid-connected systems that have to cover the annual energy demand of the owner, the annual received global solar energy on the PV modules surface has to be estimated and the design interval will be one year. Considering stand-alone systems the design interval is usually selected corresponding to the month with the lowest received solar energy.

The solar energy can be also estimated using different online or offline software tools as PVGIS (Photovoltaic Geographical Information System), PVWatts, Meteonorm, etc., that support the import of horizon/shadowing information and the estimation of the optimum tilt angle along with information on the daily, monthly and annual electrical energy expected to be produced.

Currently, many countries are considering actions to generate the solar cadastre aiming at supporting a larger implementation of solar energy conversion systems in the build environment, e.g. Switzerland [22, 89] or USA [35]. These enable the owners to easily check if a specific location, building or part of a building is suitable for installing solar energy conversion systems and to estimate its renewable energy potential.

Based on the received solar energy (E_{Gn}) during the design interval, the average daily number of equivalent hours (N_{hS}) with solar irradiance of 1 Sun (1000 W/m^2) can be calculated using

$$N_{hS} = \frac{E_{Gn}}{1000 \cdot N_d} \tag{3.7}$$

where N_d is number of days of the design interval.

This indicates the average daily duration when the PV modules work in conditions similar to STC, excluding the variable temperature of the modules.

If the parameters of the PV module are also indicated in Nominal Operating Cell Temperature (NOCT) conditions the average daily number of equivalent hours with solar irradiance of 800 W/m^2 can be calculated and further used instead of N_{hS} along with all PV parameters that correspond to these conditions.

Electrical utility grids are usually available in the built environment because conventional generators in small grids may already power even remote communities; therefore, the PV systems installed here are in most cases grid-connected. When the utility grid is not available, the PV systems are designed as stand-alone; however, the high installation and maintenance costs of the energy storage system (batteries) and the over-sizing of the PV array for meeting the energy demand even during the month with the lowest solar irradiance in the year should be considered.

A. **Design of the grid-connected PV systems**

The design of a grid-connected PV system allows an optimal sizing of the PV array and of the inverter. It addresses the nominal power required for the PV system, the area and the number of PV modules and their serial and/or parallel connection in order to be managed by the inverter [58].

Step 3.a. Estimation of the required PV power

The nominal power of the entire PV system ($P_{\mathrm{PV,s}}$) is commonly expressed in Watt peak (Wp) and can be estimated using

$$P_{\mathrm{PV,s}} = \frac{E_{\mathrm{d}}}{N_{\mathrm{hS}}} \tag{3.8}$$

The local PV market may provide different module types, with different dimensions, prices and output parameters. There are different types of PV modules recommended for grid-connected and for stand-alone systems, e.g. modules with higher MPP voltage, of about 36 V, should be selected for grid-connected systems. Various other criteria can be used by the PV system designer (or owner) to select a certain type of module.

The main parameters that should be considered for the selected module during the design phase are the MPP nominal power (P_{MPP}), voltage (V_{MPP}) and current (I_{MPP}), the open-circuit voltage (V_{OC}) and the short-circuit current (I_{SC}), the photovoltaic conversion efficiency (η_0) and the dimensions (l, w).

The number of required PV modules (N_{PV}) is the upper integer value of the ratio

$$N_{\mathrm{PV}} = \frac{P_{\mathrm{PV,s}}}{P_{\mathrm{MPP}}} \tag{3.9}$$

Step 4.a. Estimation of the required PV surface

The required surface area (A_{PV}) of the PV modules can be estimated using

$$A_{\mathrm{PV}} = N_{\mathrm{PV}} \cdot l \cdot w \tag{3.10}$$

If this area is higher than the available implementation surface area another type of module, with higher efficiency, should be selected and the process is repeated until the available implementation area is large enough for the required PV area.

Step 5.a. Selection of the grid-connected solar inverter

Further on, the selection of the optimum, grid-connected solar inverter able to convert the entire power produced by the PV system has to be done. In normal conditions, the rated power of the inverter should be the same as the nominal power of all PV modules. However, in practice and especially in the built environment, smaller inverters with up to 20% may be selected (sizing factor of 1.2). Using this sizing factor, during periods when the PV power is higher than the rated power of the inverter, the inverter will limit the output power, phenomenon known as *inverter clipping*. This can be accepted considering the PV modules aging and the fact that they will rarely operate under STC. However, *under no circumstances should the PV array voltage exceed the maxim acceptable voltage of the inverter* as outlined by Eq. (3.11).

The main parameters that have to be considered for the selected inverter are the rated AC power, the voltage range of the maximum power point tracking ($V_{MPPT,min} \dots V_{MPPT,max}$), the maximum admissible input voltage (V_{max}), the maximum admissible (short-circuit) input current (I_{max}) and the DC-AC conversion efficiency. Moreover, it should be noticed that grid-connected inverters considerably differ from their stand-alone counterparts.

Step 6.a. Selection of the interconnection type of the PV modules

The number of PV modules serially connected in a PV string (N_{SM}) has to be an integer value that satisfies the conditions expressed by Eq. (3.11), where the lowest limit supports an optimal conversion efficiency of the inverter, while the upper limit protects the inverter from overvoltage damages

$$\frac{V_{MPPT,min}}{V_{MPP,min}} < N_{SM} < \frac{V_{max}}{V_{OC,max}} \tag{3.11}$$

where

- $V_{MPPT,min}$ is the minimum accepted value of the MPPT voltage range of the inverter;
- $V_{MPP,min}$ is the minimum MPP voltage of the PV module at the highest module temperature (T_{max}, is usually 70 °C in temperate climates);
- V_{max} is the maximum admissible input voltage of the inverter;
- $V_{OC,max}$ is the maximum open-circuit voltage of the module at the lowest module temperature during daytime (T_{min}, usually −10 °C in temperate climates).

The minimum MPP voltage ($V_{MPP,min}$) can be estimated using

$$V_{MPP,min} = V_{MPP} \cdot [1-\lambda \cdot (25-T_{max})] \tag{3.12}$$

where λ is the temperature coefficient that indicates the MPP voltage variation with temperature (about −0.5%/°C for crystalline silicon modules). This coefficient is usually not given in the PV modules datasheets; instead of this the temperature coefficient (γ) of P_{MPP} is given and can be used [34].

The maximum open-circuit voltage ($V_{OC,max}$) can be estimated using

$$V_{OC,max} = V_{OC} \cdot \left[1 - \beta \cdot (25 - T_{min})\right] \tag{3.13}$$

where β is the temperature coefficient that indicates the open-circuit voltage variation with temperature (about −0.35%/°C for crystalline silicon modules).

If one single PV string cannot cover the entire required PV power, several strings have to be parallelly connected to form a PV array. Their number (N_{PS}) can be calculated as the upper integer value of the ratio

$$N_{PS} = \frac{N_{PV}}{N_{SM}} \tag{3.14}$$

For inverters with a single input, the PV strings in an array have to contain an equal number of identical modules. Thus, the inverter should be accordingly chosen.

If the maximum admissible (short-circuit) current accepted by inverter (I_{max}) is also indicated in the inverter datasheet, it should be checked if the maximum short-circuit current of the entire PV array, at the maximum module temperature is lower than I_{max}

$$N_{PS} \cdot I_{SC,max} < I_{max} \tag{3.15}$$

where $I_{SC,max}$ is the maximum short circuit current of a PV module and can be calculated for T_{max} using

$$I_{SC,max} = I_{SC} \cdot [1 - \alpha \cdot (25 - T_{max})] \tag{3.16}$$

where α is the temperature coefficient that indicates the current variation with temperature (about 0.05%/°C for crystalline silicon modules).

If this condition is not met, another inverter should be selected.

B. **Design of the stand-alone PV systems**

The design of a stand-alone PV system allows an optimal sizing of the required PV array, of the battery storage system, charge controller and inverter. It addresses the optimum voltage of the stand-alone system, the number of serially and parallelly connected PV modules and batteries, the specific requirements related to the charge controller and inverter [58].

Step 3.b. Selection of the system's voltage

The first parameter to be selected for a stand-alone PV system is the optimal DC voltage (V_{sys}) according to the maximum required power (P_{max}). The recommended values to be selected are, roughly: 12 V for less than 1200 W, 24 V for 1200–2400 W and 48 V for 2400–4800 W; voltages higher than 48 V are rarely used. These values support relatively low cable losses and low cross sections of the cables.

Step 4.b. Estimation of the required storage capacity and selection of the interconnection type of the batteries

The average daily DC storage capacity (C_d) at the selected system voltage can be calculated, based on E_d (Eq. 3.3) and the DC voltage V_{sys} using

$$C_d = \frac{E_d}{V_{sys}} \tag{3.17}$$

Stand-alone PV systems may contain only DC loads ($E_d = E_{d,DC}$); this represents an advantage because an inverter is not required. However, these systems have the disadvantage of higher appliances costs and thicker cables.

For sizing the *storage system*, the number of consecutive backup days has to be selected, during which the system has to provide power to the local consumers even if very little or no solar radiation is available; an usual average backup interval is $N_{bk} = 3$ days.

The DC storage capacity required for powering the local consumers during the entire backup interval (C_{bk}) is calculated using

$$C_{bk} = C_d \cdot N_{bk} \tag{3.18}$$

The capacity of a battery depends on its temperature and on the discharge rate (C/t). The discharge rate indicates the current that completely discharges the battery (with the capacity C) during the time (t).

The nominal capacity of lead-acid batteries is usually given at 25 °C, considering a 20 h long discharge time ($C/20$). Therefore, before estimating the storage capacity of a stand-alone PV system the minimum working temperature (T_{min}) of the batteries has to be selected, according to the conditions in the implementation location. Moreover, the discharge time should be calculated based on the selected backup interval; for example a backup of 3 days means a discharge rate of $C/(24 \times 3) = C/72$.

The total capacity (C_{total}) of the storage system at the nominal temperature and discharge rate (25 °C and $C/20$) is calculated using

$$C_{total} = \frac{C_{bk}}{\text{DF} \cdot \text{DOD}} \tag{3.19}$$

where

- DF is the discharge rate factor (the ratio of the available to the rated battery capacity);
- DOD is the maximum allowed depth of discharge (the ratio of the capacity that can be withdrawn to the battery capacity).

The discharge rate factor (DF) can be estimated based on *C*/*t* and $T_{\min}$, using diagrams similar to that in Fig. 3.10 specific for each battery type.

The maximum allowed depth of discharge (DOD) indicates the discharge threshold above which discharging the battery causes irremediable damages. Lead-acid batteries commonly used in renewable energy systems usually have a DOD = 0.8 that can be further adjusted using diagrams similar to that in Fig. 3.11; this DOD prevents the battery from freezing at $T_{\min}$. The higher the DOD is the shorter the lifetime of the battery, but also the lower the required capacity and cost.

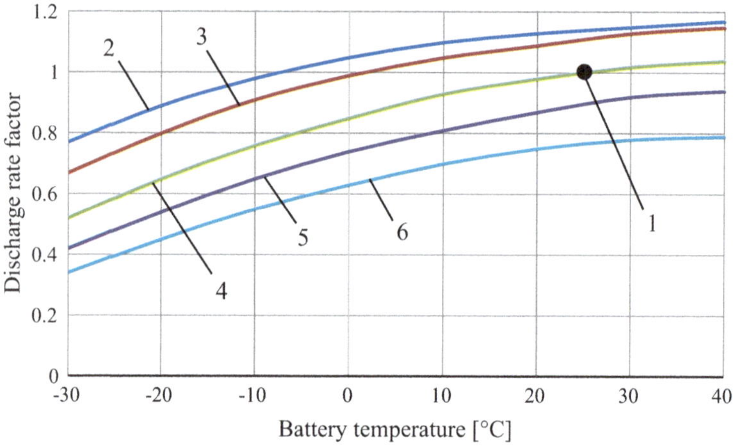

Fig. 3.10 Discharge rate factor of a lead-acid battery at different discharge rates and temperatures, compared to the rated capacity (1) at 25 °C and *C*/20 discharge rate: *C*/72 (2); C/48 (3); *C*/20 (4); *C*/10 (5); *C*/5 (6) [58]

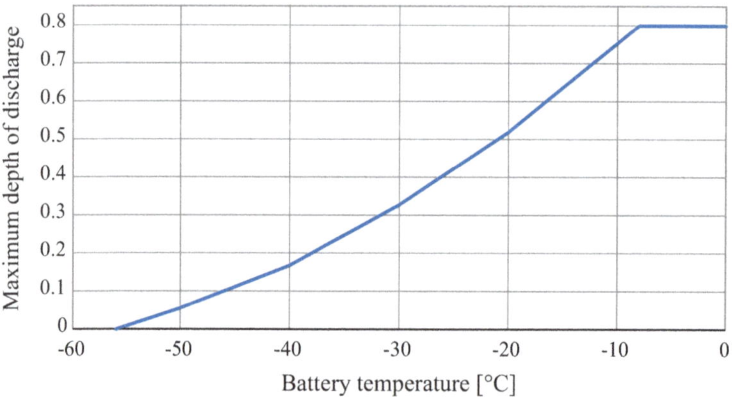

Fig. 3.11 Limits of the allowable depth of discharge of a lead-acid battery considering avoiding the electrolyte freezing (after [58])

For not excessively shortening the batteries lifetime it should be verified that the total storage capacity (C_{total}) is higher than the minimum allowed capacity, calculated if the discharge time is 5 h ($C/5$ discharge rate) at the maximum power required by the loads (P_{max})

$$C_{total} > \frac{P_{max} \cdot 5}{V_{sys} \cdot \mathrm{DF} \cdot \mathrm{DOD}} \tag{3.20}$$

The number of batteries serially connected ($N_{bat,S}$) is calculated as

$$N_{bat,S} = \frac{V_{sys}}{V_{bat}} \tag{3.21}$$

where V_{bat} is the nominal voltage of the battery.

The number of parallel strings of batteries ($N_{bat,P}$) is calculated using

$$N_{bat,P} = \frac{C_{total}}{C_{bat}} \tag{3.22}$$

where C_{bat} is the capacity of one single rechargeable battery.

Step 5.b. Estimation of the required PV modules number and selection of their interconnection type

For *sizing the photovoltaic array*, the month with the lowest solar energy has to be firstly identified, further called *design-month*.

For the design-month the *solar fraction* (F) has to be selected; the solar fraction represents the ratio between the electrical energy produced by the PV system and the total electrical energy required to power the loads. In temperate climates, a solar fraction of 0.5 during the worst month in the year roughly corresponds to an annual solar fraction of 0.8. Obviously, the difference corresponding to the energy when $F = 1$ and the selected solar fraction has to be covered by other systems able to produce electrical energy.

The daily DC capacity that has to be produced by the PV array during the design-month ($C_{d,d}$) is calculated using

$$C_{d,d} = F \cdot C_d \tag{3.23}$$

Further on, the number of serially connected modules in a PV string (N_{SM}) depends on the selected system's voltage (V_{sys}) and the nominal PV voltage (V_{MPP}) and is the upper integer value of the ratio

$$N_{SM} = \frac{V_{sys}}{V_{MPP}} \tag{3.24}$$

Moreover, the number of parallel PV strings (N_{PS}) is the upper integer value of the ratio

$$N_{PS} = \frac{C_{d,d}}{C_{d,S}} \tag{3.25}$$

where $C_{d,S}$ is the capacity of a PV string during an average day in the design month

$$C_{d,S} = N_{hS} \cdot I_{MPP} \cdot \eta_{bat} \cdot k_{PV} \tag{3.26}$$

where

- η_{bat} is the Coulomb charging efficiency of the batteries (about 0.9);
- k_{PV} is the de-rating factor of the PV modules in time, about 0.9.

Step 6.b. Selection of the charge controller and inverter

The *selection of the charge regulator* has to satisfy at least the following conditions that are similar to the conditions formulated for the input parameters of grid-connected inverters:

- The PV system voltage (V_{sys}) has to be in the range of the accepted voltage values;
- The minimum MPP voltage of the PV strings has to be in the working range of the MPP tracker of the charge regulator (Eq. 3.11);
- The open-circuit voltage of the PV strings has to be lower than the maximum accepted voltage (Eq. 3.11);
- The short-circuit current of the PV array has to be lower than the maximum accepted input current (Eq. 3.15), otherwise the PV array has to be connected to more than one charge regulator.

The selection of the stand-alone inverter has to satisfy the following conditions:

- The voltage of the stand-alone system has to be among the values accepted by the inverter;
- The maximum power required to simultaneously run the local electrical loads has to be lower than the rated output power of the inverter.

Step 7. Sizing the cables and the protection equipment

After the design of the PV system was developed, the last step focuses on sizing the cables and the corresponding protection equipment (switches and fuses).

The cables and the protection devices have to be sized to support currents 1.25 times higher than the maximum currents indicated on the devices. Moreover, the DC current produced by the PV array has to be considered 1.25 times higher than the short-circuit current, to cover the PV current that can be larger for solar irradiance values above 1 Sun (1000 W/m^2) and temperatures higher than 25 °C [42, 58].

Protection is required on the DC side for each photovoltaic string and for the entire PV array before the inverter or charge controller and for the entire battery bank. On the AC side, protection is required at the inverter output regardless it is part of a grid-connected or a stand-alone PV system.

The selection of the optimal cables is important for limiting the conduction losses (c) to specific values approved by local standards. The cross section of the required cable (A_c) can be calculated according to Deutsche Gesellschaft für Sonnenenergie [23], under DC and single-phase AC conditions using

$$A_c = \frac{2 \cdot L_c \cdot I_n \cdot \cos\varphi}{c \cdot V_n \cdot \sigma} \tag{3.27}$$

where

- L_c is the one way length of the cable;
- I_n is the nominal current which flows through the cable;
- $\cos\varphi$ is the power factor, that under DC conditions is 1, while for the AC it can take values in the range of 0.8–1;
- c conduction losses approved by local standards, usually in the 0.01–0.03 range;
- V_n is the nominal voltage on the cable;
- σ is the electrical conductivity; for copper and aluminum, at 70 °C, the conductivity values are about $\sigma_{Cu} = 50$ m/(Ω mm^2) and $\sigma_{Al} = 30$ m/(Ω mm^2).

Cables used in the three-phase circuits can be sized according to Deutsche Gesellschaft für Sonnenenergie [23] using

$$A_c = \frac{\sqrt{3} \cdot L_c \cdot I_n \cdot \cos\varphi}{c \cdot V_n \cdot \sigma} \tag{3.28}$$

All or part of the steps hereby described are included in different software tools able to provide support for correctly sizing and evaluating the photovoltaic systems. One of the mostly used software tool is PV_{syst}, developed in Switzerland at the University of Geneva for architects, engineers, researchers and for teaching activities. This tool can be used for grid-connected and stand-alone PV systems and for DC grids as those used in public transportation [30].

Most of the inverters manufacturers, as SMA, Fronius, SolarEdge developed similar software that supports the end-user to optimally design a PV system using inverters provided by these manufacturers. Obviously, these are not only design tools but also a convenient marketing solution.

The optimal design of photovoltaic systems according to the owner's demand plays an important role in the implementation of the new PV systems in the built environment. If not all the design aspects are correctly solved or less durable components are used, the maintenance costs may easily double the life-cycle cost of the system and generate a negative advertising on renewable energy systems [76].

3.3 Increasing the Electrical Output of PV Systems by Using Solar Tracking

The photovoltaic systems design and implementation in the built environment face the particular challenge risen by the limited implementation area available on or near the building, that can be used to locally produce electrical energy using solar energy. Therefore, increasing the energy output of the PV systems and intensively exploiting the available implementation space represent key actions in meeting the energy demand in the built environment.

The PV modules can be mounted on the building, i.e. on the sloping roofs, on horizontal flat roof terraces and on facades, or using the available land nearby the buildings.

A first approach to increase the annual electricity production is to set the PV modules at an optimally fixed tilt angle, according to the implementation location and climatic profile data. The fixed and optimally tilted PV systems can collect up to 55–75% of the available solar energy during clear sky days [61], depending on the season and latitude thus, their electrical output is at least 25% lower than the maximum possible one. Moreover, PV modules installed on vertical facades produce less than 70% energy as compared with the optimally tilted PV modules, with a severe depreciation in low latitude locations (<30°).

Another solution for increasing the PV output in the built environment is the use of *solar tracking*. Tracking aims at *increasing the received solar energy* by orienting the PV modules so that the direct solar radiation is as close as possible to the normal to their surface thus *decreasing the incidence angle* of the sunrays on the PV surface.

Solar tracking is one of the mostly applied solutions to increase the energy output of the solar energy convertors [1, 46, 61, 93]. The PV tracking systems are still effective even if they do not accurately follow the Sun as, if the incidence angle is lower than 10°, the output energy can still be 98.5% of the energy collected using full-continuous tracking [62]. In the cloudiest and haziest locations, the gain in the yearly output using tracked PV systems can be close to 20%. In sunnier areas, the typical annual energy gain is up to 25% using single-axis tracking and up to 45% for dual-axis tracking compared to the energy produced using fixed, optimally tilted PV modules [55]. This significant increase represents an important advantage for the PV systems in the built environment, considering the limited available implementation space. However, this advantage has also several drawbacks when compared to the fixed tilted PV systems [36, 49, 69]:

- An increased initial cost due to the mobile mechanical structure and the motion command and control systems. To be economically feasible, the energy gains obtained through solar tracking must support the recovery of the additional investment as well as the installation, operation and maintenance costs of the tracking system over the PV system lifetime. These can represent an unneglectable share in the overall cost of the entire system;
- Higher weights that can be important when considering roof implemented solar projects. However, tracked PV systems are usually installed on the ground;

- PV tracked systems require a larger implementation area for the same energy production capacity, to avoid self-shading of the PV modules organized in strings or in arrays. On the other hand, under a restricted available implementation area, the PV tracked systems lead to a lower installed PV capacity than a fixed tilt system; thus, the use of high efficiency PV modules in the tracked systems is strongly recommended.

In line with the data widely reported in literature, the current tracked PV systems reduce the cost of the provided electrical energy, as they:

- Produce more energy during the year than the fixed systems with the same photovoltaic surface and PV modules type;
- Require a smaller number of PV modules to deliver the same energy output as a fixed optimally tilted system;
- Require low amounts of electrical energy to power the actuators of the tracking system as these operate only a few minutes per day;
- Are reliable and cost-effective, particularly for large PV systems.

However, these advantages have to be correlated with the latitude and the climatic profile of the implementation location of the tracked PV system, the available space on and near the buildings, the type of tracker and its cost over the entire lifecycle.

There are two main ways for increasing the electrical output of a PV module, Fig. 3.12:

- *Increasing the conversion efficiency* in real operating conditions, i.e. selecting a PV module with increased *nominal efficiency and decreasing the module's temperature*;
- *Increasing the received solar energy.*

A common way for increasing the electrical output of PV systems is to select photovoltaic modules with higher nominal efficiency; however, this approach is limited by the high production costs of these new technologies. Reducing the temperature

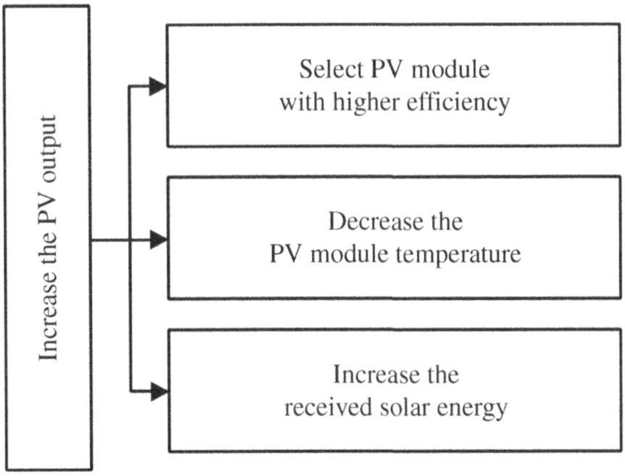

Fig. 3.12 Strategies for increasing the electrical output of PV modules

on the PV modules is another option, but it involves high-cost, active cooling systems. Increasing the amount of solar energy received by the PV modules leads to a significant gain in the electrical output and this can be reached by Neagoe et al. [66], Burduhos et al. [11]:

- *Optimizing the fixed tilt angle* of the PV modules, providing a moderate increase of the input solar energy;
- *Solar tracking* that allows receiving up to nearly 100% of the available solar radiation.

It should be noted that increasing the solar energy received by a PV module leads to an increase in the PV cell temperature and thus to a drop in the photovoltaic conversion efficiency. The decrease in the efficiency due to heating the module depends on the photovoltaic materials and on the system configuration. For a PV module cooled down using water an increase of 9–12% of the electrical yield is reported [25]. However, the efficiency decrease due to the temperature increase has a much more limited effect on the output than the increase due to tracking [66]. The main way of *increasing the PV output power* is thus the *increase of the received solar energy*.

3.3.1 Solar Angles

In the design of the solar tracking systems, the angles of the sunray vector are required. Considering an implementation location (Q) at the latitude (φ), the *solar angles* can be defined in the global reference system of the Earth ($OXYZ$) as presented in Fig. 3.13, or in the local reference system of the observer ($QX_0Y_0Z_0$), Fig. 3.14, by projecting the sunray vector on a reference plane, i.e. the *Equatorial plane* in the $OXYZ$ system and the *three orthogonal planes* in the local $QX_0Y_0Z_0$ system. As results, four pairs of the solar angles result [100]:

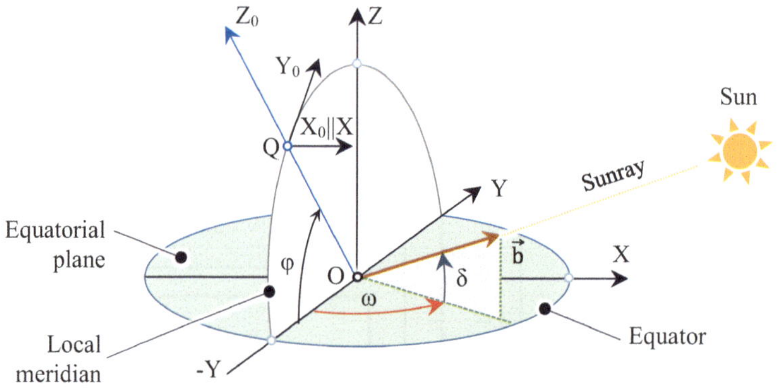

Fig. 3.13 Equatorial solar angles (ω, δ) in the global *OXYZ* reference system; the latitude angle (φ) of a locations Q on the local meridian, the zenith axis Z_0 and the sunray unit vector $\vec{b}$

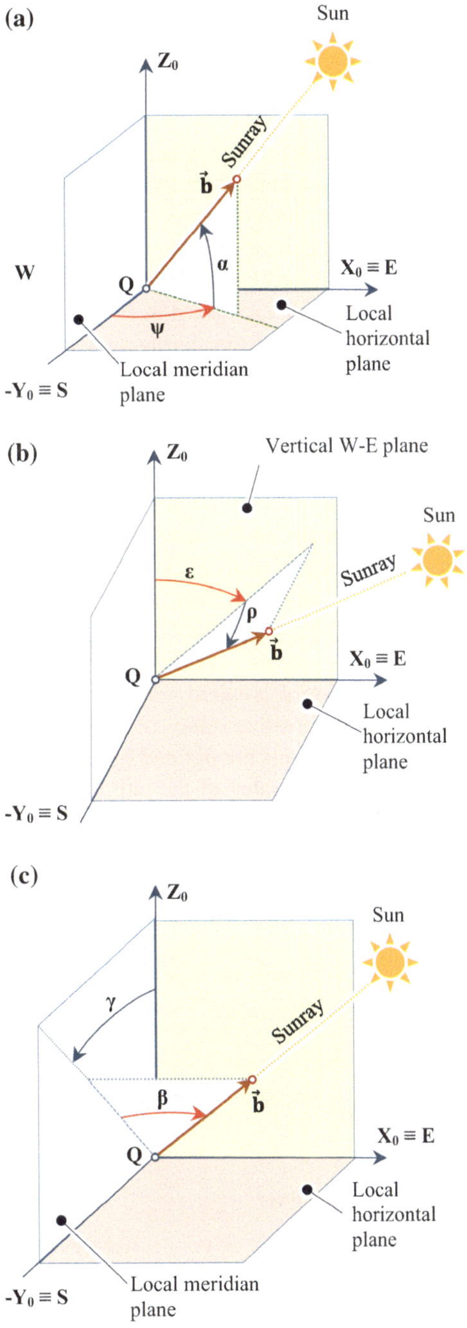

Fig. 3.14 Solar angle pairs in the local reference system: **a** azimuth pair (ψ, α); **b** pseudo-azimuth pair (ε, ρ); **c** pseudo-equatorial pair (β, γ)

- (ω, δ): *equatorial* (denoted by Eq) type solar angles (Fig. 3.13);
- (ψ, α): *azimuth* (Az) type solar angles (Fig. 3.14a);
- (ε, ρ): *pseudo-azimuth* (PAz) type solar angles (Fig. 3.14b);
- (β, γ): *pseudo-equatorial* (PEq) type solar angles (Fig. 3.14c).

In the global reference system (Fig. 3.13), the Z axis is coincident with the Earth's axis of rotation and positive to North. The Y axis is the intersection line of the local meridian plane of the observer Q with the equatorial plane. The X axis is also located in the equatorial plane and it results from the condition that $OXYZ$ is a right-handed system. The origin O is the Earth's centre.

The solar angle ω, known as *hour* angle, is measured between the $-Y$ axis and the projection line of the sunray vector $(\vec{b})$ on the equatorial plane; the *declination* angle δ is defined as the angle between this projection line and the sunray vector. The ω angle is positive towards X axis and δ towards Z axis.

The four pairs of the solar angles are expressed according to the data in Table 3.3, where the solar angles in the local reference system depend on the location latitude (φ) and are defined in relation with the hour angle (ω) and declination angle (δ) that explicitly depends on the solar time (t_s) and on the day number in the year (N).

The solar angles (ω), (ψ), (ε) and (β) are *diurnal angles* and describe the relative motion of the Sun from East (*sunrise*) towards West (*sunset*). The solar angles (δ), (α), (ρ) and (γ) are *elevation angles.*

The diurnal angles equal zero at solar noon. In the particular case of the azimuth type solar angles, ψ is called *azimuth angle* and α *altitude angle*. The daylight interval corresponds to positive values of the altitude angle $(\alpha > 0°)$ and the sunrise (t_{sr}) and sunset (t_{ss}) moments are defined by $\alpha = 0°$.

The maximum value of the altitude angle (α) during a day, for a location at the latitude (φ), is registered at solar noon [100]

$$\alpha_{noon} = 90° - \varphi + \delta \tag{3.29}$$

Table 3.3 Expressions of the solar angles [100]

Type of the solar angle pair	Relations
Equatorial (Eq)	$\delta = 23.45 \cdot \sin \frac{360\cdot(N-80)}{365}$ $\omega = 15 \cdot (12 - t_s)$
Azimuth (Az)	$\alpha = \arcsin(\cos\omega \cdot \cos\delta \cdot \cos\varphi + \sin\delta \cdot \sin\varphi)$ $\psi = \mathrm{sgn}(\omega) \cdot \arccos\left(\frac{\cos\omega\cdot\cos\delta\cdot\sin\varphi - \sin\delta\cdot\cos\varphi}{\cos\alpha}\right)$
Pseudo-Azimuth (PAz)	$\rho = \arcsin(\cos\omega \cdot \cos\delta \cdot \sin\varphi - \sin\delta \cdot \cos\varphi)$ $\varepsilon = \mathrm{sgn}(\omega) \cdot \arccos\left(\frac{\cos\omega\cdot\cos\delta\cdot\cos\varphi + \sin\delta\cdot\sin\varphi}{\cos\rho}\right)$
Pseudo-Equatorial (PEq)	$\gamma = \mathrm{arctg}\left(\frac{\cos\omega\cdot\cos\delta\cdot\sin\varphi - \sin\delta\cdot\cos\varphi}{\cos\omega\cdot\cos\delta\cdot\cos\varphi + \sin\delta\cdot\sin\varphi}\right)$ $\beta = \mathrm{sgn}(\omega) \cdot \arccos\left(\frac{\cos\omega\cdot\cos\delta\cdot\cos\varphi + \sin\delta\cdot\sin\varphi}{\cos\gamma}\right)$

The declination angle (δ) ranges yearly between −23.45 and +23.45°, as the Earth axis is tilted at 23° 26′ 21″ from the normal vector on its orbit plane (Fig. 2.1); its variation during 24 h is maximum 0.4° thus, for practical applications, the declination angle (δ) is considered constant over one day.

The solstices and equinoxes are defined in direct relation with the declination angle (δ) values:

- $\delta = 0°$: at spring and autumn equinoxes;
- $\delta = 23.45°$: at summer solstice in the northern hemisphere and winter solstice in the southern hemisphere;
- $\delta = -23.45°$: at winter solstice in the northern hemisphere and summer solstice in the southern hemisphere.

The Sun crosses the local meridian at solar noon, i.e. when $t_s = 12$ it results $\omega = 0$.

The local time (t) of a given location is defined according to its time zone. Based on the relations presented by Duffie and Beckman [26], the local time and the solar time (t_s) are correlated using

$$t = t_s - \frac{[4 \cdot (\lambda_{loc} - \lambda_{sm}) + \text{EOT}]}{60} + k \tag{3.30}$$

where

- λ_{loc} is the longitude of the investigated location, expressed in degrees East;
- λ_{sm} is the longitude of the standard meridian for the local time zone, expressed in degrees East;
- $k = 1$ for daylight saving time and $k = 0$ otherwise;
- EOT is the equation of time [90]

$$\text{EOT} = 0.0172 + 0.4281 \cdot \cos x - 7.352 \cdot \sin x - 3.3498 \cdot \cos 2x - 9.372 \cdot \sin 2x \tag{3.31}$$

where x is calculated using

$$x = \frac{360}{365} \cdot (N - 1) \tag{3.32}$$

where N is the day number in the year.

The conversion between the solar time (t_s) and the local time (t) is required for the tracking system controller that calculates the solar angles according to the solar time and controls the motion considering the local time, according to its internal clock.

The incidence angle (ν) of the sunray on a plane surface represents the angle between the unit vector $\vec{b}$ of the sunray and the normal unit vector $\vec{n}$ associated to the surface (Fig. 3.15).

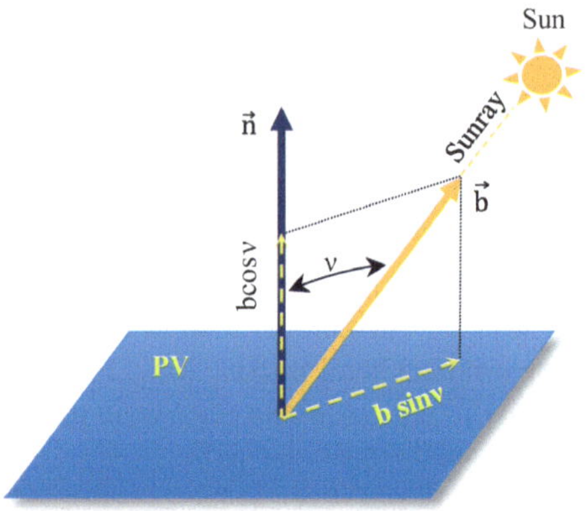

Fig. 3.15 Incidence angle of the sunray on the PV surface

Table 3.4 Equations for the incidence angle [100]

	$\cos\nu = \vec{b} \cdot \vec{n} =$
Equatorial system (ω, δ)	$\cos\delta \cdot \cos\delta_n \cdot \cos(\omega - \omega_n) + \sin\delta \cdot \sin\delta_n$
Azimuth system (ψ, α)	$\cos\alpha \cdot \cos\alpha_n \cdot \cos(\psi - \psi_n) + \sin\alpha \cdot \sin\alpha_n$
Pseudo-Azimuth system (ε, ρ)	$\cos\rho \cdot \cos\rho_n \cdot \cos(\varepsilon - \varepsilon_n) + \sin\rho \cdot \sin\rho_n$
Pseudo-Equatorial system (β, γ)	$\cos\beta \cdot \cos\beta_n \cdot \cos(\gamma - \gamma_n) + \sin\beta \cdot \sin\beta_n$

The incidence angle is of particular importance in evaluating the direct solar radiation received by the PV module and can be calculated based on the solar angles as presented in Table 3.4, where the subscript n is used for the angles of the unit vector $\vec{n}$ (i.e. for the PV module angles).

The maximum values of the solar angles in Table 3.4 and, particularly, of the diurnal angles are important in the design of the optimal tracking system suitable for a given location. The variations of the solar angles during representative days (winter solstice, equinoxes, and summer solstice), in various locations at different latitudes are outlined in Figs. 3.16, 3.17 and 3.18.

At 30°N latitude, the four diurnal solar angles have almost similar variations during the day (Fig. 3.16), except the azimuth angle (2) in Fig. 3.16c. The maximum range of the diurnal angles is observed at this latitude during the summer solstice. However, at the same latitude, the elevation angles α (azimuth, Az), δ (equatorial, Eq), ρ (pseudo-azimuth, PAz) and γ (pseudo-equatorial, PEq) have significant daily variations, as Fig. 3.17 shows, depending on the day number in the year.

The difference between the largest and smallest annual values (i.e. the maximum variation range) of the solar angles and particularly of the diurnal angles are important in the design of the optimal tracking system suitable for a given implementation location. The dependence of this range on the latitude is presented in Fig. 3.18.

The maximum variation ranges of the PAz and PEq diurnal angles do not depend on latitude (180°, curves (3) and (4) in Fig. 3.18a); oppositely, the variation ranges of the Eq and Az diurnal angles significantly increase with the latitude as outlined

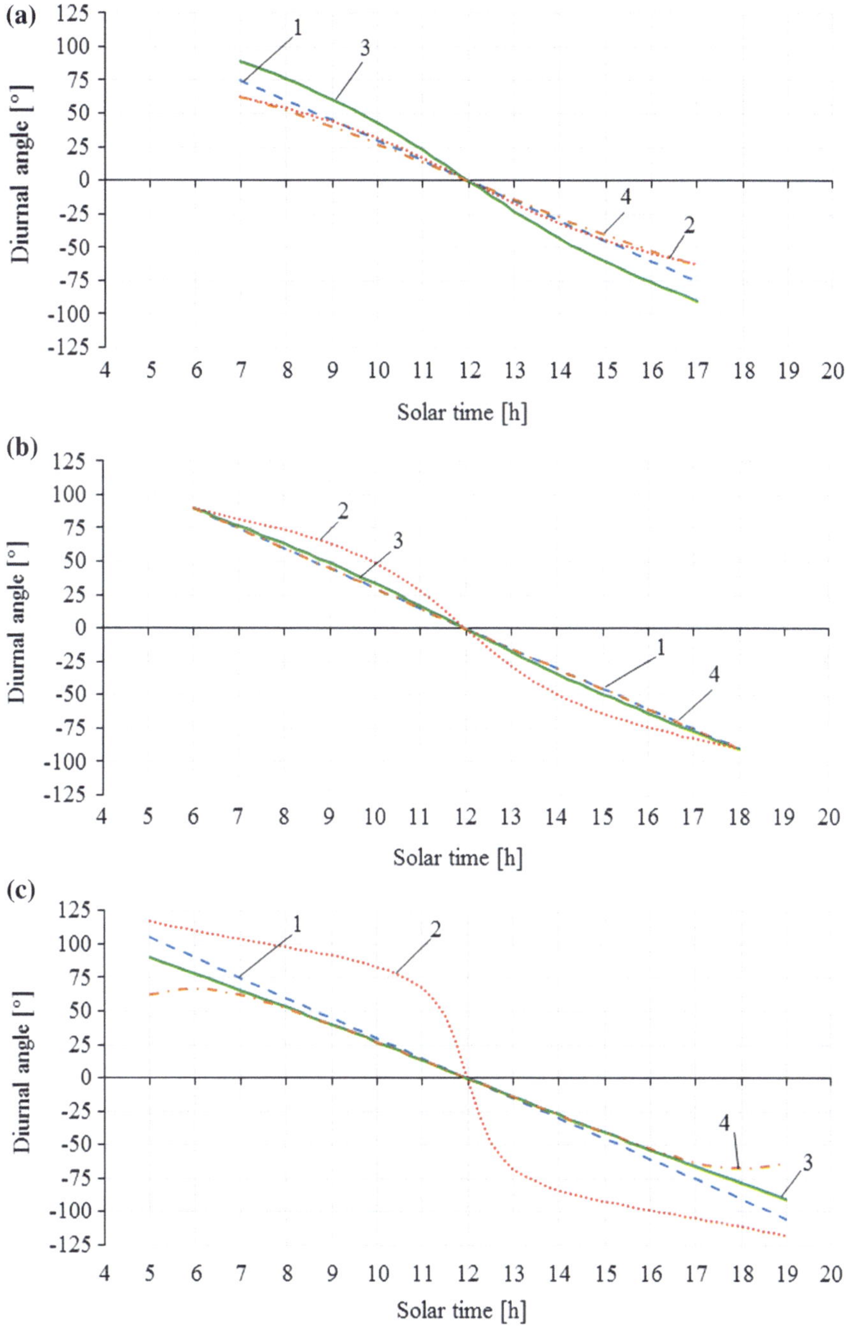

Fig. 3.16 Variation in time of the diurnal solar angles of equatorial type ω (1), azimuth type ψ (2), pseudo-azimuth type ε (3) and pseudo-equatorial type β (4) at the 30°N latitude during: **a** winter solstice; **b** equinoxes; **c** summer solstice

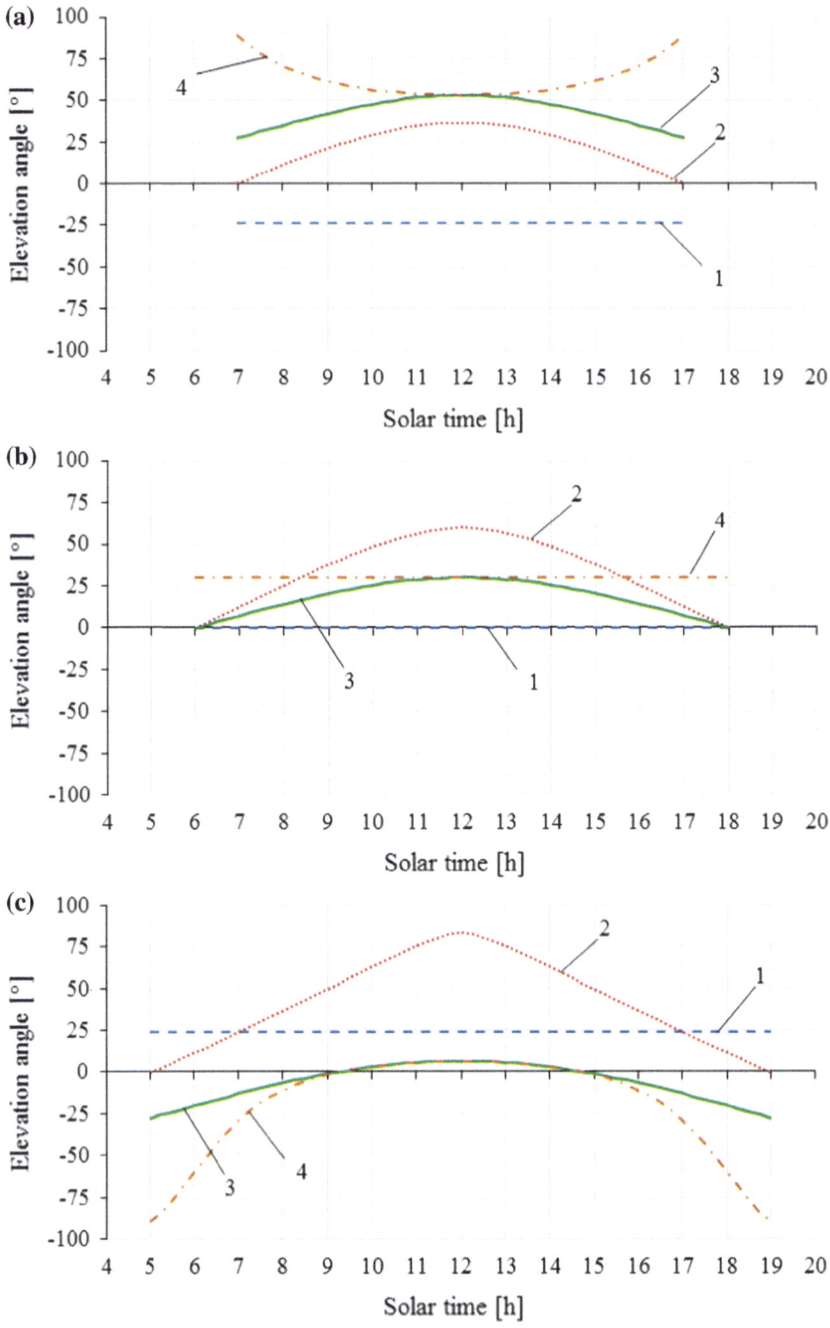

Fig. 3.17 Variation in time of the elevation solar angles of equatorial type δ (1), azimuth type α (2), pseudo-azimuth type ρ (3) and pseudo-equatorial type γ (4) at the 30°N latitude during: **a** winter solstice; **b** equinoxes; **c** summer solstice

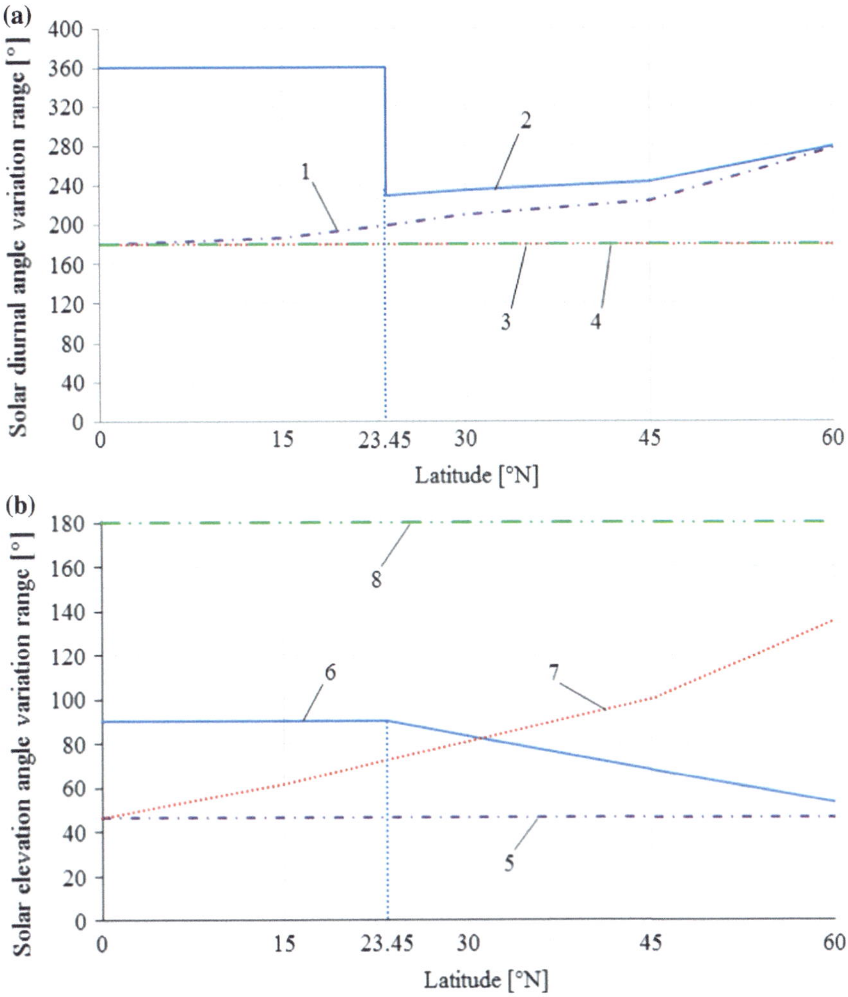

Fig. 3.18 The annual variation range of the solar angles at different latitudes in the northern hemisphere: **a** diurnal angles: hour angle ω (1), azimuth angle ψ (2), PAz angle ε (3), PEq angle β (4); (b) elevation angles: declination angle δ (5), altitude angle α (6), PAz angle ρ (7), PEq angle γ (8)

in Fig. 3.18a, curves (1) and (2). Similarly, the maximum variation ranges of the Eq and PEq elevation angles are constant over the latitudes (curves (5) and (8) in Fig. 3.18b), while the Az and PAz elevation angles are decreasing and respectively increasing with the latitude (curves (6) and (7) in Fig. 3.18b). This information is useful in the selection and the design of the tracking mechanisms for PV modules and of their stepwise motions with angular strokes usually smaller than the annual variation ranges of the solar angles.

3.3.2 Solar Tracking Systems

According to the literature data, there are four types of tracking systems defined in direct relation with the four pairs of solar angles, as presented in Fig. 3.19, with their main features synthetically presented in Table 3.5:

- *Equatorial* (or *polar*) *tracking systems*, with the diurnal axis parallel to the Earth polar axis [20, 53, 67];
- *Azimuth* (or *azimuth-altitude*) *tracking systems*, where the diurnal axis is parallel to the zenith (the local vertical axis) [50, 53, 59, 67, 98];
- *Pseudo-Equatorial tracking systems*, with a fixed horizontal rotation axis parallel to the East–West direction [9, 24, 98, 100];
- *Pseudo-Azimuth tracking systems*, with a fixed horizontal rotation axis parallel to the North–South direction [9, 24, 98, 100].

The tracking systems can be of dual-axis or single-axis type. The dual-axis systems are used to develop two rotational (diurnal and elevation) motions of the solar energy convertor, while the single-axis systems control only one rotational motion, typically the diurnal one.

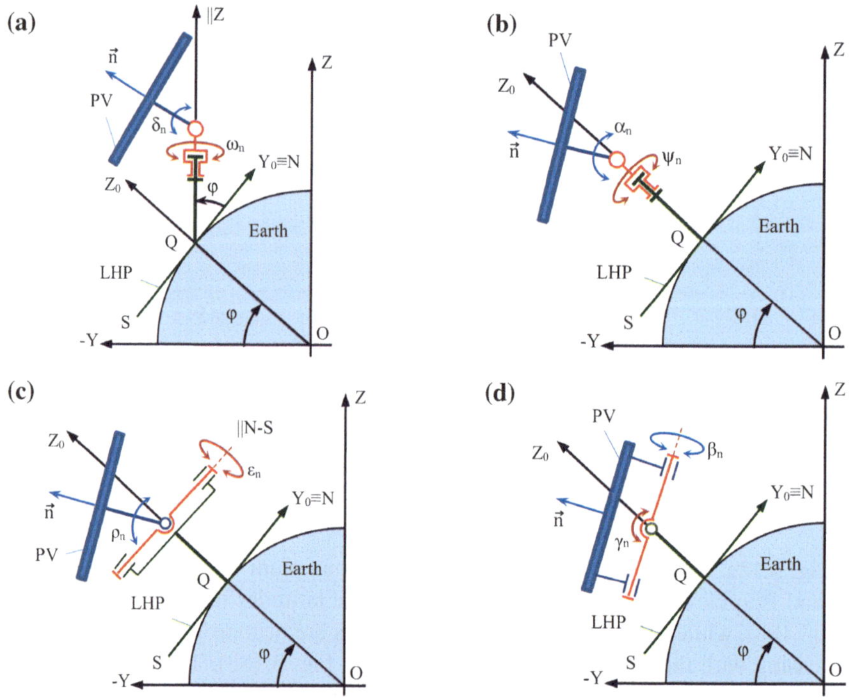

Fig. 3.19 Dual-axis tracking systems: **a** Equatorial (Eq) type; **b** Azimuth (Az) type; **c** Pseudo-Azimuth (PAz) type; **d** Pseudo-Equatorial (PEq) type; LHP—local horizontal plane

Table 3.5 Specific features of the solar tracking systems

Features	Tracking system type			
	Equatorial (Eq)	Azimuth (Az)	Pseudo-Azimuth (PAz)	Pseudo-Equatorial (PEq)
Solar angle pair (diurnal, elevation)	(ω, δ)	(ψ, α)	(ε, ρ)	(β, γ)
Diurnal tracking angle and its limit values	ω_n −139 to +139° (at 60° lat.)	ψ_n −180 to + 180° (between Tropics)	ε_n −90 to +90° (at all latitudes)	β_n −90 to +90° (at all latitudes)
Elevation tracking angle and its limit values	δ_n −23.45 to + 23.45° (at all latitudes)	α_n 0–90° (between tropics)	ρ_n −53 to +83° (at 60°N lat.)	γ_n −90 to +90° (all latitudes)
Primary axis position	parallel to the polar axis (tilted at φ)	Vertical	Horizontal N–S	Horizontal E–W
Rotation around the primary axis	Diurnal	Diurnal	Diurnal	Elevation
Secondary axis position	Perpendicular to the primary axis			
Rotation around the secondary axis	Elevation	Elevation	Elevation	Diurnal

Conceptually, the dual-axis tracking systems contain two revolute joints, where the first (*primary*) axis is fixed relative to the ground, and the other (*secondary*) is perpendicular to the first axis. The primary axis is used for diurnal motion, and the secondary axis for the elevation, except the pseudo-equatorial system where the role of the two motion axes is reversed.

In the general case of a tracker that can be used at any latitude, the Eq and Az tracking systems have the largest diurnal strokes (360°) and the smallest elevation strokes, while the PAz and PEq tracking systems require 180° strokes for both motions. In a particular location, the strokes of the four tracking systems types have specific values, as presented in Fig. 3.18.

The two motions of the dual-axis tracking systems can be developed by serial or parallel type of mechanisms, driven by linear or rotary actuators [99, 102].

Several tracking systems implemented in the built environment are presented in Fig. 3.20. These systems are located on the building's terrace (Fig. 3.20a, b, d–f, h) and nearby (Fig. 3.20c, g) as diurnal single-axis (Fig. 3.20a, f, g—P_4 and P_5, *h*) and dual-axis (Fig. 3.20b–e, g—P_1, P_2 and P_3) systems.

At a lower cost than the rotary actuators, the broad majority of these systems use linear actuators to control the elevation motion (in the case of dual-axis systems, Fig. 3.20b–d, f, g) and the diurnal motion (Fig. 3.20d–f, h).

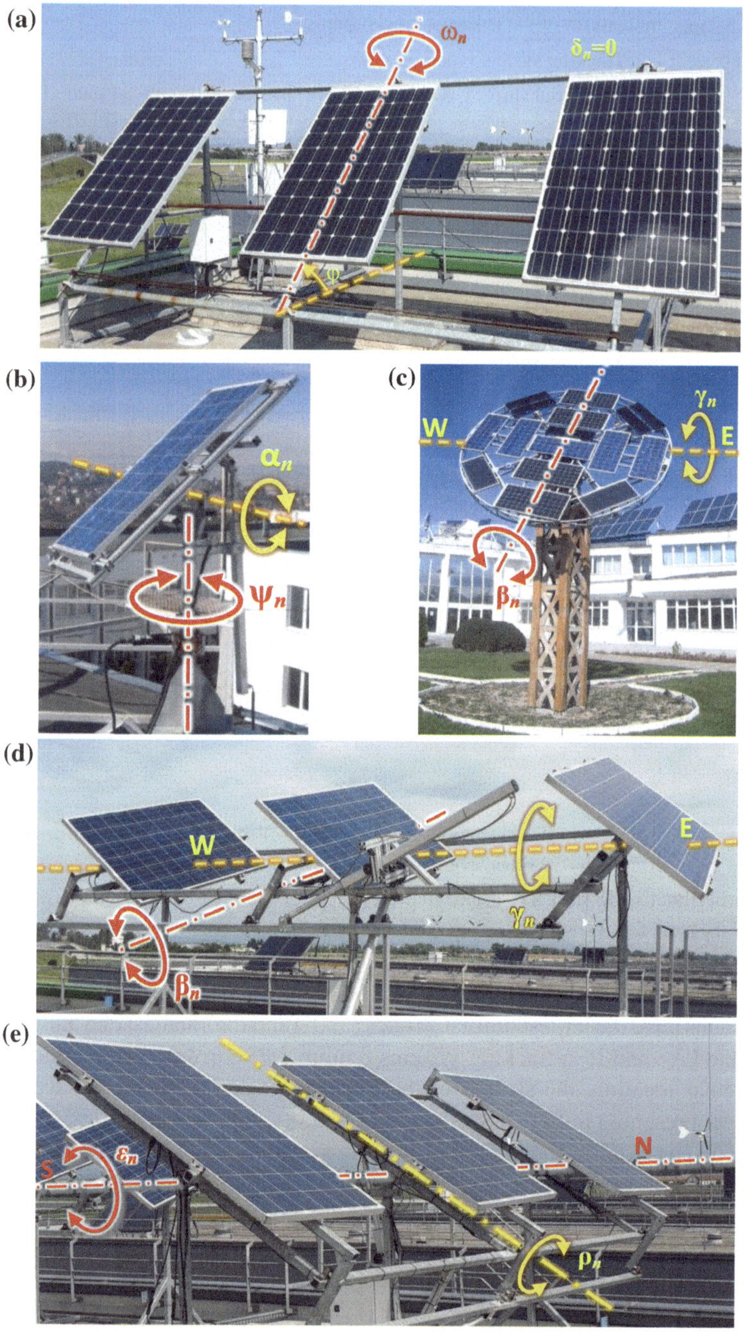

Fig. 3.20 PV tracking systems installed in the Transilvania University of Brasov, Romania: **a** single-axis of Eq type; **b** dual-axis of Az type; **c** dual-axis of PEq type; **d** dual-axis of PEq type; **e** dual-axis of PAz type; **f** single-axis of PAz type; **g** row of five PV platforms with dual-axis of Az type (P_1–P_3) and single-axis (P_4, P_5) systems; **h1** single-axis of PAz type and **h2** its actuation mechanism

(f)
(g)
(h1)
(h2)

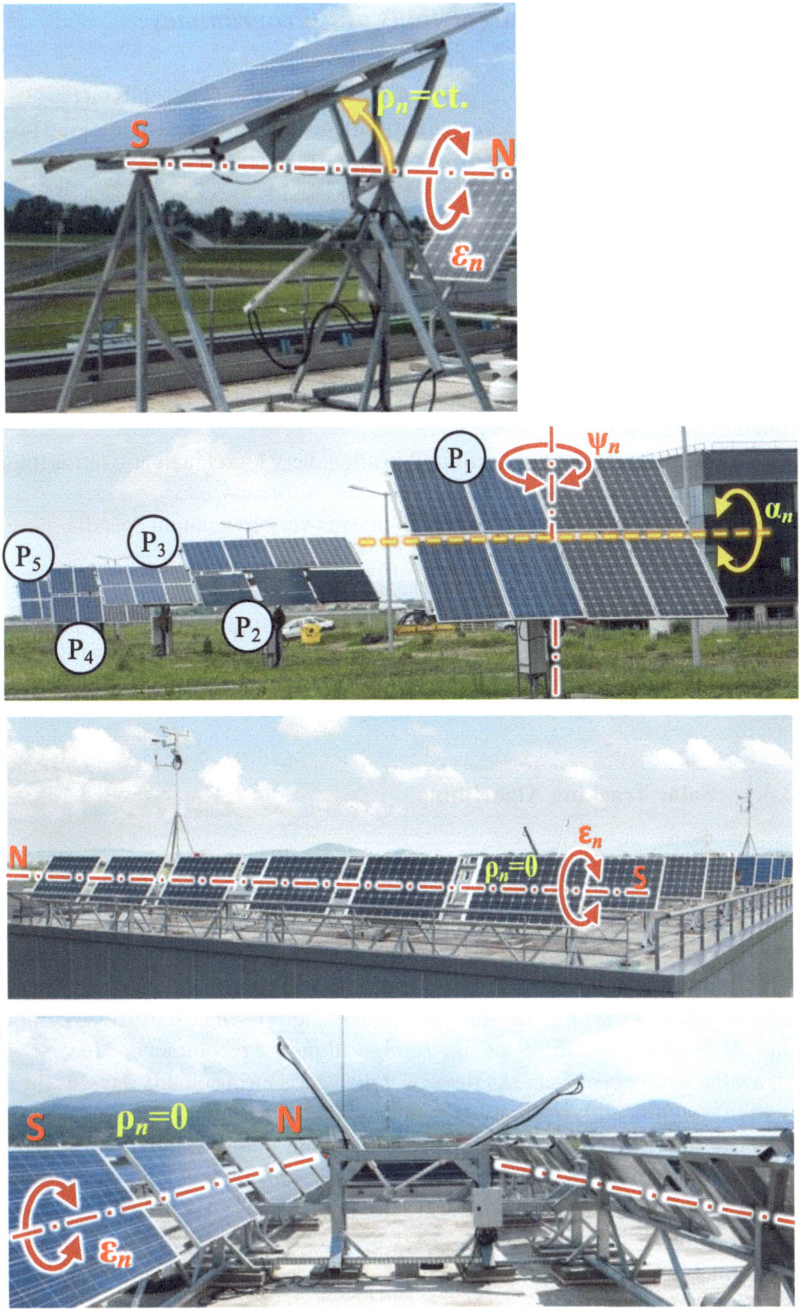

Fig. 3.20 (continued)

3.3.3 Solar Tracking Algorithms and Programmes

The amount of solar energy received by a tracked PV system depends on the available solar radiation, on the used tracking algorithm, on the implemented tracking program that controls the daily motions of the tracking mechanism and on the diurnal and elevation angular strokes.

Continuous tracking motions are not feasible because of technical and economic reasons [100], as:

- The permanent operation of the actuators may have a negative impact on the actuator's lifetime and thus on the system's maintenance;
- The energy consumption for operating the actuators may be significant, when compared with the energy gained by tracking using as reference an optimally fixed tilt system;
- Complex speed reducers are required to allow very high kinematic ratios thus, low mechanical efficiencies;
- The high price of the actuators able to deliver very low output speeds.

The tracking algorithms are mainly of two types, aiming at increasing:

- The received direct solar irradiance (B_n) by neglecting the diffuse radiation;
- The received direct (B_n) or diffuse (D_n) solar irradiance, under variable environmental conditions [5, 47, 54].

3.3.3.1 Solar Tracking Algorithms

Since continuous tracking does not represent a fully feasible alternative, stepwise tracking algorithms are typically used in practice. These *solar tracking algorithms* are designed to increase the amount of the received solar energy by using a *stepwise* motion.

The mostly used tracking algorithms aim at increasing the received direct solar irradiance (B_n) by tracking the Sun on its path in the sky. Since the diffuse component of the solar radiation is neglected, the B_n algorithms are recommended in sunny areas with a large numbers of clear sky days over the year. During cloudy days, as the direct radiation is significantly decreased (it can reach zero), the efficiency of a B_n algorithm is reduced and the output may be even lower than that of a horizontally fixed PV module, used to maximize the received D_n. As results, a tracking algorithm able to switch between two functional modes (increasing B_n or D_n) is useful in regions with frequent extreme weather changes.

The tracking algorithms recommended during various weather periods are [66]:

- *Clear sky periods*: the classical method of increasing B_n;
- *Cloudy sky periods*: fixed horizontal positioning the PV module is recommended to increase the received diffuse solar radiation;

- *Partially clouded* (*sunny-cloudy*) *periods*: an algorithm able to switch between increasing B_n (during sunny periods) or D_n (horizontally positioning during cloudy periods). Such an algorithm accesses live data on the available diffuse and direct or global radiation [66].

A general design scheme of the tracking systems and algorithms is presented in Fig. 3.21.

According to the information in Fig. 3.21, a designer should face in the PV system design the following challenges:

- Which is the optimal tracking system type (Eq, Az, PAz, PEq)?
- Is there a dual-axis solar tracking required (motions on both the diurnal and elevation axes), or a single-axis tracking or even the optimally fixed and tilted system is good enough?
- Which would be the optimal tracking algorithm and how complex the tracking program should be?

An answer to these questions is further on proposed.

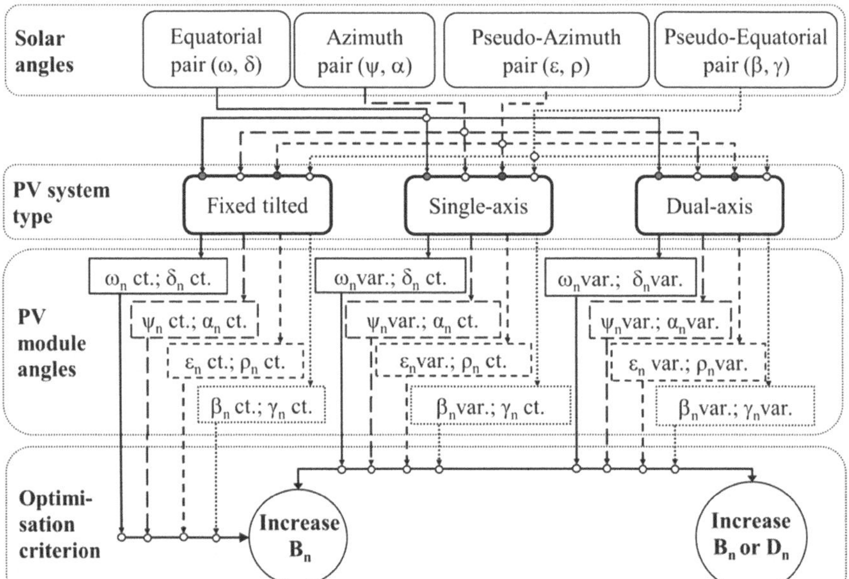

Fig. 3.21 Synthesis on the solar tracking systems and algorithms (ct. = constant value, var. = time variable value)

3.3.3.2 Solar Tracking Programs

For a given application the steps of the tracking algorithm are implemented into a tracking program to control the tracking mechanism motions.

The tracking program is based on the following parameters that can be optimized depending on the purpose of the solar energy collection [66, 100]:

- Number and duration of the tracking seasons (the period when a tracking program remains unchanged, e.g. one month, one season);
- The number of steps during one day;
- The type of the steps (constant or with variable duration);
- The angular strokes of the diurnal and of the elevation motions.

The yearly received solar energy increases as the number of tracking seasons is higher, the number of daily steps is higher, steps with variable optimized durations are used and the angular strokes of the tracking system are close to the entire variation ranges of the solar angles.

The values of these parameters can be the result of an optimization problem of the tracking system, as higher values lead to an increased complexity of the tracking mechanism (e.g. for angular strokes larger than 120°) and of the command and control system. Tracking programs with constant steps (e.g. 1 h) are easy to use under specific conditions. However, the implementation of variable steps allows a better solar energy collection by reducing the steps duration at solar noon, when the irradiance reaches the maximum value and increasing the steps duration for the rest of the day. Changing the tracking program each month and a 120° diurnal stoke represent feasible options for most photovoltaic applications.

3.3.3.3 Numerical Simulations

A comparative study of the four types of tracking systems (Az, Eq, PAz, PEq) is further presented. The simulation results are detailed for the 45°N latitude and are further presented for locations at different latitudes in the northern hemisphere (Equator 0°, 15°N, 30°N and 60°N), located on the 25.5°E meridian, considering only the effect of the direct solar radiation (the B_n tracking algorithm). The monthly values of the turbidity factor (T_R), required to calculate the direct solar radiation (B) according to Eq. (2.4), were obtained for each location using the SoDa database [86].

The following two tracking programs versus fixed tilted orientation used as reference are analysed:

- *Scenario* 1: fixed PV module South oriented and tilted at an angle equal to the latitude;
- *Scenario* 2: single-axis diurnal tracked PV module tilted at an angle equal to the latitude;
- *Scenario* 3: dual-axis stepwise tracking with four unequal steps.

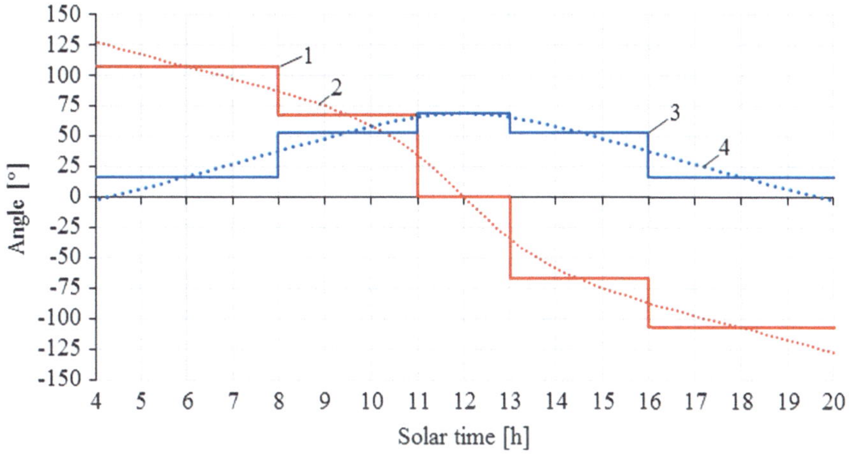

Fig. 3.22 Variation of the solar (2 and 4) and tracking (1 and 3) angles for a dual-axis tracking program with four unequal steps (scenario 3) in the case of the Az type tracking system in mid-June at 45°N latitude: the angles ψ_n (1), ψ (2), α_n (3), α (4)

A four-steps tracking program, with unequal steps for an Az system in June at 45°N latitude, is presented in Fig. 3.22.

This tracking program (Fig. 3.22) contains four fast motions (2 steps in the morning and 2 steps in the afternoon). The solar time and the value of these angular displacements are defined as follows:

- Steps with shorter duration are required close to the solar noon and longer duration steps can be otherwise set;
- The solar angle curves (2) and (4) split into roughly equal segments in the tracking angle lines (1) and (3), respectively.

Dual-axis tracking programs with 2 steps and 6 steps were also investigated: the 4-steps program leads to a tracking efficiency increase of about 3–4% relative to a 2-steps approach, while a 6-steps program proved only a slight increase (<0.5%). As results, a dual-axis 4-steps tracking program (scenario 3) represents a suitable option, particularly in the case of limited available space on buildings.

According to Figs. 3.23 and 3.24, the variation of the *direct solar energy receiving share* (i.e. the percentage of the available solar energy received by the PV module, defined by $E(B_n)/E(B)\cdot 100$ where E is daily energy) at the 45°N latitude allows to formulate the following conclusions:

- The fixed PV modules tilted at an angle equal to the latitude (Scenario 1) variably receive the direct solar energy and the receiving share decreases in summer from Equator (about 70%) to high latitudes (about 50% at 60°N), Fig. 3.23. This trend is reversed in winter (Fig. 3.24) when the receiving rate reaches about 85% at the 60°N latitude. As result, the solar energy is received throughout the year almost constant at Equator and with large variations at high latitudes;

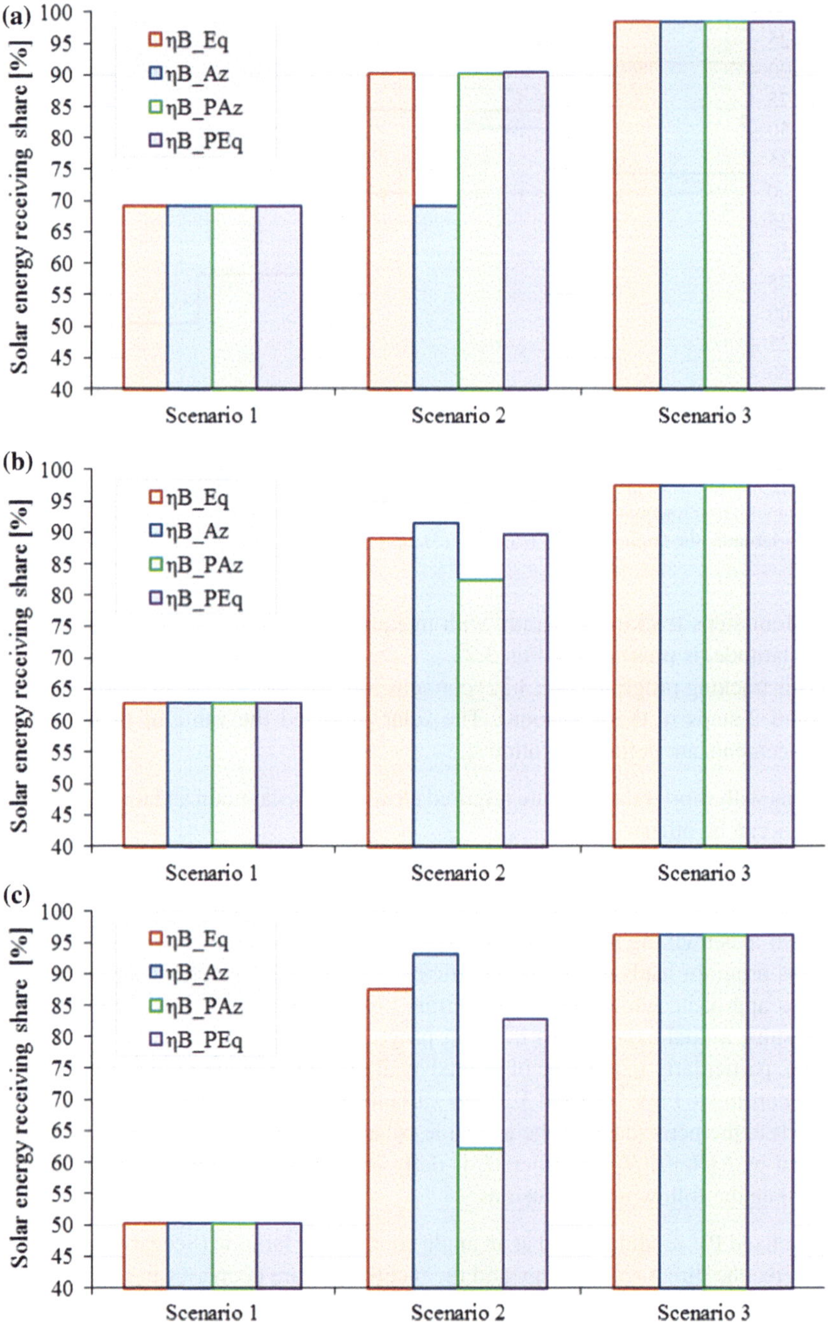

Fig. 3.23 Direct solar energy receiving share in mid-June for Scenarios 1–3 at various latitudes: **a** 0° (equator); **b** 30°N; **c** 60°N

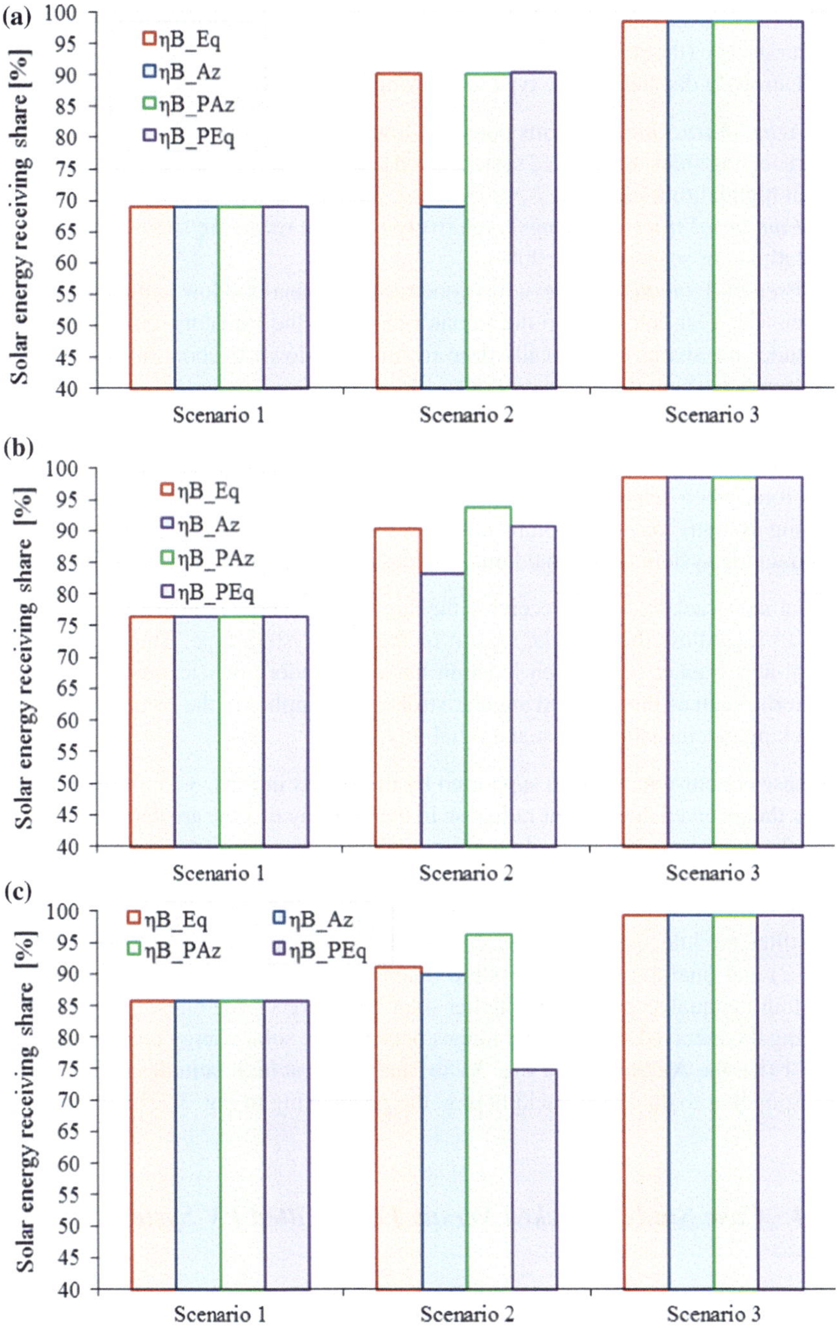

Fig. 3.24 Direct solar energy receiving share in mid-December for Scenarios 1–3 at various latitudes: **a** 0° (equator); **b** 30°N; **c** 60°N

- The single-axis tracking (Scenario 2) brings a significant gain to the received direct solar energy (the receiving rate can increase up to 95%) compared to Scenario 1, but strongly depends on the type of tracking:
  - *Azimuth tracking*: performs poorly at low latitudes (at Equator it has the same receiving share as the fixed system tilted at 0°), but it becomes the most efficient at high latitudes;
  - *Equatorial tracking*: brings a relatively constant receiving share regardless of latitude or season (about 90%);
  - *Pseudo-Azimuth tracking*: it has good receiving shares at low latitudes throughout the year (identical to the Equatorial one at the Equator), but at high latitudes the shares dramatically drop in summer (down to about 60%) and grow significantly in winter (over 95%);
  - *Pseudo-Equatorial tracking*: ensures good results at low and medium latitudes, regardless the season.

Therefore, when selecting single-axis tracking systems, the use of the Eq or PAz tracking systems may be recommended at low latitudes (between the Tropics), the PEq tracking system type at medium latitudes and the Az type at high latitudes;

- Dual-axis tracking allows receiving the direct solar energy with higher rates (over 96%) regardless the latitude, season or tracking system type. Thus, selecting a dual-axis tracker for a given location has to consider other technical-economic criteria, such as the required angular strokes and implicitly the complexity of the tracking mechanism, its cost and reliability, etc.

These conclusions are also supported by the results in Figs. 3.25, 3.26 and 3.27, where the received direct solar radiation in the mid-day of June are depicted considering the three previously stated scenarios and the four tracking types (Az, Eq, PAz, PEq) at three latitudes of 0°, 30°N and 60°N. The variation of the solar irradiance clearly shows the relative position of the curves associated with the three scenarios: fixed tilted module, used as reference (Scenario 1), single-axis tracked module (Scenario 2) and dual-axis tracked module (Scenario 3). While the dual-axis tracking with four unequal steps supports higher solar energy receiving rates, for single-axis tracking the received solar energy ranges between the solar energy received in Scenario 1 (i.e. the Az tracking in Fig. 3.25b) and tends, at high latitudes, to that value corresponding to dual-axis tracking (e.g. the Az tracking in Fig. 3.27b).

3.3.4 Case Study: Tracked Versus Fixed Tilted PV Systems

The implementation location represents a key factor in selecting an optimal tracking solution for the PV surfaces, i.e. the tracking system type and the tracking program. The complexity of this problem increases when switching from the clear to the real sky assumption(s), by considering the variability of the meteorological conditions in the implementation location. Experimental tests in the outdoor environment for

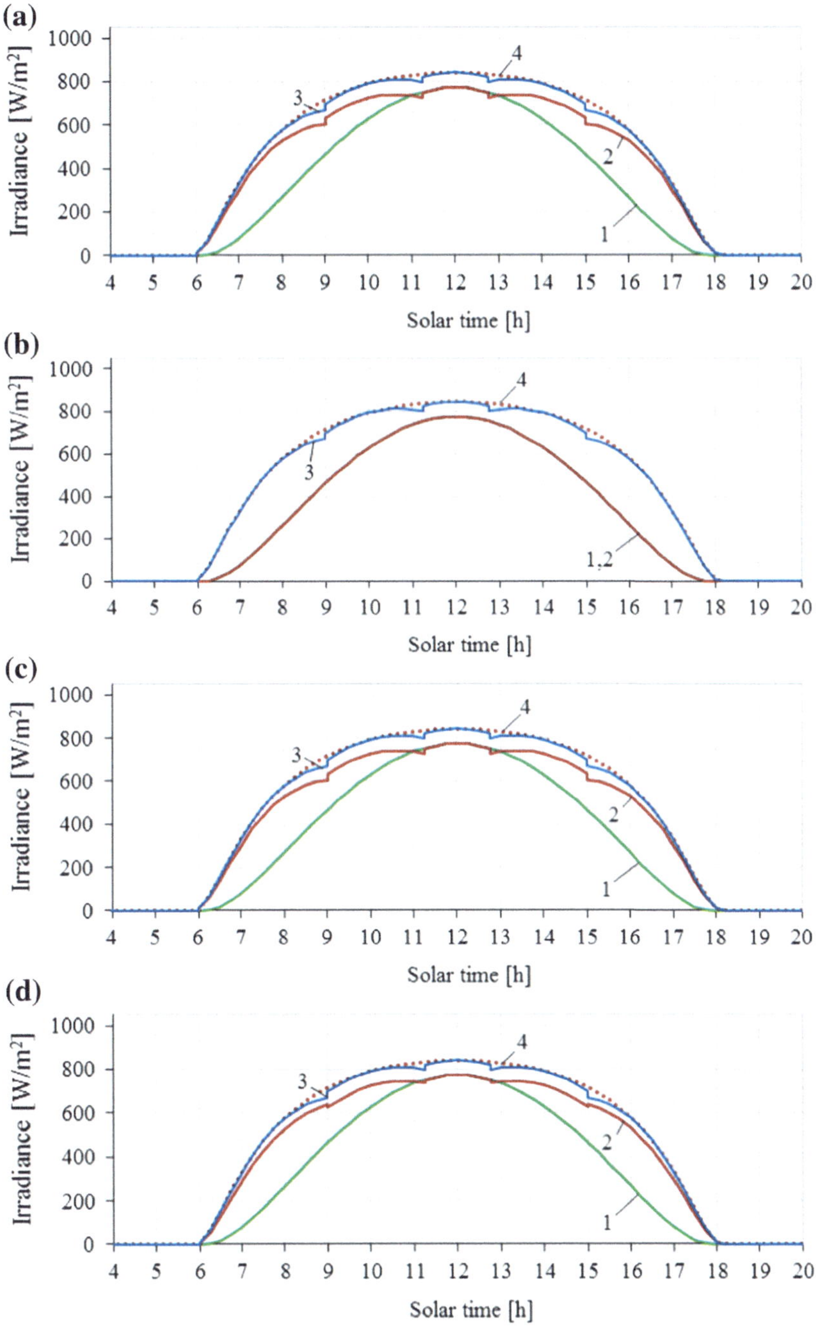

Fig. 3.25 Irradiance variation of the available (4) and of the received direct solar radiation in mid-June at 0°latitude (Equator) for the reference Scenario 1 (1) and for the tracking programs according to Scenario 2 (2) and to Scenario 3 (3) for a tracking system of: **a** Eq type; **b** Az type; **c** PAz type; **d** PEq type

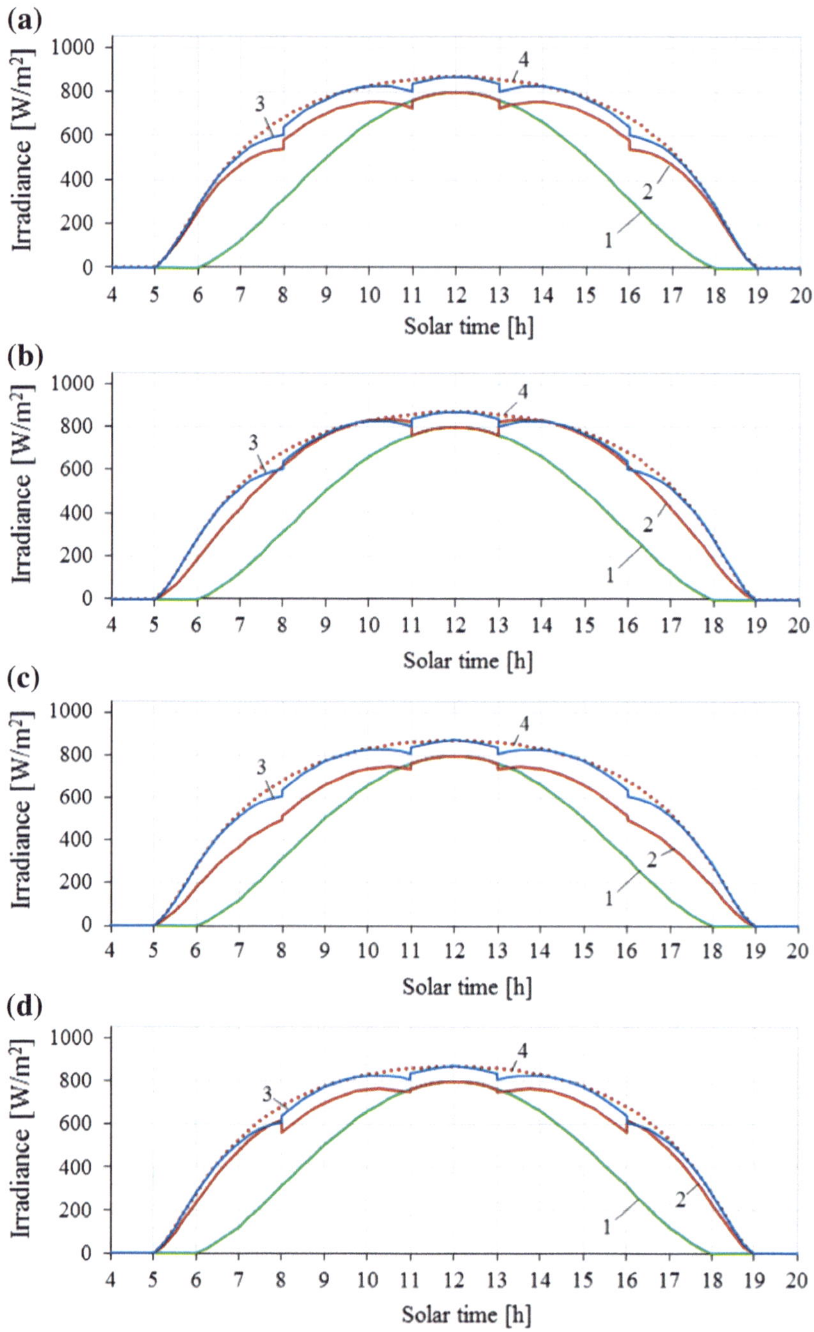

Fig. 3.26 Irradiance variation of the available (4) and of the received direct solar radiation in mid-June at 30°N latitude for the reference Scenario 1 (1) and for the tracking programs according to Scenario 2 (2) and Scenario 3 (3) for a tracking system of: **a** Eq type; **b** Az type; **c** PAz type; **d** PEq type

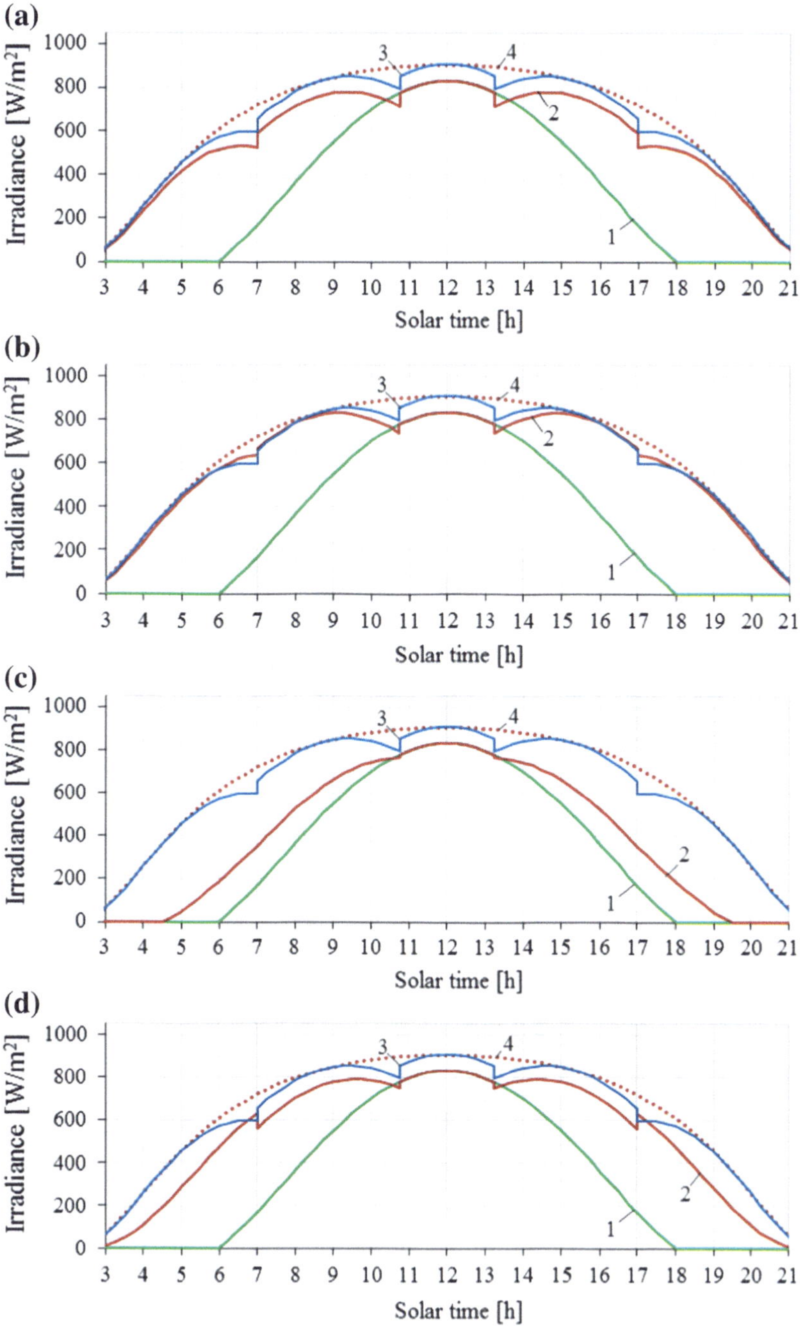

Fig. 3.27 Irradiance variation of the available (4) and of the received direct solar radiation in mid-June at 60°N latitude for the reference Scenario 1 (1) and for the tracking programs according to the Scenario 2 and to the Scenario 3 (3) for a tracking system of: **a** Eq type; **b** Az type; **c** PAz type; **d** PEq type

various tracking solutions are recommended as an important step in the optimal design of the PV systems.

The experimental results presented in Figs. 3.28 and 3.29 allow comparing dual-axis tracked versus fixed tilted PV systems. The data were recorded during one year (June 2017–May 2018) in Brasov, Romania using an outdoor testing rig that has five PV platforms (P_1–P_5, Fig. 3.20g). Among these, two platforms (P_1 and P_4) were selected for comparative testing:

- P_1: a platform with a dual-axis tracked system of Az type with the diurnal stroke of 220° (−110 to +110°) and the elevation stroke of 90° (0–90°);
- P_4: fixed and tilted at 45° (close to the latitude angle of the implementation location).

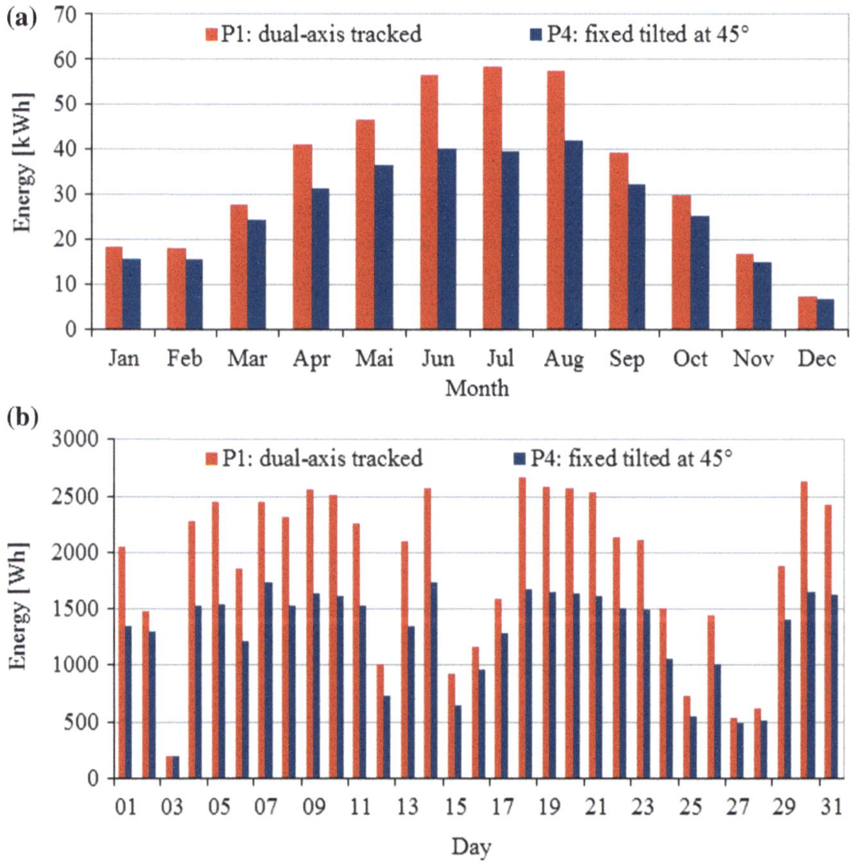

Fig. 3.28 Experimental data on the electrical energy produced by a dual-axis tracked (P_1) versus fixed and optimally tilted (P_4) PV platform in Brasov, Romania: **a** monthly produced electrical energy; **b** daily electrical energy produced during July 2017

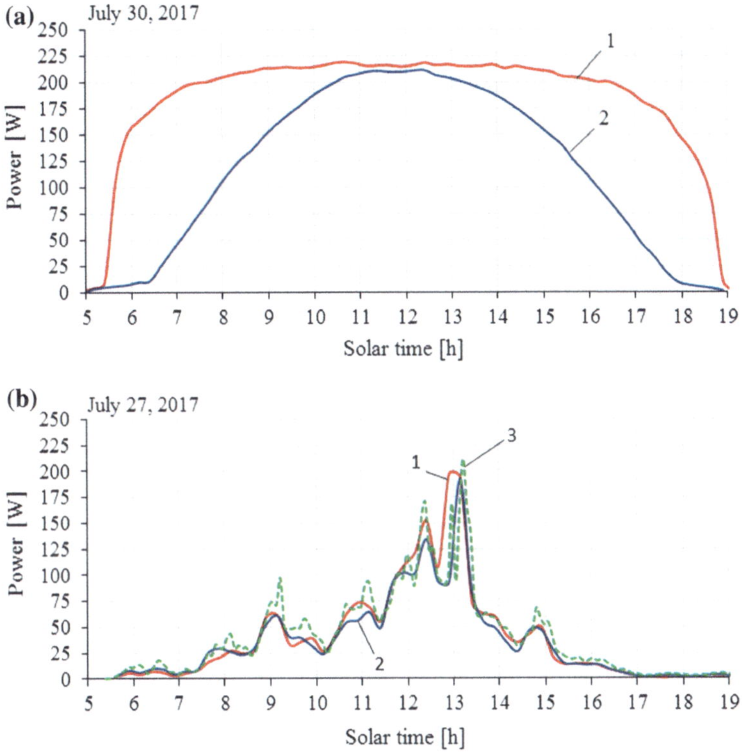

Fig. 3.29 Experimental output power of tracked versus fixed PVs in July 2017 (Brasov, Romania) during: **a** a clear sky day; **b** a clouded sky day; platform P_1: dual-axis tracked PV modules (1); platform P_4: fixed and tilted PV modules at 45° (2) versus horizontally fixed PV module(s) (3)

Another PV array (Fig. 3.20h1, h2) was horizontally positioned during the testing period and used as reference during cloudy days. All the PV modules are of polycrystalline silicone (p-Si) type with the nominal power of 250 W and the nominal efficiency of 15.32%. The output power of only one PV module from each platform was further considered.

According to the experimental data presented in Fig. 3.28a, the dual-axis tracked platform (P_1) is highly efficient during the summer season while tracking has a low contribution (less than 10%) during the winter months.

Throughout the year, the electricity additional gain is of 28.7% due to tracking. During July, the month with the highest electrical output, a consistent gain in the electrical output during clear sky days (e.g. during July 18–21, 2017) was observed while the tracking contribution is almost insignificant during cloudy days (e.g. July 03 and July 27, 2017). Thus, tracking gets more effective, compared to the fixed tilted PV systems, during sunny days, as confirmed by the experimental data on the electrical output presented in Fig. 3.29. These results allow formulating the following conclusions:

- During a summer sunny day, the dual-axis stepwise tracking (platform P_1) adds an extra 50% to the electrical energy that was produced by an identical fixed tilted platform (reference platform P_4), as the results in Fig. 3.29a show. Moreover, tracking is also recommended during the sunny periods of mixed (sunny-cloudy) days;
- During cloudy days, it is recommended to horizontally position the PV modules, to receive the predominantly diffuse solar radiation and thus to increase the electrical output (about 20% of that registered during clear sky days), Fig. 3.29b;
- Moreover, during cloudy days tracking leads to an insignificant energy gain as shown in Fig. 3.29b curve (1), compared with the fixed tilted PV modules on platform P_4, curve (2), or even to a slightly lower energy output as compared with the horizontally fixed PV module, curve (3).

In conclusion, compared to an optimally fixed tilted PV module, the tracked module shows about 50% increase in the energy output on a clear sky day, being effective even on mixed days. However, during clouded sky, tracking becomes inefficient, and it is recommended to keep the PV modules fixed, in the horizontal plane. The horizontal position is also favourable in the case of high speed wind that frequently occurs during cloudy periods.

As solar tracking increases the received amount of solar energy on the PV module surface, thus the electrical energy production, identifying the optimal PV tracking system depends on certain parameters, such as the available implementation area and the energy demand, the cost of the PV modules and of the tracking system, the installation, operation and maintenance costs, i.e. the cost of the entire PV system over its lifetime.

According to the study developed by Greentech Media [36], considering the current price of the PV modules (about 0.2–0.4 €/Wp), single-axis tracking systems remain an attractive option for sunnier areas, supported by the recent significant decrease in the cost of single-axis solar trackers. Nevertheless, dual-axis diurnal tracking with the optimum tilt angle monthly adjusted can also be considered an affordable option as it allows up to 8% more solar radiation to be received on the PV module surface at low and middle latitudes, compared with the single-axis tracking systems with a fixed tilt angle all over the year.

Considering the current trends on decreasing the cost/W and increasing the nominal efficiency of the PV modules, it can be concluded that tracked PV systems remain a viable option in the future. This in mainly true for applications with severe space constraints (as in the built environment) or when a high share of energy produced using renewable energy sources represents a clear target.

3.4 Exploitation and Maintenance of the PV Systems

After the PV system is installed and all verification procedures indicate a well-functioning system, the end-user has to be informed about the limits of the system and about specific, regular check-ups that have to be done during its exploitation. These will support an increase of the advantages brought by the use of photovoltaic systems [68].

Whether a PV system is grid-connected or stand-alone, installed in the built environment or on a land surface in the outskirts, the exploitation and maintenance activities mainly target the same directions and follow the same steps. Similar to the case of any conventional power system, the human intervention targets:

- The correct *exploitation* of the photovoltaic system so that its components work in normal conditions and allow the best use of the produced electrical energy, according to the needs;
- The regular *maintenance* of all the equipment in the PV system, to maximize the electricity production.

3.4.1 Exploitation of the PV Systems in the Built Environment

The correct exploitation of any PV system focuses on aspects the end-user has to consider and that are further on outlined.

Regular analysis of the electrical energy produced by the PV system is required to verify if the system is working properly. This analysis allows to compare the electrical output with the design expectations, with various other PV configurations (different PV types, different locations) and to detect possible faults.

The analysis can be either manually done on a weekly/monthly/annual basis by reading the electrical utility meter or the electricity bill that indicates the amount of electricity produced [72], or automatically if an autonomous energy monitoring system is part of the PV system. The monitoring system can save the data on a local storage device or send them to an online monitoring platform for remote access and analysis (Fig. 3.30).

The produced electrical energy allows to calculate the *final yield* (Y_f). The final yield indicates the *time during which the PV system would operate at the nominal power to provide the energy produced during the reported interval* [43] and can be determined [57] using

$$Y_f = \frac{E_{AC}}{P_{DC}} \tag{3.33}$$

where

- E_{AC} is the produced AC energy during a specific time interval;
- P_{DC} is the nominal DC power of the PV system.

For a 10 years old PV system with a nominal DC power of 10.6 kW, small values of the final yield are indicated in Fig. 3.31 during the winter months as result of the

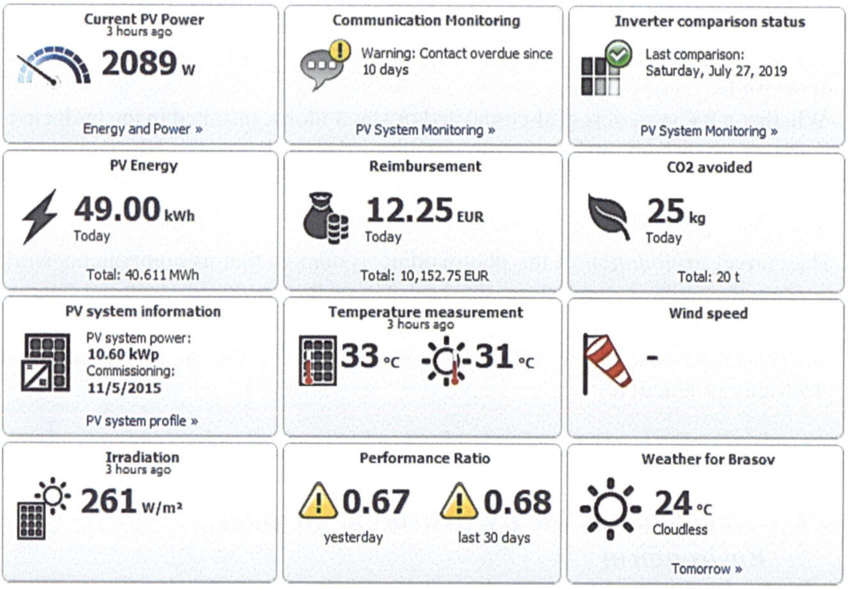

Fig. 3.30 Online SMA Sunny Portal used to monitor and analyze the performance of the PV systems installed at the Transilvania University of Brasov, Romania

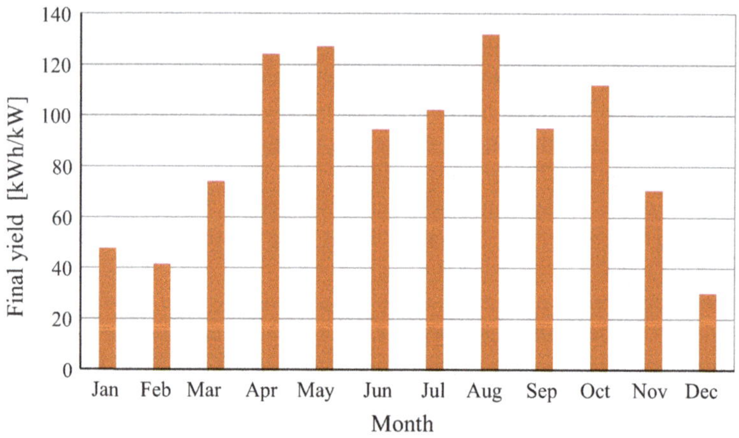

Fig. 3.31 Monthly final yield values in 2018 of a PV system, installed fixed tilted at 45° in the Transilvania University of Brasov, Romania

low available solar irradiance, while the spring and autumn months show the highest yields because of the higher solar irradiance and the relatively low temperature of the PV modules. The summer months, June and July, show average yield values during 2018 because of the relatively unstable weather conditions and because of the higher temperature of the PV modules. The largest yield corresponds to August that was in 2018 a month with predominantly clear sky days thus with a significant amount of available solar energy.

An advanced analysis of the PV systems can be developed if the irradiance of the received solar radiation on the PV modules is known. This requires a pyranometer installed in the same plane as the photovoltaic modules, able to automatically record data every several minutes. Using these data, the *performance ratio* (PR), of the PV system can be calculated using

$$\mathrm{PR} = \frac{E_{\mathrm{AC}}}{E_{\mathrm{DC}}} \tag{3.34}$$

where E_{DC} is the calculated nominal DC energy produced by the PV system during the same time interval (t_1, t_2) of interest as E_{AC}

$$E_{\mathrm{DC}} = \int_{t_1}^{t_2} P_{\mathrm{PV}} \mathrm{d}t; \quad P_{\mathrm{PV}} = G_n \cdot \eta_0 \cdot A \tag{3.35}$$

where

- P_{PV} is the power produced by the PV array;
- G_n is the received solar irradiance in the plane of the PV array;
- η_0 is the nominal conversion efficiency of the PV modules in STC;
- A is the area of the entire PV array.

The PR represents a more accurate indicator, independent of the PV system location, that shows the overall effect of losses (thermal losses due PV modules heating, electrical conduction losses, efficiency of the PV module/inverter/charge regulator, etc. and possible defects of the components) on the system output [85].

The monthly performance ratio of standard PV systems ranges between 0.6 and 0.75 (as presented in Fig. 3.32), while for high performance PV systems the value can reach 0.8.

If the analysis of the performance parameters indicates unusual variations of the PV system's final yield or performance ratios lower than 0.5, further investigations are required, focusing on the PV system components. In the case of an automated analysis, based on online platforms, the individual modules can be checked if the DC power-optimizers (Fig. 3.33) or micro-inverters are used, and different alarm signals can be triggered by sending an email, SMS, etc. to the end-user with additional information.

When receiving various alarms, the end-user should first do a visual inspection of all the equipment in the PV system. This allows to observe minor problems, such as objects that are responsible for shadowing the PV modules (grown grass or trees, newly installed masts or cables, bird droppings, snow, soiling due to dust or pollen (Fig. 3.34), broken/commuted overvoltage or overcurrent fuses, loose cable connections, stopped devices, etc.

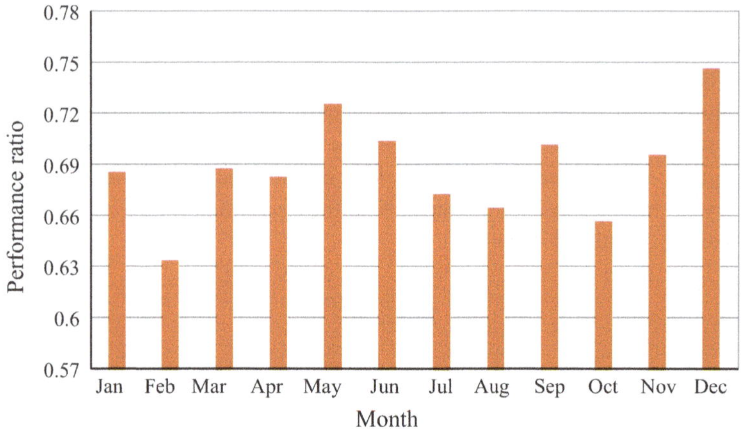

Fig. 3.32 Monthly performance ratios during 2018 of a 10.6 kW, 10 years old PV system, installed fixed tilted at 45° in the Transilvania University of Brasov, Romania

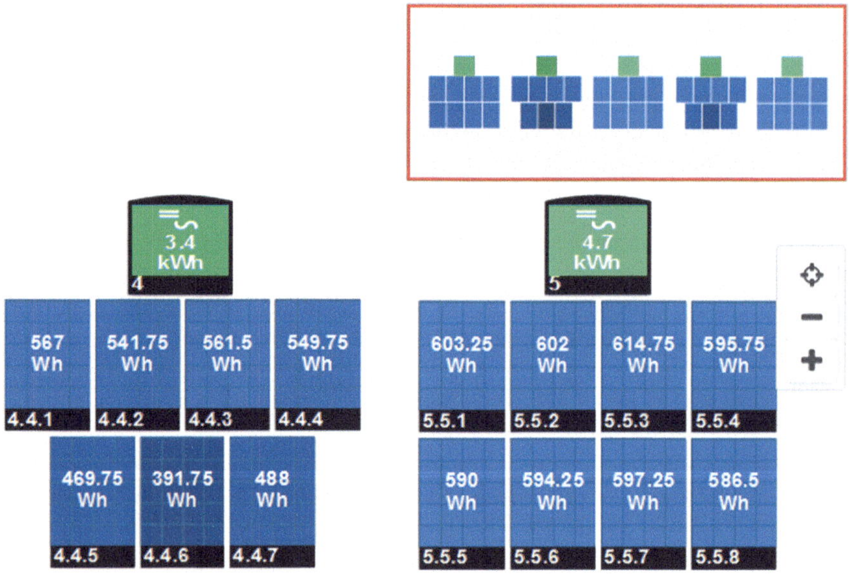

Fig. 3.33 SolarEdge monitoring portal that enables the performance analysis of the individual PV modules in tracked PV systems installed in the Transilvania University of Brasov, Romania

These minor problems can be easily managed by the PV owners by eliminating the shadow sources (leaves, branches, grass, dirt, etc.) or by reconnecting the fuses, tightening the connections or restarting the PV equipment. Major problems, such as broken or burnt out PV modules, malfunction of the inverter or charge regulator, broken or overheated batteries need to be announced to professionals and will be

Fig. 3.34 Shadow of nearby object on a PV platform installed in the Transilvania University of Brasov, Romania

solved by qualified personnel by repairing or replacing them, as part of the PV systems maintenance.

For the optimal exploitation of the PV systems, the end-users have to adjust their electrical energy consumption pattern according to the solar radiation availability to directly consume the electrical energy produced, thus decreasing the size of the storage devices. This can be either manually done, paying attention to what electrical consumers are best to be powered at a given moment or by using smart management systems able to decide which consumers are optimum/required at each moment during the day.

3.4.2 Maintenance of the PV Systems Implemented in the Built Environment

The maintenance of the photovoltaic systems is rather easy to do. However, different operations have to be performed on a regular basis and require an easy access to all the equipment that builds up the PV system, in order to check and, if necessary, to repair or replace the damaged components.

Frequent maintenance supports an uninterrupted and secure operation of the PV system during its entire lifetime. Therefore, specialized service providers that are able to verify all the aspects according to the implementation location and to the system type should do the maintenance.

Maintenance can be run based on *preventive (*thus *scheduled)* or *failure-related* operations in order to correct/repair various damages.

3.4.2.1 Preventive Maintenance

Preventive maintenance is part of the correct PV system exploitation and should be scheduled every several months, at least once per year. It is essentially a visual inspection of all main PV components, using a checklist provided by the manufactures [72]. These preventive operations and their recurrence depends on the system's size, on the PV design and complexity and on the environment. It should occur at an interval that maximizes the ratio between the financial benefits of the PV system and the cost of this scheduled maintenance. Preventive maintenance consists of general operations, as the ones described below, but also other specific operations that depend on the PV system type.

Cleaning the PV surface might be necessary due to the winter precipitations or because of other different types of dirt on the glazed surface of the modules. Snow and ice cleaning is not common because during intervals with snow coverage usually little solar radiation is available and because at higher slopes than 30° snow usually slides on the PV modules. The energy production losses due to snow/ice coverage are between 3% in temperate climates and 15% in very snowy climates [65].

Other types of dirt on the PV surface are generally referred as soiling and include dust, pollen, bird droppings and grime. Soiling can be responsible for up to 15% of the annual losses in dusty areas. Dust and pollen soiling can be avoided by water washing either manually using a standard garden hose (in case of small PV systems) or using an automatic spraying system as the ones produced e.g. by Heliotex or Ruth-agri Co. for large PV systems. According to Naeem [64] who analyzed the soiling rates and cleaning costs, water cleaning is not very cost-effective, with costs between \$0.25/m^2 for large systems and \$1/m^2 for small residential systems, and water consumptions of about 1 L/m^2.

Dust soiling mainly occurs in areas where the water reserves and supplies are scarce. This is why most PV owners rely on rain for cleaning or use other technologies that do not imply the use of water, such as Ecoppia's E4 robotic solution based on a soft, micro-fiber-rotating device that glides over the surface of the PV modules on a moving frame or the Ecovacs RayBot vacuum cleaner robot [71].

Bird droppings should be carefully removed without damaging the surface of the modules using high-pressure water or non-abrasive brushes.

Relatively simple solutions can be applied to remove the results of soiling described above; however, more effort is required when grime is formed on the surface of the PV modules, composed of a mixture of fuel exhaust gases and particles, or other specific emissions in a dirty urban environment as, e.g. the atmospheric dust. This usually requires periodic cleaning using water with different chemicals able to dissolve the grime [27].

Corrosion inspection of the PV structure and cables, connectors and terminals has to focus on the AC switches and protection devices and on the electrical combiner boxes also checking for intrusion of small animals. Maintenance includes checking the operational indicators of the inverters, the inspection of the AC transformer used in large PV systems, the inspection of the monitoring instruments for measuring the electrical and meteorological parameters and the inspection and update, if required, of the monitoring software.

Inspection of the batteries, as part of the PV system, requires checking for low electrolyte level in the battery compartments (that asks for adding deionized water). Moreover, the battery inspection also focuses on terminal corrosion (that requires cleaning and re-tightening the screws) and on sulfating that appears when batteries are not fully charged for a long period of time and may indicate a failure of the charge controller.

DC testing the PV strings can be manually done by measuring the open-circuit voltage and the short-circuit current, using instruments as the Profitest PV Grossen Metra Watt (Fig. 3.35) able to also generate the *I–V* curves of the entire PV array. This solution can indicate damaged modules and strings but an accurate comparison of the arrays can hardly be done because of the different time intervals when each array is tested as large PV systems may require days or even weeks for testing all the strings.

A quicker inspection of a PV system can be done using infrared cameras, generating thermal images of the individual modules (and cells), of the PV arrays (Fig. 3.36a) or even of the entire PV system if aerial high-resolution cameras are used, transported by drones or other flying devices and remotely managed, as presented in Fig. 3.36b.

Fig. 3.35 Testing a PV array installed on the buildings of the Transilvania University of Brasov, Romania, with a Profitest PV Grossen Metra Watt instrument

(a)
(b)

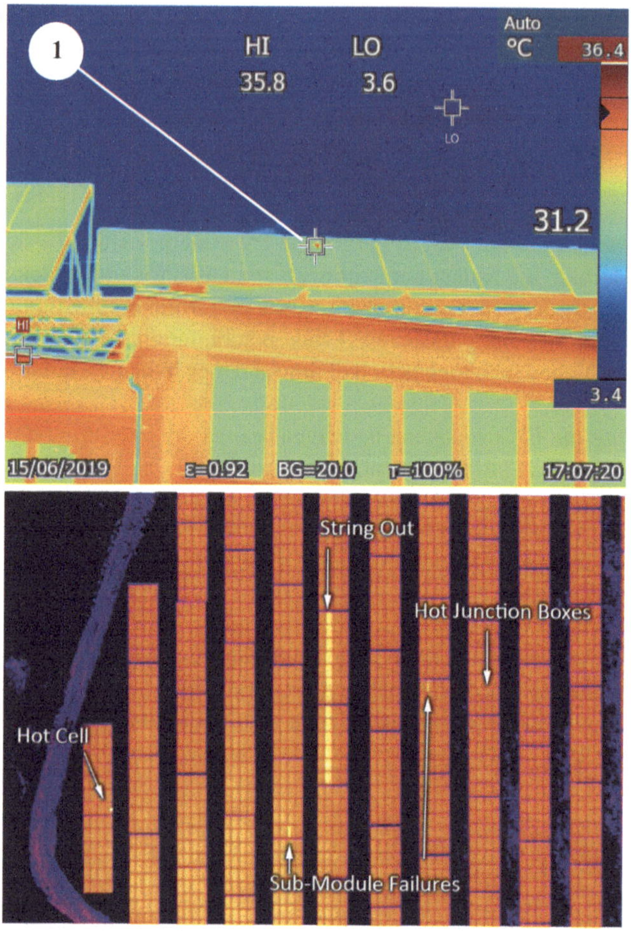

Fig. 3.36 **a** Hotspot (1) on the PV system installed in the Colina Campus at the Transilvania University of Brasov, identified with an infrared camera; **b** Aerial high-resolution infrared image used for identifying failures of PV arrays, modules and cells (Andrews [3], National Renewable Energy Laboratory [65], Courtesy of Heliolytics)

These images can indicate the problems on the entire string, on the modules, on parts of the modules and even hotspots within individual PV modules.

The preventive-scheduled maintenance operations should also include an extra inspection of the roof- or façade-mounted PV system focusing on the support structure to identify possible damages. The roof permeability/insulation and the proper drainage of rainfall have also to be verified, to avoid short-circuits due to water contact with unprotected electrical components and cables.

Ground-mounted PV systems require more extensive cleaning because they are closer to the dust and pollen sources. The vegetation in the area where the PV system is installed requires special attention and can be managed through grass mowing/trimming, the use of herbicides or the removal of trees; minimizing vegetation management costs can also be reached by replanting the land with special low-growth and pollinator-friendly plants. Another problem is the wind-blown snow that can accumulate on the access roads or between the arrays and might damage the electrical components (cables, electrical boxes or inverters).

The tracked PV systems require even more preventive operations that include: the inspection of the electrical motors/controllers, cables, connections, cooling fans, fan-filters, etc.; the inspection of all sensors used during tracking (e.g. anemometers for an optimal alignment during high-speed winds, position sensors of the tracking system, limiting switches). The software installed on the controller and used for tracking the Sun may also require a fine-tuning or an update. Another important maintenance operation of the tracked PV systems is the inspection of the tracking mechanism, checking if it is well lubricated and verifying the integrity of the joints.

3.4.2.2 Failure-Related Maintenance

The second type of maintenance operations are *failure-related* and involve repairing and/or replacing different PV components due to low values of the performance indicators or as result of irremediable defects. The use of less reliable components and of disordered cabling in the PV system causes most of these failures. These can lead to high maintenance costs that can be even double as compared to the systems that are correctly implemented. The most frequent failures are the result of loose electrical connections, due to oxidation or too short wiring, failed circuit breakers or protections that were not rated for DC applications. Issues that are more serious may occur, as when the charge controllers and inverters get damaged [76]. The *failure-related* operations are commonly covered by traditional or extended warranties of the components and are usually solved by an operation and maintenance provider [27]. This type of companies are able to replace any damaged component in a short period, as most of the used spare parts are available on stock at the providers, thus allowing a fast PV system repair.

3.4.2.3 PV System Warranty

The exploitation and maintenance costs increase with the age of the PV system while the performance in the electrical energy production decreases. Life-cycle cost standards usually mention 25 years as the lifetime of the electrical equipment, but the Energy Independence and Security Act [94] indicates that 40 years periods may also be considered.

A clear distinction should be done between the *product warranty* and the *performance warranty*. Product warranty only covers manufacturing defects of the materials and of the PV products, while the performance warranty refers to a power output guaranteed by the manufacturer over the product's lifetime; the decrease in this power output occurs following the natural degradation of the materials used in the PV system.

The product warranty of most crystalline silicon PV modules currently on the market is about 10–12 years. Their performance warranty ranges from 20 to 25 years that represents an exception considering the electric components, with degradation rates of about 0.5%/year [45].

Other electric and electronic devices have lower warranty periods and expected lifetimes. For example, the currently mostly sold inverters have a 10 years product warranty (or even longer if a 10-year service was also contracted), while older inverters need to be replaced every 10–12 years. Charge controllers have similar lifetime as inverters while the rechargeable batteries used in stand-alone PV systems have to be replaced every 5–10 years [65].

3.5 Photovoltaic—Wind Energy Mixes in the Built Environment

Photovoltaic, wind or hydro power systems individually connected to an electricity grid represent a viable solution for reaching a certain percentage of renewable energy during a broad timeframe. In these grids, during time intervals when the renewable energy source is not available due to missing solar radiation or wind, other conventional energy production systems have to compensate the energy demand.

Stand-alone systems based on photovoltaic modules or wind turbines are used in locations where only one source of renewable (solar or wind) energy is mainly available. They are able to produce the electrical energy at the designed value, if well sized.

For intervals when the renewable energy source is missing, storage systems are used. These store the excess energy during the production intervals and provide it when required. For even longer intervals electrical energy generators based on internal combustion engines are used, that may not represent a convenient solution due to the possible unavailability and high costs related to the fossil fuel use, or due to the need for heat and noise damping solutions.

In order to decrease the costs related to the storage system and combustion generator, stand-alone systems installed in locations with more than one renewable energy source are recommended to contain additional renewable energy systems as wind or hydroelectric turbines that also increase the reliability and lifetime of the system. This involves merging the systems and working together in a *hybrid system* of at least two renewable energy technologies whose energy sources are intermittent and complementary, in order to compensate the lack of one source with the other.

3.5.1 Examples of Photovoltaic—Wind Hybrid Systems

Most common applications for hybrid systems are in remote locations, as islands [77] or rural settlements [8]; these are not connected to the central electricity grid, thus the electrical energy has to be supplied to the inhabitants using local sources. Other photovoltaic-wind turbine energy systems can be used to power remote houses or chalets [81]. Photovoltaic-wind systems are also used in the built environment for improving the city aspect with functional and aesthetically pleasing structures. These structures are, for example, remote charging stations for electric cars, bicycles or any other vehicles [97] or shadowed playgrounds for children.

One example is presented in Fig. 3.37, a wooden structure with a stand-alone hybrid wind-photovoltaic power system designed by Innoventum AB, Sweden, called Giraffe 2. It occupies 25 m^2 and, according to the producers, it stores over 4 tons of CO_2 in the renewable materials that were used and is able to generate about 13,000 kWh/year electrical energy.

One of the common photovoltaic—wind hybrid systems that improves the aspect of the implementation location but also has a functional role in the built environment

Fig. 3.37 Stand-alone photovoltaic—wind hybrid power system, Giraffe 2, designed by InnoVentum AB in Sweden (InnoVentum [41], Courtesy of InnoVentum)

(a) (b)

Fig. 3.38 Example of streetlights powered by photovoltaic—wind hybrid systems with: **a** horizontal axis wind turbines; **b** vertical axis wind turbines (StreetlightCN [92], Courtesy of StreetlightCN)

are the systems used for street lighting, installed inside the settlements, at crossroads, on country roads or even on highways [7]. They have the advantage that intensive connection operations (cable trench excavation, cable laying, backfilling, road digging, etc.) to the local electricity grid are not required as the correct setup of the hybrid photovoltaic – wind structure in the implementation location is enough. Moreover, in the case of individual failure of one system, the other streetlights are not affected (Fig. 3.38).

These hybrid systems are commonly used in locations where the owner asks for an alternative solution to fossil fuel generators or when the connection to the local grid is too difficult to be done.

3.5.2 Sizing the PV-Wind Hybrid Systems

Sizing the hybrid systems with several power sources depends on the systems' type and is usually done based on simulations of the systems' components and of the entire hybrid system aiming at reaching the lowest cost for a given amount of produced electrical energy [28].

To simulate the hybrid PV-wind system, mathematical models are required for:

- The power produced using the photovoltaic modules. This model has to consider the PV module type and at least the variation of the solar radiation, ambient temperature and wind speed in the implementation location as these are the parameters with the highest influence on the photovoltaic conversion efficiency;

- The power generated by the wind turbines; in this case the model has to be able to cope with the wind turbine type, the available wind speeds and the height at which the turbine is installed;
- The charging state of the batteries (for the electrical energy storage) considering their type, temperature, input and output currents and voltages;
- The power of the electrical backup generator if such an equipment is part of the system;
- Different cost parameters to evaluate the benefits brought by the system. The mostly used parameters are: the *net present cost (NPC)*, that includes the capital, replacement and other maintenance costs of each component, part of the hybrid system [74] and the *levelized cost of energy (LCOE)*, that represents the average cost of the energy produced by the system during its entire lifetime [31].

While the power produced by the PV modules can be estimated as already described (Eq. 3.35), the power that a wind turbine can produce is estimated using

$$P_v = \frac{1}{2} \cdot \rho \cdot C_p \cdot A \cdot v^3 \tag{3.36}$$

where

- ρ is the air density;
- C_p is the power factor of the wind turbine with values in the 0.05–0.45 range [12, 38];
- A is the area of the wind turbine rotor;
- v is the wind speed.

The annual energy produced by small wind turbines, mostly suitable in the built environment, can be estimated in several ways that are ordered according to their increasing accuracy [33].

The annual energy can be roughly estimated using the diagram in Fig. 3.39 if the annual average wind speed in the implementation location and the area of the wind turbine rotor are known.

Further on, the annual energy can be estimated based on the power produced at the annual average wind speed and the total number of hours in a year

$$E = P_v \cdot 8760 \tag{3.37}$$

The disadvantage of this approach is that the power factor of wind turbines is only rarely given by manufacturers in the product datasheet.

The annual energy can be also estimated using wind speed/energy tables, usually given by the turbine producers. The disadvantage of this solution is the lack of accuracy of some turbine producers and the fact that the values are given for different wind speeds and intervals, that makes difficult to compare different turbines.

Another option for estimating the annual energy is based on the wind speed distribution (the number of annual hours for each wind speed) and the power curve

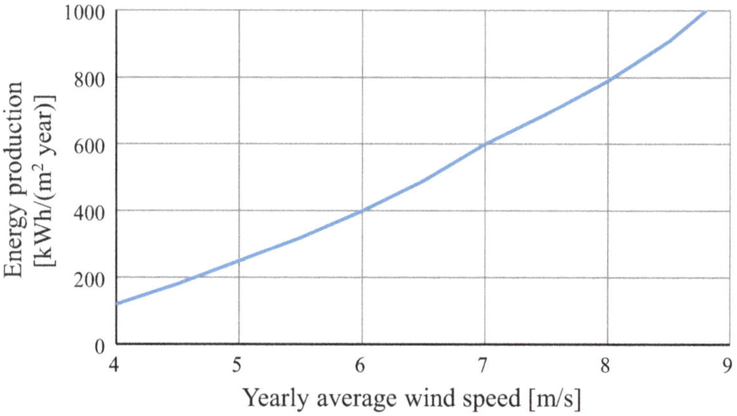

Fig. 3.39 Annual energy produced by small wind turbines according to the average wind speed available [33]

of the turbine (that indicates the correlation between the wind speed and the produced power, Fig. 3.40). This curve either is indicated by manufacturers or can be measured using controllable wind generators. The drawback of this solution is that high-frequency measurements of the wind speed are required in the implementation location for at least one entire year.

The steps to follow when simulating a PV-wind hybrid system are:

- Identifying the energy consumers and the energy demand during the operation period;
- Measuring or estimating the renewable energies potential in the implementation location;

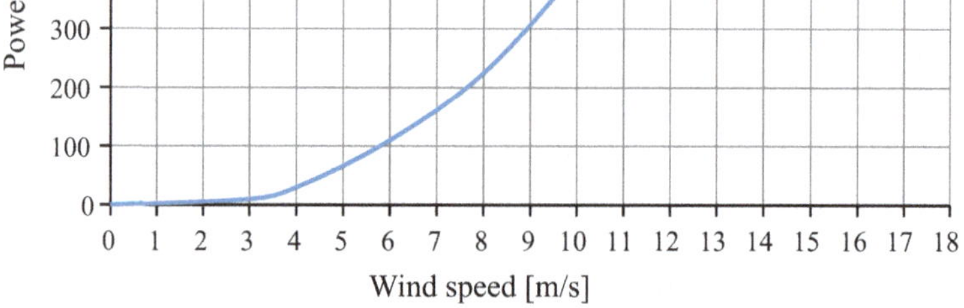

Fig. 3.40 Power curve of a LeadingEdge LE-600 wind turbine

- Selecting the PV modules type(s), the wind turbine and the batteries. These will guide the mathematical models to be used in the system design;
- Modeling the energy that can be locally produced by one unit (1 module or 1 m^2 of PV modules, 1 wind turbine) of each selected renewable energy system during the entire year or during the month with the lowest renewable energy potential;
- Selecting the renewable energy fraction, indicating how much energy is produced using renewable energy systems and how much using conventional sources;
- Selecting the optimum solution (number of PV modules, of wind turbines and batteries), usually the one with feasible cost and higher renewable energy fraction.

These models and steps can be afterwards relatively simple implemented in different spreadsheet software or general simulation software to generate the results required for implementing the hybrid systems.

An alternative option for designing the hybrid PV-wind systems is the use of dedicated software, as the ones described by Sawle et al. [79]:

- HYBRID 2, a simulation tool developed by the National Renewable Energy Laboratory (NREL) able to generate long term results considering the short term variation of the solar irradiance and wind speed; it can generate accurate technical results but it is not able to automatically optimize a system;
- HOMER, a software solution for calculating the installation and operation costs of renewable and conventional energy systems (both electrical and thermal), but also of different storage systems types connected to on-grid or to stand-alone applications; it is able to analyze any combination of available systems in order to generate the most feasible solution, with the smallest net present cost [77];
- RAPSIM, a computer aided simulator for PV modules, wind turbines, batteries and diesel generator systems where optimization can be reached based on the variation of the components sizes;
- RETScreen, a Microsoft Excel based software tool able to simulate the energy efficiency and feasibility of different renewable energy systems, but also to generate a cost analysis [84];
- SOMES, a software tool usable on an hourly basis simulation and optimization of the PV and wind hybrid systems, when searching for the lowest electricity cost at a reliability level accepted by the end-user.

3.6 Economic and Financial Aspects of the PV Systems

The large-scale implementation of the PV systems mainly depends on their economic advantages. These depend on the location where the system is installed as location-dependent elements may influence the feasibility of the renewable energy systems. Important factors to consider are:

- The solar energy potential in the implementation location;
- The existence of a local electricity grid that allows the connection of the new renewable power plant;
- The local energy prices and the incentives for renewable energy systems that are available in the implementation region, in different forms.

The lack of a local electricity grid represents an important factor when installing PV systems in the build environment. In this case, the financial aspects represent only the second part of the problem as the most important aspect is powering the specific location with electrical energy. An economic analysis is required to compare the cost of the renewable energy with the cost of the energy produced using non-renewable energy sources and systems, in most cases fossil fuel based combustion generators.

The local authorities can use various solutions that support the development of new PV systems, if renewable energy systems are planned to be an important part of the future energy scenario in the region. These include:

- Simplifying the legislation, the paper-work and the entire procedure that new photovoltaic owners have to follow in order to connect their systems to the local electricity grid and start producing;
- Making renewable energy systems of different sizes mandatory for new buildings. This is the case of California [13] and of public buildings in the European Union starting with 2019 [29];
- Increasing the taxes for the energy produced using non-renewable sources, due to the extra pollution generated when burning fossil fuels;
- Stimulating the production of renewable energy using incentives or different financial schemes that support the PV implementation in the build environment. One of the mostly used solutions is the Feed-in-Tariff that will be further discussed;
- Stimulating financing bodies (e.g. banks) to give loans with low interest rates to investors interested in the construction of PV systems. For these cases, the systems need to be accurately designed and documented so that the financing body gets warranties of a correct and complete system able to produce at its nominal capacity during its entire lifetime. These warranties are required considering the case when the investor is not able to pay back the loan and indicates a system that is known as *bankable*.

According to Armansperg et al. [4], the photovoltaic bankability should increase the trust of investors, financers and insurance companies in a PV system. It represents an active quality management process where all the stakeholders involved in a PV project are trying to identify all types of risks (legal, technical and economic) that may appear during the entire project lifecycle. The identified risks have to be quantitatively and qualitatively assessed, managed and controlled and at least mitigation measures should be proposed.

To reduce the payback time of the investment in a photovoltaic system, the owner can consider different financing schemes and funds from various sources, along with selling the electrical energy:

- Feed-in Tariff;
- Renewable Energy Certificates for the produced PV energy;
- Power Purchase Agreements (PPA);
- Member in a virtual power plant;
- Net-Metering;
- Solar coins for the produced PV energy.

The mostly used solution is the *Feed-in Tariff* (FIT). This represents an economic policy mechanism able to accelerate the investment in producing and installing renewable energy systems. It is a long-term agreement or contract, usually in the range of 5–25 years, between the energy producer using renewable sources (household/farm/business/private investor) and the electricity grid administrator or electrical system/market operator.

The FIT guarantees that the producer receives over a long timeframe a fixed price for each energy unit produced, thus protecting the producer from the typical risks associated to the renewable energy. The FIT value may depend on several factors:

- The technology used and in this way various renewable technologies can be encouraged by governments [17, 18];
- The size of the renewable energy system, being known that small or medium-sized systems have a higher cost per unit of installed capacity;
- The renewable source of energy that may differ with the implementation location (e.g. solar energy potential, average wind speed), as outlined by Couture et al. [17].

FITs usually have a variable value over time: this either can increase, to compensate the local inflation or maintenance costs, or is usually decreasing, to track and stimulate the reduction of the technology specific costs [75]. The decrease rate can be either fixed, predetermined and calculated on a regular interval basis or it can be variable, according to the market and technology development.

FITs were for the first time used in 1978, in the USA, by the Carter administration who wanted to promote solutions for energy savings and development of new technologies that would rely on renewable energy sources [39]. Since then, the mechanism was used in various countries all over the world as the data in Table 3.6 show, among which notable results were obtained in Germany, Italy and Spain where photovoltaic systems were intensively promoted until mid of the 2010s.

In Germany, the large number of PV systems generated a 40% price reduction during the early afternoon peak hours as result of the decreased demand for energy produced by high-cost power plants using fossil fuels. This supported consumers' savings between €520 and €840 million that meant losses for the major electricity companies who lobbied in 2012 for the reduction of the feed-in tariffs [70]. Similar reductions in the electricity prices were registered in Spain [21] and in Denmark [63].

An alternative to the feed-in tariff, also used in various countries, is the *Renewable Energy Certificates* mechanism. Local authorities issue the certificates to the renewable energy producers according to the amount of energy produced, usually

Table 3.6 Feed-in tariff rates for the photovoltaic energy produced and used all over the world

Country	Capacity	Year	FIT rate (currency/kWh)	Currency	References
Australia	<10 kW	2013	7.5	¢ AUD	ActewAGL [2]
Canada		2017	19.2–31.1	¢ CAD	Independent Electricity System Operator [40]
China		2011	15	¢ USD	Coco [16]
Egypt		2016	5.7–7.88	¢ USD	Shahid Law Firm [82]
Germany		2012	3.592	¢ EUR	Lang and Lang [52]
Japan	<10 kW	2012	52.5	¢ USD	Choudhury [15]
US-OR		2013	18.1–39	¢ USD	Database of State Incentives for Renewables & Efficiency [19]

for each MWh. Moreover, local authorities ask that a specified fraction of the energy sold by the electricity supply companies to be renewable. To reach this fraction, the electricity companies can buy the certificates from the renewable energy producers and prove their compliance to the local regulations. The extra cost of the certificates is then distributed by the supply company to the final electricity consumers. After this step, the certificates lose their validity and cannot be sold a second time, to avoid double benefits. The income generated to producers by selling the certificates adds to the price of the produced electricity and represents an extra stimulation for the investment in renewable energy systems.

The disadvantages of FITs and of other incentive schemes are mainly related to the difficult path that has to be followed for formulating or adjusting them to make the specific renewable technologies attractive but not to over-compensate the energy producers. Moreover, as the renewable energy producers are not encouraged to consider the dynamic variability of the electricity market, these schemes do not allow an effective implementation of the renewable energy systems in the market. Therefore, after the encouraged technologies and market are mature, no incentives should be further considered.

To get a constant revenue from their systems, large renewable energy producers usually decide to sign a long-term contract (usually between 5 and 20 years) with an entity interested in buying their energy (in general utility companies, big power consumers/corporates or power traders) even before the system is installed. The buyers have thus a warranty that for the contracted timeframe the purchased energy is renewable. This long-term contract is known as a *power purchase agreement* (PPA) and indicates all commercial terms between the seller and the buyer regarding the contracted electrical energy. It represents a simple solution, usually regulated by national energy commissions, through which an entity can directly buy energy from the producer and not from the local utility company.

The owners of small residential renewable energy systems have rarely the possibility to access incentive schemes that support renewable energy production due to bureaucratic reasons. Therefore, a common solution for obtaining revenue from their systems is joining of several small producers to form a *virtual power plant* under the umbrella of a company able to receive the incentives and divide these to the group members. To increase their profits, the virtual power plant operators can also include power consumers and manage the peak loads by power trading or balancing the production–demand of the electrical grid.

For small renewable energy systems, especially for the photovoltaic systems installed in the build environment, another solution is available for selling the produced energy, if the local electrical utility company allows it. It is called *net-metering* and involves the local use of the produced electrical energy or the injection of the excess energy in the local electricity grid, when the electric meter rotates in the opposite direction and builds up a credit. During periods when the demand is higher than the energy production, especially during the night or in intervals with low solar irradiance, the local grid provides the required power. At the end of a specified time interval, the customer, which in this case is also a producer, called *prosumer*, is receiving a bill only for the energy that could not be provided by the renewable energy system. This is the most convenient solution for the energy producer (after the feed-in tariff) since the produced energy has the same value as the consumed one and is equal to the retail price required by the utility company. There are three types of net-metering according to the retail rates locally used by the utility company:

- Net-metering at standard residential rates that may be the same all over the year or may seasonally differ, or are based on the consumed amount;
- Net-metering using time-of-use rates, which are lower during off-peak hours and higher during on-peak hours; these type of rates can be adjusted when renewable energy systems are used on a large scale, as it is the case of Italy where on-peak times were shifted from daytime to the evening hours [91];
- Net-metering using market rates that are dynamically adjusted according to the instantaneous wholesale electricity prices and energy demand and supply.

If at the end of a time interval (day, month, year), the produced energy exceeds the demand in a location, different options are available. The excess energy generated is not payed (ideal from the utility point of view), alternatively, it can be compensated at the costs spared by the utility or compensated at the retail rate (ideal from the prosumer point of view). Best practice net-metering solutions recommend a one-year interval for calculating the difference between the local energy demand versus consumption and the transfer of the eventual excess energy to future intervals.

The local electricity company may not allow net-metering and, in this case, the installation of renewable energy systems in the built environment will be slowed down. The usual explanations focus on the instability caused to the local grid by the small, distributed generation systems. Another explanation, usually not mentioned by the companies, is that the small-distributed energy systems can generate profit losses for the utility companies if they use traditional business models.

In these cases, other financing options may be available for grid-connected prosumers that allow different rates for the produced and consumed energy. The electrical grid can be used for injecting the excess energy but the received rate only refers to the price of the electrical energy, being usually less than half of the retail price which includes energy transportation costs and different other taxes. The best economic option for the prosumer would be in this case to locally consume the energy when produced, thus avoiding, as much as possible, the demand for the grid. This requires the use of a smart control system able to decide which consumer should be powered and when, based on the instantaneous renewable energy potential.

Among these options used to give value to the renewable energy, FITs are the most convenient for the producer, with compensations usually higher than the retail energy price and gradually decreasing. The second best option is the use of net-metering with compensation at the retail price. The last option is the use of PPA, for which the compensation is usually lower than the retail price, but can be higher in the case of PV energy that is usually generated during peak daytime hours.

Another income source, mainly for the electrical energy produced using PVs, is the use of *solar coins*. The SolarCoin (SLR) is a new type of digital cryptocurrency dedicated to accelerating the transition to clean energy by giving incentives to solar energy producers [87]. One SolarCoin is the reward that a producer receives for each produced MWh using solar energy. In May 2019, the value of a SolarCoin was still very low, around 5 ¢ USD and this does not represent a real incentive. If the value will increase, the SolarCoin will represent a good compensation along the local subsidies and other types of incentives.

For stand-alone renewable energy systems, the economic analysis is rather simple and is in fact a comparison between the price of renewable energy and the price of the energy produced using fossil fuel-based technologies. The average price of the electrical energy produced using photovoltaics during the system's lifetime can be roughly estimated using different software solutions, e.g. the online tool PVGIS.

Two economic parameters are used for a rough evaluation of the financial benefits of the renewable energy systems: the *simple payback time (SPT)* and the *simple (initial) rate of return (SRR)*. These parameters indicate if a specific project is of interest and are generally used in a more detailed economic analysis.

The *simple payback time* represents the time required to recover a specific investment considering the generated savings; it is calculated using

$$\mathrm{SPT} = \frac{C}{S} \tag{3.38}$$

where

- C is the initial cost of the system;
- S is the savings during a specific time interval (e.g. one year).

The simple rate of return (SRR) represents the inverse of the simple payback time and indicates how efficient an investment is

$$\mathrm{SRR} = \frac{S}{C} \tag{3.39}$$

3.7 Integrating the Renewable Energy Systems in the Urban Electrical Distribution and Transmission Infrastructure

The energy system is usually structured on several grid layers with different voltage levels: ultra and extra high voltage (UHV and EHV), high voltage (HV), medium voltage (MV) and low voltage (LV) (Table 3.7).

Traditional power systems have a hierarchical structure with a uni-directional top-down power flow from the energy producers to the energy consumers. These systems include three different sections: the *power generation* using large-scale power stations, the *power transmission* using the grids and the *distribution* to the final consumers. Newer power systems include more types of power producers that can be located on several grid levels. This aspect generates a fluctuating bidirectional power flow at each grid level and between different levels.

Following grid types are commonly available in the built environment: the *low-voltage grid* (for powering residential buildings or small businesses) and *the medium voltage grid* (for powering the local industry buildings). These grids are also available for connecting different renewable energy systems. Small PV systems or wind turbines up to 20 kW are usually located in the build environment and are generally connected to the low-voltage grids, while larger power plants, PV or wind-based, with powers up to 1 MW are usually located at the built environment outskirts and are connected to medium voltage grids.

General protection requirements, functional requirements and monitoring requirements have to be fulfilled for being allowed to connect renewable energy systems to the power grid and these are further detailed.

Table 3.7 Voltage levels of the grids used for electrical energy transmission all over the world [6, 56]

Voltage level	Voltage range	Type of system	Name of the section	Approx. range
Low voltage	<1000 V	AC	Secondary distribution	0–3 km
Medium voltage	1–70 kV	AC	Primary distribution	3–120 km
High voltage	70–110 kV	AC	Secondary transmission	120–220 km
Extra high voltage	110–800 kV	AC, DC	Primary transmission	220–1500 km
Ultra high voltage	800–1000 kV	AC, DC	Primary transmission	>1500 km
	>1000 kV	HVDC		

3.7.1 General Protection Requirements

The internal electrical protection against over-voltage and over-current, using circuit breakers or electrical fuses installed on each part and on the entire renewable energy system (PV modules/strings/arrays/plant, wind turbines, etc.), as presented in Fig. 3.41.

Before connecting the renewable energy system to the local electricity grid, two separate devices are required: a *distribution panel* for dividing the electricity between the local consumers and the local grid, and an electricity *meter*.

A utility disconnect protection instrument has to be available between the local electricity grid and the electrical power converters that adjust the produced electrical energy to the grid parameters; this instrument protects the converters/local consumers from grid faults and the grid from inverter or local consumers faults.

3.7.2 Functional Requirements

The quality of the electrical energy produced using the renewable energy systems has to be similar to the energy available on the local electricity grid to avoid grid mismatches or problems to the nearby consumers. The most important parameters that need to be synchronized are *the frequency* and *the voltage*. Local utility companies may accept different thresholds for these parameters.

Distributed renewable energy systems have to be able to avoid *electrical islanding*; islanding corresponds to the case when a generator continues to produce electrical power even when the utility grid is missing. Such a situation may pose a high injury risk to utility personnel that is working on the grid while knowing that a specific part of the grid is disconnected.

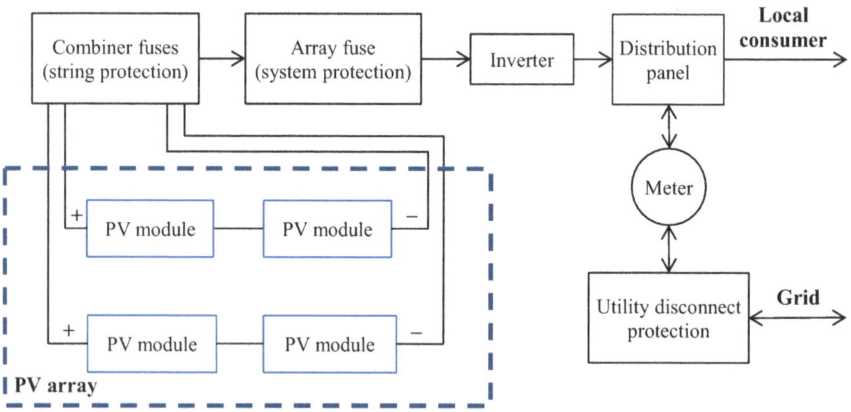

Fig. 3.41 Electrical protection at different levels for a photovoltaic system

Electrical connections to the grid are characterized by the maximum nominal power limit that can be drawn from or given to the grid and this defines the protection components and transformers used by the utility companies in that part of the grid. Considering this limit, the distributed energy generators have to take all the measures to avoid exceeding the Upper Feed-in Limit allowed. This may occur during periods with high solar irradiance levels, when a PV generator was over-dimensioned for reaching high profitability and works most of the time at the Upper Feed-in Limit. This may also occur when the local consumers of the produced electricity are not continuously working, e.g. an industrial consumer that is not working during weekends [88].

3.7.3 Monitoring Requirements

The most common monitoring solution of the distributed renewable energy systems is the use of *double-sense/bidirectional electrical meters* (Fig. 3.42). This allows the electrical utility company to have an overview, on a regular basis, on the amount of energy taken from the grid and of the energy injected into the grid; based on these types of data, the company is able to generate the electrical bills and to reward the producer. Bidirectional metering also allows the benefit from FIT or PPAs incentive schemes previously described.

Currently, grid distributors accurately monitor the medium voltage grids but they do not usually monitor the low-voltage grids in detail, where many residential PV systems installed in the build environment feed in their output. This is why during periods with high solar irradiance and low local electricity use, the renewable energy systems installed in a certain area may generate more electrical power than the transformer of that area was designed to cope with. This can damage an insufficiently protected transformer, increasing the maintenance costs and leaving the entire area without electricity.

Thus, the utility grid operators must install protection devices for transformers and must be able to remotely dispatch the renewable energy generators connected

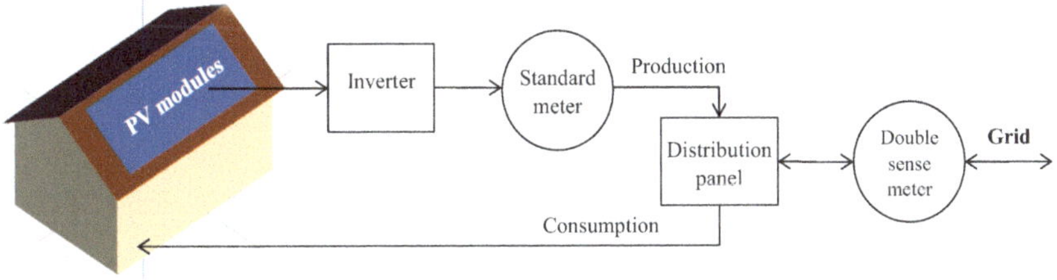

Fig. 3.42 Metering solution used for measuring the produced photovoltaic energy, the grid injected energy and the locally consumed energy

to the grid. The usual minimum allowed dispatch levels are at 10% of the installed system power; in this way the electrical utility operator is able to protect the grids from failures by decreasing or stopping certain generators. Further on, as low-level voltage grids monitoring will improve in time, the ability to dispatch a distributed generator will increase in importance and will be mandatory for all grid-connected generators.

To check if at least the previously mentioned requirements are met, the local electrical utility companies generally have the possibility to validate the products or the companies that produce electrical converters and components used for connecting the renewable energy systems to the grid, thus insuring that no dysfunctionalities occur in the electrical grid or to the consumers.

Another problem of the distributed renewable energy systems is their unequal, discontinuous power generation that depends on the day/night succession (partly predictable) and on several meteorological factors (not accurately predictable before a certain period). This makes their integration in the electricity distribution systems rather difficult. As mentioned, during excess production periods, the power can be limited by the grid operator, while for insufficient production, high-price electricity has to be acquired from the grid. To limit this inconvenience, storage systems are usually installed. These storage systems are locally installed, at the prosumer, where deep-discharge lead-acid or lithium-ion batteries are deployed to allow the use of locally produced power during intervals with no renewable energy source. This type of systems represents a combination between a grid-connected and a stand-alone system and requires a home energy management system (HEMS) to feasibly use the locally produced electrical energy. This HEMS has the role to optimally power the consumers according to the renewable power availability [83].

Storage systems can also be installed at community level. Currently, under testing is the use of electrical vehicles for storing the energy during peak production periods and giving it back to the grid during time intervals with small or no production [96]. One example is the district of Lombok in Utrecht, The Netherlands, where a small company uses 30 shared Renault Zoe vehicles, charged by 25 bidirectional public charging stations powered with locally produced PV energy [73].

At regional or national levels, usable storage systems are the bi-directional hydropower plants able to use the excess power for pumping water and to generate electricity using a hydroturbine during periods when more energy is required. Such a hydro storage solution in connection with large wind turbine power plants is able to support the 100% renewable energy use in El Hierro, Canary Island [32].

References

1. Abdallah S (2004) The effect of using sun tracking systems on the voltage–current characteristics and power generation of flat plate photovoltaics. Energy Convers Manag 45:1671–1679

2. ActewAGL (2013) Changes to ActewAGL solar buyback scheme 1 July 2013. https://web.archive.org/web/20130623054251/http://www.actewagl.com.au/Product-and-services/Offers-and-prices/Prices/Residential/ACT/Feed-in-schemes/ActewAGL-Solar-buyback-scheme/Transition-to-net-metering.aspx. Accessed 20 May 2019
3. Andrews R (2016) DC assessment proposed best practice. Heliolytics Inc.
4. Armansperg M, Oechslin D, Schweneke M (2017) Solar bankability, horizon 2020 project, No 649997, PV investment—technical risk management—best practice guidelines for risk identification, assessment and mitigation. Deliverable D5.8 20/02/2017. http://www.solarbankability.org/fileadmin/sites/www/files/documents/D4.3_Technical_Bankability_Guidelines_Final-SB_Website_170215.pdf. Accessed 25 May 2019
5. Armstrong S, Hurley WG (2010) A new methodology to optimise solar energy extraction under cloudy conditions. Renew Energy 35:780–787
6. Asplund G (2007) Ultra high voltage transmission—alternative scenarios for long distance bulk power transmission—800 kV HVDC and 1000 kV HVAC. ABB Review 2. https://pdfs.semanticscholar.org/63d2/5dc1653b8c31ef79e0b977e962404b3f5b08.pdf. Accessed 15 May 2019
7. Beck M (2019) Hybrid street lighting systems market chain analysis 2019: Global market studies, surveys, outlook, trends analysis and forecast 2024. Market Research News. https://themarketresearchnews.com/2019/04/20/hybrid-street-lighting-systems-market-chain-analysis-2019-global-market-studies-surveys-outlook-trends-analysis-and-forecast2024. Accessed 15 May 2019
8. Bekele G, Boneya G (2012) Design of a photovoltaic-wind hybrid power generation system for Ethiopian remote area. Energy Proc 14:1760–1765
9. Burduhos BG, Toma C, Neagoe M, Moldovan M (2011) Pseudo-equatorial tracking optimization for small photovoltaic platforms from Toronto/Canada. Environ Eng Manag J 10(8):1059–1068
10. Burduhos BG, Visa I, Duta A, Neagoe M (2018) Analysis of the conversion efficiency of 5 types of photovoltaic modules during high relative humidity time periods. J Photovolt 8(6):1716–1724
11. Burduhos BG, Visa I, Neagoe M, Badea M (2015) Modeling and optimization of the global solar irradiance collecting efficiency. Int J Green Energy 12(7):743–755
12. Burton T, Sharpe D, Jenkins N, Bossanyi E (2001) Wind energy handbook. Wiley, West Sussex
13. California Energy Commission (2019) https://www.energy.ca.gov. Accessed 25 May 2019
14. Chan J (2012) Westmill solar park becomes world's largest co-operatively owned solar farm, Solar Power Portal. https://www.solarpowerportal.co.uk/news/westmill_solar_park_becomes_ worlds_largest_co_operatively_owned_solar_farm. Accessed 10 May 2019
15. Choudhury N (2012) Japan announces preliminary FiT rates. PVTECH. https://www.pv-tech.org/news/japan_announces_preliminary_fit_rates. Accessed 20 May 2019
16. Coco L (2011) China uses feed-in tariff to build domestic solar market. New York Times. https://archive.nytimes.com/www.nytimes.com/cwire/2011/09/14/14climatewire-china-uses-feed-in-tariff-to-build-domestic-25559.html. Accessed 20 May 2019
17. Couture T, Cory K, Kreycik C, Williams E (2010) Policymaker's guide to feed-in tariff policy design, NREL—National Renewable Energy Laboratory. US Department of Energy. https://www.nrel.gov/docs/fy10osti/44849.pdf. Accessed 25 May 2019
18. Couture T, Gagnon Y (2010) An analysis of feed-in tariff remuneration models: Implications for renewable energy investment. Energy Policy 38(2):955–965
19. Database of State Incentives for Renewables & Efficiency (2013) Solar volumetric incentive and payments program. https://web.archive.org/web/20131104060604/. http://www.dsireusa.org/incentives/incentive.cfm?Incentive_Code=OR134F&re=1&ee=1. Accessed 20 May 2019
20. Davies PA (1993) Sun-tracking mechanism using equatorial and ecliptic axes. Sol Energy 50(6):487–489
21. de Miera GS, González PR, Vizcaino I (2008) Analysing the impact of renewable electricity support schemes on power prices: the case of wind electricity in Spain. Energy Policy 36(9):3345–3359

22. Desthieux G, Carneiro V, Camponovo R, Ineichen P, Morello E, Boulmier A, Abdennadher N, Dervey S, Ellert C (2018) Solar energy potential assessment on rooftops and facades in large built environments based on LiDAR data, image processing, and cloud computing. Methodological background, application, and validation in Geneva (solar cadaster). Front Built Environ. https://doi.org/10.3389/fbuil.2018.00014
23. Deutsche Gesellschaft für Sonnenenergie (2010) Planning & installing solar thermal systems—a guide for installers, architects and engineers. Earthscan, London
24. Diaconescu DV, Vatasescu MM (2008) Two new pairs of local solar angles and their corresponding tracking systems. Bull Trans Univ Brasov 1(50):113–120
25. Dubey S, Sarvaiya JN, Seshadri B (2013) Temperature dependent photovoltaic (PV) efficiency and its effect on PV production in the world—a review. Energy Proc 33:311–321
26. Duffie JA, Beckman WA (2006) Solar engineering of thermal processes, 3rd edn. Wiley, New Jersey
27. Eiffert P, Kiss GJ (2000) Building-integrated photovoltaic designs for commercial and institutional structures: a sourcebook for architects. DIANE Publishing. https://www.nrel.gov/docs/fy00osti/25272.pdf. Accessed 15 July 2019
28. Eltamaly AM, Mohamed MA (2015) Sizing and techno-economic analysis of stand-alone hybrid photovoltaic/wind/diesel/battery power generation systems. J Renew Sustain Energy 7:063128
29. European Parliament and Council (2010) Directive 2010/31/EU on the Energy Performances of Buildings (Recast). http://www.buildup.eu/en/practices/publications/directive-201031eu-energy-performance-buildings-recast-19-may-2010. Accessed 08 Jun 2019
30. Fara L, Craciunescu D (2017) Output analysis of stand-alone PV systems: modeling, simulation and control. Energy Proc 112:595–605
31. Fulzele JB, Daigavane MB (2018) Design and optimization of hybrid PV-wind renewable energy system. Mater Today Proc 5(1–1):810–818
32. Gils HC, Simon S (2017) Carbon neutral archipelago—100% renewable energy supply for the Canary Islands. Appl Energy 188:342–355
33. Gipe P (2019) Fundamentals of wind energy. http://www.wind-works.org/cms/fileadmin/user_upload/Files/presentations/Gipe_ZNE-SCE-Wind_02_Sizing.pdf. Accessed 15 May 2019
34. Global Sustainable Energy Solutions (2019) Grid-connected PV systems design & installation, 8th edn. https://www.gses.com.au/product/grid-connected-pv-systems-design-and-installation-8th-edition. Accessed July 2019
35. Google Project Sunroof (2019) https://www.google.com/get/sunroof. Accessed 10 May 2019
36. Greentech Media (2019) Solar balance of system. To track or not to track. https://www.greentechmedia.com/articles/read/solar-balance-of-system-to-track-or-not-to-track-part-I. Accessed May 2019
37. Guevara-Stone L (2014) Solar co-ops bring affordable green power to the people. GreenBiz. https://www.greenbiz.com/blog/2014/04/28/solar-co-ops-bring-affordable-green-power-people. Accessed 10 May 2019
38. Hau E (2006) Wind turbines: fundamentals, technologies, application, economics. Springer, Germany
39. Hirsh RF (1999) PURPA: the spur to competition and utility restructuring. Electr J 12(7):60–72
40. Independent Electricity System Operator (2017) Ontario FIT/microFIT price schedule, 1 Jan 2017. www.ieso.ca/-/media/files/ieso/document-library/fit/2017-fit-price-schedule.pdf. Accessed 20 May 2019
41. InnoVentum (2019) Projects. http://trends.archiexpo.com/innoventum/project-79012-223088.html. Accessed 15 May 2019
42. International Electrotechnical Commission (2017) International standard. Low voltage electrical installations 60364 Part 7-712: requirements for special installations or locations—solar photovoltaic (PV) power supply systems. https://www.lsp-international.com/iec-60364-7-712-2017-requirements-for-special-installations-or-locations-solar-photovoltaic-pv-power-supply-systems. Accessed 15 July 2019

43. International Electrotechnical Commission (2019) 61724 standard series. Photovoltaic system performance. https://www.hukseflux.com/applications/solar-energy-pv-system-performance-monitoring/how-to-calculate-pv-performance-ratio. Accessed 15 April 2019
44. International Energy Agency (2018) Trends 2018 in photovoltaic applications—survey report of selected IEA countries between 1992 and 2017. http://www.iea-pvps.org/fileadmin/dam/intranet/task1/IEA_PVPS_Trends_2018_in_Photovoltaic_Applications_03.pdf. Accessed 17 June 2019
45. Jordan D, Kurtz S (2012) Photovoltaic degradation rates—an analytical review. National Renewable Energy Laboratory NREL/JA-5200-51664. http://www.nrel.gov/docs/fy12osti/51664.pdf, Accessed 15 April 2019
46. Karimov KS, Saqib MA, Akhter P, Ahmed MM, Chattha JA, Yousafzai SA (2005) A simple photo-voltaic tracking system. Sol Energy Mat Sol C 87:49–59
47. Kelly N, Gibson T (2011) Increasing the solar photovoltaic energy capture on sunny and cloudy days. Sol Energy 85:111–125
48. Kesler S, Kivrak S, Gurleyen H, Dincer F, Yilmaz S, Ozcalik HR (2015) A low cost shading analyzer and site evaluator design to determine solar power system installation area. Int J Photoenergy. Article ID 126373. https://dx.doi.org/10.1155/2015/126373
49. Kiewit (2019) Fixed-tilt vs. axis tracker solar panels. https://www.kiewit.com/plant-insider/current-issue/fixed-tilt-vs-axis-tracker-solar-panels. Accessed 15 May 2019
50. Koussaa M, Haddadib M, Saheba D, Maleka A, Hadjic S (2012) Sun tracking mechanism effects on flat plate photovoltaic system performances for different step time and main parameters affecting the obtained gains: case of North Africa and Mediterranean site. Energy Proced 18:817–838
51. Krawczynski M, Strobel MB, Betts TR, Gottschalg R (2011) Spectral irradiance measurements for photovoltaic systems in maritime climates. In: Proceedings of 7th photovoltaic science applications and technology conference (PVSAT-7). Heriot-Watt University, Edinburgh, pp 277–280
52. Lang M, Lang A (2012) 2012 EEG surcharge increases slightly to 3.592 ct/kWh. https://web.archive.org/web/20160803082319/http://www.germanenergyblog.de/?p=7526. Accessed 20 May 2019
53. Lee CY, Chou PC, Chiang CM, Lin CF (2009) Sun tracking systems: a review. Sensors-Basel 9:3875–3890
54. Liu GY, Ren Y, Nguang SK (2012) Simulated photovoltaic array systems under a changing environment for temporal performance. Int J Green Energy 9(7):673–684
55. Lo S, Cheng F, Chang V, Liu WDJ, Chang L, Adurodija OF, Chou E, Lung C, Liu J, Lu T, Su J, Cheng E (2015) Design, operation, and performance evaluation of a cable-drawn dual-axis solar tracker compared to a fixed-tilted system. Energy Sci Eng 3(6):549–557
56. Mandal A (2014) Power evacuation from generating station with fault level calculation & single line diagram. World of Electrical Engineering. https://kalyan07.wordpress.com/tag/ror-plant. Accessed 20 May 2019
57. Marion B, Adelstein J, Boyle K, Hayden H, Hammond B, Fletcher T, Canada B, Narang D, Shugar D, Wenger H, Kimber A, Mitchell L, Rich G Townsend T (2005) Performance parameters for grid-connected PV systems. In: 31st IEEE photovoltaics specialists conference and exhibition lake Buena Vista, Florida, 3–7 Jan, NREL/CP-520-37358
58. Masters GM (2004) Renewable and efficient electric power systems. Wiley, New Jersey
59. Miloudi L, Acheli D, Chaib A (2013) Solar tracking with photovoltaic panel. Energy Proc 42:103–112
60. Mittal AK, Passow K, Lee M (2019) Precipitable water effect on spectral shift and solar PV module energy. https://www.solarquarter.com/index.php/analysis1/1294-precipitable-water-effect-on-spectral-shift-and-solar-pv-module-energy. Accessed 10 May 2019
61. Moldovan MD, Visa I, Burduhos BG (2011) Energetic autonomy for a solar house. Environ Eng Manag J 10(9):1283–1290
62. Moradi H, Abtahi A, Messenger R (2016) Annual performance comparison between tracking and fixed photovoltaic arrays. In: IEEE 43rd photovoltaic specialists conference (PVSC), Portland, USA, 5–10 June 2016, pp 3179–3183. https://doi.org/10.1109/5.7750252

63. Munksgaard J, Morthorst PE (2008) Wind power in the Danish liberalized power market—policy measures, price impact and investor incentives. Energy Policy 36:3940–3947
64. Naeem MH (2014) Soiling of photovoltaic modules: modelling and validation of location-specific cleaning frequency optimization. Master's thesis, Arizona State University
65. National Renewable Energy Laboratory (2016) Best practices in photovoltaic system operations and maintenance, 2nd edn. NREL/Sandia/Sunspec Alliance, SuNLaMP PV O&M Working Group
66. Neagoe M, Visa I, Burduhos BG (2014) Increasing the tracking efficiency of photovoltaic systems. In: Visa I (ed) Sustainable energy in the built environment—steps towards nZEB. Springer, Heidelberg, p 443
67. Nsengiyumva W, Chen SG, Hu L, Chen X (2018) Recent advancements and challenges in solar tracking systems (STS): a review. Renew Sustain Energy Rev 81:250–279
68. Olariu N, Mantescu G (2013) Study long-term behavior of building integrated photovoltaic panels—thermal effect. In: 28th European photovoltaic solar energy conference and exhibition, pp 4369–4372
69. Paris E, Beisner K (2018) Fixed-tilt vs. tracker: why a one-size-fits-all approach can limit solar production. SPW, 19 Feb 2018. https://www.solarpowerworldonline.com/2018/02/fixed-tilt-vs-tracker-one-size-fits-approach-can-limit-solar-production. Accessed May 2019
70. Parkinson G (2012) Euro utilities declare war on solar PV. https://reneweconomy.com.au/euro-utilities-declare-war-on-solar-pv-57935. Accessed 25 May 2019
71. Pickerel K (2015) Fighting dirty: manual washing vs. automatic cleaning of solar modules. Solar Power World. https://www.solarpowerworldonline.com/2015/02/fighting-dirty-manual-washing-vs-automatic-cleaning-of-solar-modules. Accessed 15 April 2019
72. Prasad DK, Snow M (2014) Designing with solar power: a source book for building integrated photovoltaics. Routledge, Australia
73. PVeurope (2018) We drive solar—utrecht goes ahead. https://www.pveurope.eu/News/E-Mobility/We-drive-solar-Utrecht-goes-ahead. Accessed May 2019
74. Rawat R, Chandel SS (2013) Simulation and optimization of solar photovoltaic-wind stand alone hybrid system in hilly terrain of India. Int J Renew Energy Res 3(3):595–604
75. Rickerson W, Grace RC (2007) The debate over fixed price incentives for renewable electricity in Europe and the United States: fallout and future direction. Heinrich Boll Foundation, Washington, DC
76. Roman E, Lopez JR, Alves L, Eisenschmid I, Melo P, Rousek J, Tsoutsos TD (2008) BIPV potential and benefits—intelligent energy Europe, PURE project. http://pvtrin.eu/assets/media/PDF/Publications/Other%20Publications/38.BIPV%20POTENTIAL%20AND%20BENEFITS.pdf. Accessed 10 May 2019
77. Sajid A, Choon-Man J (2019) Evaluation of PV-wind hybrid energy system for a small island. In: Wind solar hybrid renewable energy system. IntechOpen. https://dx.doi.org/10.5772/intechopen.85221
78. Sarver T, Al-Qaraghuli A, Kazmerski L (2013) A comprehensive review of the impact of dust on the use of solar energy: history, investigations, results, literature, and mitigation approaches. Renew Sustain Energy Rev 22:698–733
79. Sawle Y, Gupta SC, Bohre AK (2016) PV-wind hybrid system: a review with case study. Cogent Eng 3:1189305
80. Schweiger M, Herrmann W (2015) Comparison of energy yield data of fifteen PV module technologies. In: 42nd photovoltaic specialists IEEE conference, New Orleans, USA, 14–19 June 2015. https://doi.org/10.1109/PVSC.2015.7356123
81. Shady A (2010) Zero energy retrofit case study of a chalet in Ain-Sukhna, Egypt. In: Proceedings of 39th American solar energy society conference, Phoenix, Arizona
82. Shahid Law Firm (2016) Egyptian feed-in tariff programme update on 06 Sept 2016. http://shahidlaw.com/2016/09/06/egyptian-feed-in-tariff-programme-update-on-06-september-2016. Accessed 20 May 2019
83. Shakeri M, Shayestegan M, Reza SMS, Yahya I, Bais B, Akhtaruzzaman Md, Sopian K, Amin N (2018) Implementation of a novel home energy management system (HEMS) architecture with solar photovoltaic system as supplementary source. Renew Energy 125:102–120

84. Sinha S, Chandel SS (2015) Review of recent trends in optimization techniques for solar photovoltaic–wind based hybrid energy systems. Renew Sustain Energy Rev 50:755–769
85. SMA (2019) Performance ratio, quality factor for the PV plant, Technical information, version 1.1. http://files.sma.de/dl/7680/Perfratio-TI-en-11.pdf. Accessed April 2019
86. SoDa (2019) Solar radiation data. http://www.soda-pro.com. Accessed 25 May 2019
87. Solarcoin (2019) https://solarcoin.org. Accessed 25 May 2019
88. Solar-Log (2019) Solar-log guarantees the upper feed-in limit. https://www.pes.eu.com/renewable-news/solar-log-guarantees-the-upper-feed-in-limit. Accessed 20 May 2019
89. Sonnendach (2019) https://www.uvek-gis.admin.ch/BFE/sonnendach. Accessed 10 May 2019
90. Spencer JW (1971) Fourier series representation of the position of the Sun. Search 2(5):162–172
91. Stagnaro C (2012) How solar subsidies can distort the power market: the case of Italy. MPRA paper 44853, University Library of Munich, Germany. https://mpra.ub.uni-muenchen.de/44853. Accessed 25 May 2019
92. StreetlightCN (2019) https://www.streetlightcn.com/led-module/wind-pv-hybrid-street-light.html. Accessed 15 May 2019
93. Sungur C (2007) Sun-tracking system with PLC control for photo-voltaic panels. Int J Green Energy 4(6):635–643
94. United States Congress (2007) Energy Independence and Security Act. Public Law 110–140, 110th Congress, 121 stat. 1492, 19 Dec 2007
95. United States Department of Energy (2003) The history of solar. https://www1.eere.energy.gov/solar/pdfs/solar_timeline.pdf. Accessed 16 June 2019
96. van der Kam M, van Sark W (2015) Smart charging of electric vehicles with photovoltaic power and vehicle-to-grid technology in a microgrid; a case study. Appl Energy 152:20–30
97. Vermaak HJ, Kusakana K (2014) Design of a photovoltaic–wind charging station for small electric Tuk–tuk in D.R. Congo. Renew Energy 67:40–45
98. Visa I (2014) Renewable energy systems: case study—solar energy conversion systems. In: Visa I (ed) The 11th IFToMM international symposium on science of mechanisms and machines. Mechanisms and machine science, vol 18. Springer, Cham, p 31
99. Visa I, Moldovan M (2019) Solar tracking linkage RSSR for all latitudes. Mech Mach Sci 73:3541–3550
100. Visa I, Jaliu C, Duta A, Neagoe M, Comsit M, Moldovan M, Ciobanu D, Burduhos B, Saulescu R (2015) The role of mechanisms in sustainable energy systems. Transilvania University Pub. House, Brasov
101. Visa I, Burduhos BG, Neagoe M, Moldovan MD, Duta A (2016) Comparative analysis of the infield response of five types of photovoltaic modules. Renew Energy 95:178–190
102. Visa I, Neagoe M, Moldovan M, Comsit M (2018) Solar tracking parallel linkage applicable for all latitudes. Mech Mach Sci 57:3–11. https://doi.org/10.1007/978-3-030-20131-9_349
103. World Energy Council (2013) World energy resources: solar. https://www.worldenergy.org/wp-content/uploads/2013/10/WER_2013_8_Solar_revised.pdf. Accessed 16 Jul 2019

Chapter 4
Increasing the Solar Share for Domestic Hot Water, Heating and Cooling in the Built Environment

4.1 Solar Thermal Systems in Buildings and at Community Level

The future development scenarios have to consider the population growth along with an increased energy demand in the residential area. Moreover, due to environmental protection strategies, this energy demand has to be fulfilled using renewable energy sources to a large extent, as these are not responsible for greenhouse gas emissions.

Among the renewable energy sources available in the built environment, solar energy is highly investigated to provide electrical and thermal energy. However, the direct use of solar energy in solar energy conversion systems has to be optimized considering that the periods with available solar radiation do not fully match the periods with a high energy demand, especially for thermal energy production. The extent of this problem depends on the geographical location and is more important in temperate and cold climates, where the thermal energy for heating is required during short daylight periods and with less amount of solar energy, e.g. during the cold season. Therefore, the solar thermal systems implemented in the built environment have to be specifically designed considering the climatic profile of the implementation location along with the energy demand to be covered. The design has to consider the development of efficient thermal energy storage systems and the review developed by Parameshwaran et al. [68] outlined that the use of phase-change materials represents a viable path.

According to the data provided by the International Energy Agency [38], the solar heating and cooling sector were the largest solar sector worldwide with 472 GW (thermal power) installed at the end of 2017, followed by photovoltaics (402 GW installed by the end of 2017) and concentrated solar power (5 GW). The same report outlines that the major end-users of the energy produced using solar radiation were from the industrial sector followed by the solar district heating sector.

I. Visa et al., *Solar Energy Conversion Systems in the Built Environment*, Green Energy and Technology, https://doi.org/10.1007/978-3-030-34829-8_4

As mentioned by the International Energy Agency [38], heating accounts for 47% of the world's energy demand, which is higher than the combined share of electricity (17%) and transport (27%). However, as a result of the low prices of fossil fuels and of the use of other renewables, the added solar thermal capacity is declined by 4.2% in 2017. This declining percentage is lower than the values observed in 2016 and in 2015 and mainly corresponds to the markets in Europe and in China. Oppositely, three countries registered significant market growth: India (26%), Mexico (7%) and Turkey (4%). Consequently, India and Turkey have the lowest solar hot water prices (0.02–0.03 €/kWh), as compared to the highest costs of solar hot water registered in France (0.19 €/kWh for small systems and 0.14 €/kWh for multifamily houses).

According to Weiss and Spörk-Dür [99], the largest solar district heating system is in Denmark and has an output of 110 MW insured by 156,694 m^2 of flat plate solar thermal collectors that started to produce in 2016. One of the largest solar thermal district heating systems in China was installed in 2013 in the Hebei University of Economics and Business in Shijiazhuang to supply space heating and domestic hot water (DHW) for the students' dormitories. The system has a capacity of 8.1 MW (11,592 m^2) connected to a 20,000 m^3 heat storage system. In 2017, four large-scale solar thermal plants were installed in China, two with flat plate collectors and two with concentrating collectors. Besides the concentrating systems installed in China, another one was mounted in 2017 in Denmark.

According to the International Energy Agency [38], solar district heating systems were installed in 2017 in France, Serbia, Australia and Kyrgyzstan. A total of 15 large-scale solar thermal systems (with an installed area of over 500 m^2/system, corresponding to 350 kW) were installed in 2017 mainly in Austria, China, Denmark, Germany and Sweden. By the end of 2017, over 300 large-scale solar thermal systems connected to district heating systems were in operation all over the world, with an installed capacity of 1140 MW insured by a 1,630,415 m^2 collectors' surface, excluding the concentrators' surface that adds 110,929 m^2.

Depending on the size and on the implementation location, the costs of such a system vary between 0.03 €/kWh (Brazil) and 0.185 €/kWh (Denmark). The lifetime considered for the systems also influences the cost level. A lifetime of 15 years is considered in China while 25 years is considered in all the other countries.

The top ten countries that installed solar thermal systems in 2016 are synthesized in Table 4.1.

As the values in Table 4.1 show, there are countries with a rather large number of solar thermal systems earlier installed and that did not install too many such systems during the analysed period, as, e.g., Greece. Moreover, according to the study published by Martinopoulos [50] even in Greece, a solar thermal system could only cover 68.2% of the entire thermal energy demand of a building in Athens. On the other hand, the population number in the implementation location significantly influences these values as countries with low population reach their installation limit faster than countries with a larger population. Therefore, reference values of installed capacity per 1000 inhabitants are relevant. Considering this indicator, Table 4.2 presents the data delivered by the International Energy Agency for 2016.

Table 4.1 Top ten countries with installed solar thermal systems [38]

Country	New installations in 2016 (MW)	Country	Total installations by 2016 (MW)
China	27,664	China	324,506
Turkey	1296	USA	17,565
Brazil	913	Turkey	14,993
India	841	Germany	13,535
USA	682	Brazil	9555
Germany	536	India	6673
Australia	381	Australia	6181
Denmark	335	Austria	3645
Israel	295	Israel	3244
Mexico	256	Greece	3148

Table 4.2 Top ten countries with installed solar thermal systems per 1000 inhabitants in 2016 [38]

Country	New installations in 2016 (kW)	Country	Total installations by 2016 (kW)
Denmark	60	Barbados	515
Cyprus	36.1	Austria	418
Israel	36	Cyprus	399
Barbados	27	Israel	397
China	20	Greece	292
Greece	18	Palestinian Territories	289
Australia	17	Australia	269
Turkey	16	China	236
Austria	9	Denmark	204
Palestinian Territories	8	Turkey	186

The analysis of these data outlines that countries with a large density of solar thermal systems mainly have a warm climatic profile. However, solar thermal systems were recently installed also in countries with a colder climate (as Denmark or Austria). It can be also observed that China is part of all four types of ranking hereby discussed, despite the fact that it has the largest population in the world, thus proving its development strategy towards a clean future.

According to Weiss and Spörk-Dür [101] the cumulated solar thermal capacity in operation by the end of 2018 was 480 GW that outlines a yearly growth factor of 7.7 compared to 2000. The corresponding annual solar thermal energy yield in 2018 amounted 396 TWh that indicates 42.6 million tons of oil savings and a reduction in the CO_2 emissions of 137.5 million tons. By the end of 2017, the output of the

Table 4.3 Share of various solar thermal technologies in the newly installed capacities in the world in 2017 [101]

Solar thermal technology	Evacuated tube collectors	Flat plate collectors	Unglazed water collectors	Unglazed air collectors
Share, in 2017 (%)	71.0	22.6	6.1	0.3

installed solar thermal systems was 375 TWh, thus saving over 40 million tons of oil and 130 million tons of CO_2. It is to mention that in 2016, for the first time, solar thermal heating technologies based on parabolic trough collectors were implemented in district heating systems. However, the evacuated tube collectors were mostly used, representing a large share of the newly installed capacity, most of these collectors being delivered by the Chinese market that also had the largest share of new solar thermal installed systems in 2017 (70.6%). Table 4.3 shows the share of each solar thermal technology in 2017.

Integrating solar thermal systems on buildings was done since over 50 years and is still done mainly to produce domestic hot water (DHW). However, if this trend will be extended and the production would, at least partially, also cover the heating and/or the cooling energy demand, careful urban planning needs to be developed, considering from the initial design stage the installation of the solar thermal systems. This is required for, e.g., avoiding the shading of the solar thermal collectors by a newly built construction.

Following the analysis developed by Lobaccaro et al. [48] on existing and on new urban areas and on solar landscapes, it can be observed that there are specific solutions for each analysed case study, according to the urban context (the social–cultural value of the urban area and the visibility of the building surface). Further on, the implementation of the solar energy conversion systems (photovoltaic and solar thermal systems) should consider a pattern that well fits the landscape and the ecological aspects. For example, in newly designed urban areas the building's architecture considered the integration of solar thermal collectors, as in the case of the *Residential Plot B45* (China), where evacuated tube solar thermal collectors were installed on the roofs of the high-raise building blocks. The *Drake Landing Solar Community*, in Okotoks, Canada, where roof-installed solar thermal collectors support the district heating system that also includes a community storage system, represents another example. This district system was built over ten years (2003–2012) based on a feasibility study that considered the solar thermal systems installed on the garage and on the buildings' roofs and a borehole for long-term, seasonal thermal energy storage that works correlated with a short-term thermal energy storage systems using storage tanks. Overall, developing a solar thermal community system to support the district heating has firstly to solve the thermal energy storage as, considering the currently known solutions, long-term storage in large-sized systems requires a significant land area and this is usually not affordable at the city level. Thus, new storage systems along with public commitment and acceptance are required to support large solar district heating systems. The development of combined solar thermal–heat pump systems represents a possible path that has to be well designed, also considering the electrical energy required for the heat pump operation.

Thus, when assessing the solar energy potential at community/district/neighbourhood level, one should consider the urban morphology and density as some of the most important factors along with shading [1]. Following the way the district unit is defined, the energy performance of that unit needs to be assessed based on a set of relevant indicators that are expected to be formulated and harmonized.

The trends identified by the International Energy Agency [38] for the near future also consider the implementation of the solar energy conversion systems (including the solar collectors) in the building's envelope, particularly in/on the suitably oriented facades. The collectors can be mounted on the facades (building-added solar thermal systems, BAST) or new solutions for integrating the collectors in the building parts, e.g. in the walls (building-integrated solar thermal systems, BIST) are considered, targeting a reduction in the duration of the construction process. Moreover, integrating solar thermal collectors on the building's façade can bring an additional advantage during the warm periods of the year [46]. This advantage is represented by the role of BAST mounted at a well-optimized distance from the solid façade to effectively remove the solar heat gain and support the internal thermal comfort. As the authors also show, the facades can be more efficiently used when integrating photovoltaic thermal (PVT) systems to produce DHW and provide the electrical energy required by the building operation.

Solar thermal systems can also be used for cooling the buildings, thus replacing the traditional vapour compression air-conditioning systems. As Jani et al. [40] showed in their review, the solar-powered solid desiccant–vapour compression hybrid represents a good candidate for space cooling. The best option, considering the existing technologies, is represented by the solar-powered, desiccant-based hybrid air-conditioning system.

Integrating solar thermal systems in or on the buildings requires architectural acceptance [24] particularly for the systems that are mounted on the visible parts of the building, e.g. on the facades. Thus, *the visual amplitude* is considered as a new indicator that would support the sizing and the positioning of the solar energy convertors on the building's facade.

Moreover, as the existing built environment has to be further used and improved, the renovation step may represent a good opportunity to install solar thermal systems on the building, as proposed in the project UrbanSol+ [86]. Considering this option, the architectural integration is discussed as for example using the VELUX solar systems that combine the attic windows with the collectors in a harmonized product. As examples of solar thermal systems integrated with retrofitted buildings, there are mentioned the *Campo Pequeno Building* in Lisbon, Portugal, that is a multiapartment building with collectors installed on the southern and western facades and on the roof. The excess thermal energy produced during the summer is sent to the condominium inner pool or to the thermo ventilator (outside mounted). Other examples are *the Sanctuary of the Blessed Virgin Mary*, Redemptorist Monastery in Poland, *the Königlich Sächsische Schrotfabrik*, Freiberg, Germany, *the Palace of Bishops* in Poland, *the Caixa Geral de Depósitos* building in Lisbon, Portugal, etc.

These are old and representative buildings in the community and have solar thermal systems installed mainly in less visible places (as on the roof). Significant is that for each such a project the full use of the solar heat is considered for the primary application (DHW or heating) along with secondary applications as for air-conditioning.

Solutions to increase the solar thermal energy conversion efficiency and the architectural acceptance of these systems are further discussed in this chapter as prerequisites in an extended implementation of the solar thermal systems in the built environment.

4.2 Design of the Solar Thermal Systems Implemented in the Built Environment

The design of the solar thermal systems implemented in the built environment is structured according to each functionality (DHW preparation, space heating and space cooling) on the following main steps:

1. Estimation of the thermal energy demand;
2. Estimation of the received solar energy;
3. Selection of the solar thermal collector type;
4. Estimation of the number of solar thermal collectors;
5. Estimation of the storage capacity;
6. Interconnection of the solar thermal collectors;
7. Design of the piping system;
8. Selection of the hydraulic pump;
9. Estimation of the expansion vessel capacity

Step 1. **Estimation of the thermal energy demand**

The thermal energy demand for domestic hot water preparation (E_{DHW}) is evaluated on a daily basis using

$$E_{\mathrm{DHW}} = V_{\mathrm{DHW}} \cdot \rho_{\mathrm{w}} \cdot c_{\mathrm{w}} \cdot (T_{\mathrm{HW}} - T_{\mathrm{CW}}) \tag{4.1}$$

where

- V_{DHW} is the volume of domestic hot water that must be heated;
- ρ_{w} is the water density;
- c_{w} is the specific heat capacity of water;
- T_{HW} is the hot water temperature;
- T_{CW} is the cold water temperature.

Table 4.4 Specific DHW consumption in Romania [54]

Building type	Specific DHW consumption [L/(user day)]
Residential building	40–110
Office building, school	5
Restaurant	10
Hotel	60–100

The domestic hot water volume (V_{DHW}) is obtained by multiplying the number of users with the specific consumption expressed in *L*/(user day), as given in specific building codes formulated for each region; selected values from the Romanian Standard I9-2015 are included in Table 4.4.

For space heating and cooling, the thermal energy demand depends on the buildings' geometry and orientation towards the cardinal points, on the occupancy schedule, on the thermal characteristics of the materials embedded in the buildings' envelope (walls, slabs, roofs, exterior doors and windows) and on the meteorological profile of the implementation location. Thus, the assessment of the thermal energy demand for space heating and cooling is a complex procedure, and dedicated software (e.g. TRNSYS, EnergyPlus, DOE, PHPP) was developed to support this task.

An assessment example of the thermal energy demand is further presented, using the TRNSYS software, for a residential building of 100 m^2, with 4 users, located in a temperate continental climate (Brasov, Romania). A multizone building was developed using the TRNSYS 17 Simulation Studio with the structure presented in Fig. 4.1.

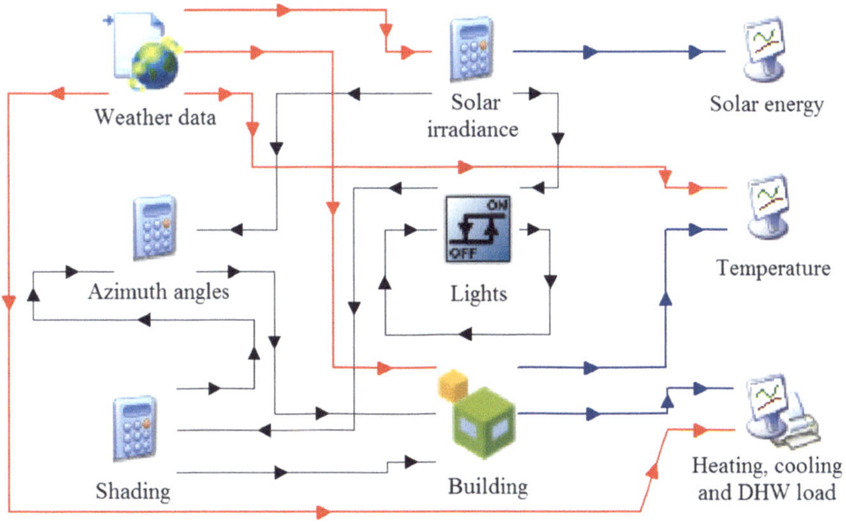

Fig. 4.1 TRNSYS 17 model used for the assessment of the thermal energy demand

Here, the building file was imported as a Type56 unit, on-site measured meteorological data were imported into the Type 15-2 Weather data unit, the equation cards were used to calculate the azimuth angles of the building's envelope components and the associated solar irradiance. A Type 2d Lights unit was used as a controller for the lighting system functionality. Output data (the solar energy received on each building's envelope component; the outdoor and indoor air temperatures; the heating and cooling loads for each thermal zone of the building) were plotted and exported in *.csv files through the Type 65a online plotter units.

The input data divided into two categories (site and building characteristics) are further presented.

The site characteristics mainly refer to the meteorological data, and the best option is to use on-site measured values; if these are not available, dedicated applications (e.g. Meteonorm), interpolating data from the closest weather stations, can be used. In the analysed case study, a local weather station registered each minute the outdoor air temperature and its relative humidity, the wind speed and direction, the global and diffuse solar irradiance values measured in the horizontal plane and the direct solar irradiance measured on the direction of the Sun. These values were hourly averaged and are plotted in Fig. 4.2 for the outdoor air temperature (1) and for the global solar irradiance in the horizontal plane (2).

The building's characteristics refer to the building's envelope components features and are evaluated using TRNBuild based on the input data considered for the layers: 15 cm concrete slabs thermally insulated with 20 cm of polystyrene for the ground floor and for the rooftop, 30 cm light concrete masonry thermally insulated with 10 cm of polystyrene for the exterior walls. A thermal resistance of 5.62 m^2K/W resulted for the ground floor, 5.52 m^2K/W for the rooftop, 4.15 m^2K/W for the exterior walls and 1.23 m^2K/W for the exterior windows.

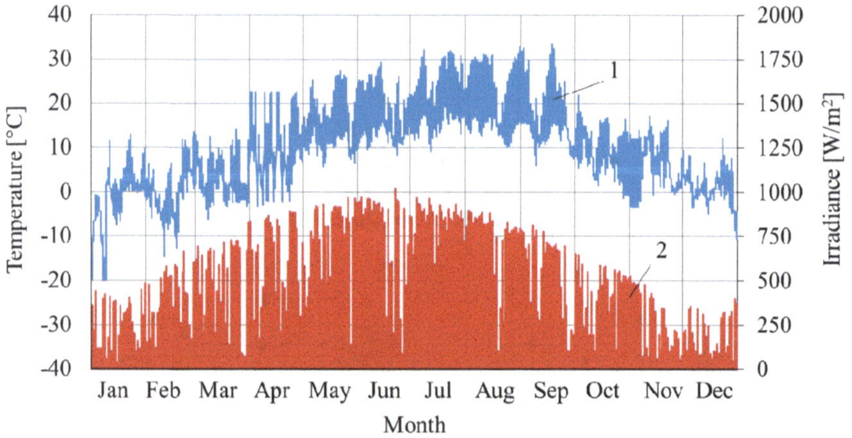

Fig. 4.2 Hourly values of the outdoor air temperature (1) and the global solar irradiance in the horizontal plane (2) on-site measured in Brasov, Romania, in 2015

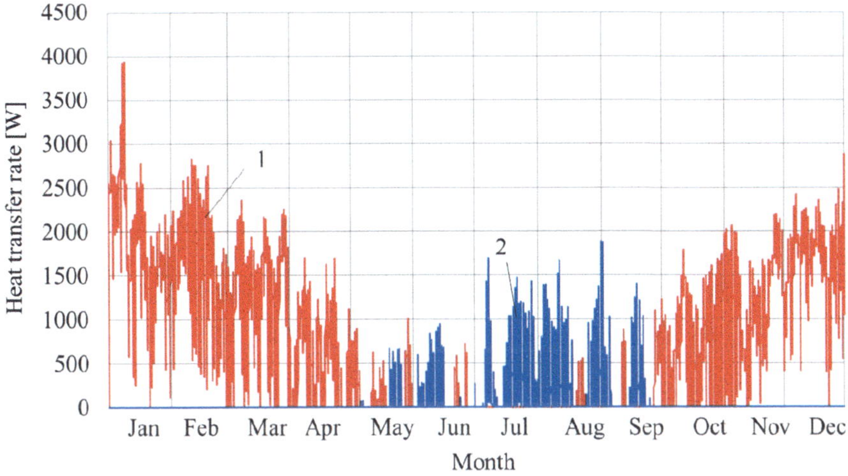

Fig. 4.3 Hourly heat transfer rates for space heating (1) and for space cooling (2)

Based on the TRNSYS numerical simulations, the hourly values of the heat transfer rates for space heating (1) and for space cooling (2) are plotted in Fig. 4.3. As expected, the highest heat transfer rate value (3940 W) results in the heating season during the coldest month (January) when the outdoor air temperature decreases down to −20 °C. This maximum transfer rate for space heating is important in the design of the heating system, imposing the peak power value of the heating system (heat pump, biomass boiler, etc.).

It is important to notice the long heating period (from September till May) with some cool days even in June due to the mountain climate in the building's location. During the summer of 2015, the highest cooling heat transfer rate (1896 W) occurs on the first day of September due to the combined effect of outdoor air temperature (31.8 °C) and horizontal global solar irradiance (714 W/m^2). The highest temperature of the outdoor air (33.4 °C) was recorded on 18 September, at 16:00. However, the horizontal global solar irradiance was of only 327 W/m^2 at that hour, corresponding to a cooling heat transfer rate of 1164 W. The highest horizontal global solar irradiance, of 1020 W/m^2, occurred in 26 June, when the outdoor air temperature was of only 16.9 °C; thus, in that day, both the heating and the cooling heat transfer rate were zero. Based on these types of data, it may be concluded that, in the design process of the space cooling system, the combined effect of outdoor air temperature and solar irradiance has to be considered; otherwise, the cooling capacity could be oversized.

The thermal energy demand is calculated by integrating the assessed hourly heat transfer rates over the period of interest (e.g. one day, week, month, season or year) as presented in Fig. 4.4 for the monthly thermal energy demand for space heating (E_{H}) and for space cooling (E_{C}). The thermal energy demand for the domestic hot water preparation (E_{DHW}) calculated using Eq. (4.1) is obtained considering the number of days in each month. Typical for a mountain temperate continental climate, the

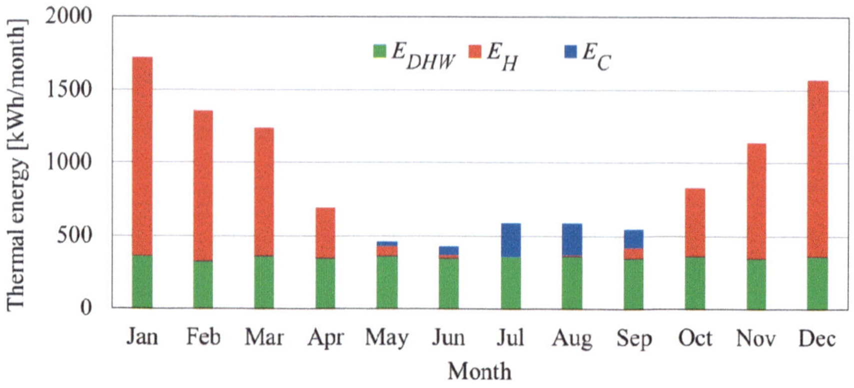

Fig. 4.4 Monthly thermal energy demand for domestic hot water preparation (E_{DHW}), space heating (E_{H}) and space cooling (E_{C})

monthly thermal heating energy demand has the highest variability, with four times higher values in winter than during summer, due to the seasonality of the heating and cooling energy demand. However, the thermal energy demand for DHW preparation only slightly varies due to the number of days in each month.

When a higher energy demand for the DHW preparation is expected (multifamily residential buildings, wellness centres, industrial buildings, etc.), the slight variation of the cold water temperature or the daily/seasonally consumption variability should be also considered. In the investigated case study, during the transition months (May, September) a small energy demand occurs both for heating and cooling, making these months good candidates for the use of solar energy for these functionalities.

By summing up the monthly values of the thermal energy demand, the yearly values are obtained (Table 4.5) and these values must be compared with the standard values recommended for low energy buildings (LEB) in each country (e.g. 50–150 kWh/(m^2 year) for the member states in the European Union).

For new buildings, due to the in-force building codes that impose energy efficiency measures, the thermal energy demand generally complies with the low energy building standards (111.56 kWh/(m^2 year) in the analysed case study). However, new buildings still represent a small share in the entire building stock. In the case of old buildings, the specific energy demand usually overpasses the standards and specific measures should be considered before implementing renewable energy systems. These measures consist of reducing the amount of wasted DHW, increasing the thermal insulation of the building's envelope, increasing the airtightness of the building, recovering the thermal energy from the exhausted air, etc. Only after applying these measures to reduce the thermal energy demand down to the LEB standards, the next steps in implementing renewable energy systems should be followed.

Step 2. **Estimation of the received solar energy**

The global solar irradiance received in the plane of the solar thermal collectors (G_n) has to be assessed over the entire period of one year, in the implementation location,

Table 4.5 Monthly breakdown of the total thermal energy demand for domestic hot water preparation (E_{DHW}), space heating (E_H) and space cooling (E_C)

Month	January	February	March	April	May	June	July	August	September	October	November	December	Year
E_H (kWh)	1359	1032	879	345	72	22	0	10	76	468	789	1213	6265
E_C (kWh)	0	0	0	0	26	59	229	219	118	0	0	0	652
E_{DHW} (kWh)	360	325	360	348	360	348	360	360	348	360	348	360	4239
Total (kWh)	1719	1357	1239	693	458	430	589	589	543	828	1138	1573	11,156

as presented in Fig. 4.5 for Brasov, Romania. Due to the optimum tilt angle selected for the solar thermal collector, peak values close to 1000 W/m^2 are recorded from February till October, more frequently than in the case of the global solar irradiance measured in the horizontal plane (Fig. 4.2). However, values higher than 900 W/m^2 only occur in 8.7% out of the total hourly values of the received global solar irradiance and almost 50% of these hourly values are lower than 200 W/m^2.

By integrating the hourly values of the global solar irradiance numerically simulated in the horizontal (G_{h}) and in the collector's (G_n) planes, the monthly solar energy received in both planes was calculated and is comparatively plotted in Fig. 4.6. Except the months of May, June and July, the solar energy received in the plane of

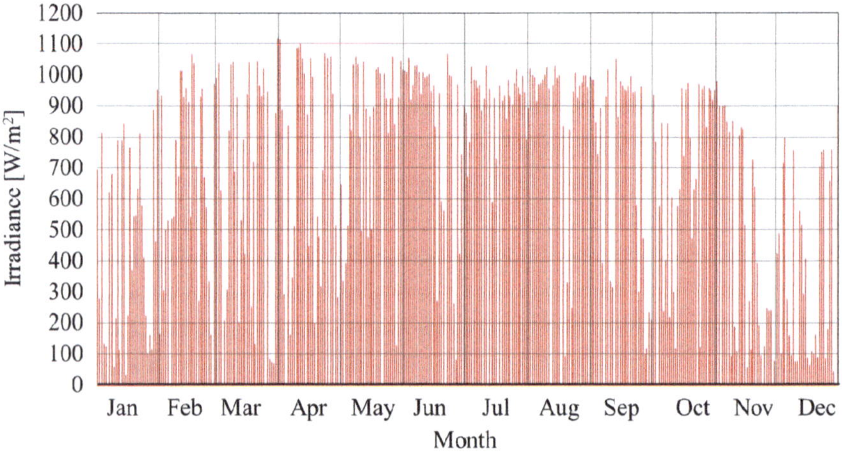

Fig. 4.5 Hourly variation of the global solar irradiance calculated in the plane of the solar thermal collector tilted at 40° in Brasov, Romania

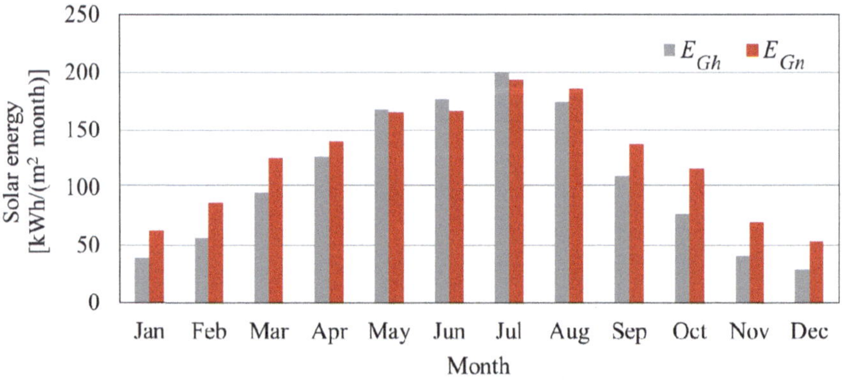

Fig. 4.6 Global solar energy received in the horizontal plane ($E_{G_{\mathrm{h}}}$) and in the plane of the solar thermal collector (E_{G_n}) tilted at 40° in Brasov, Romania

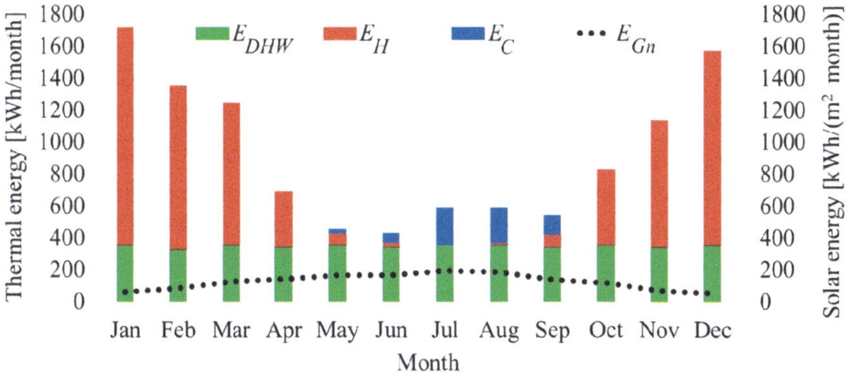

Fig. 4.7 Thermal energy demand of the building's functionalities: domestic hot water (E_{DHW}), heating (E_{H}) and cooling (E_{C}), and the received global solar energy (E_{G_n}) on the tilted plane of a solar thermal collector with a surface of 1 m^2 in Brasov, Romania

the collector is always higher than that received in the horizontal plane, reaching a total (yearly) amount of 1506 kWh/m^2 in the tilted plane and 1293 kWh/m^2 in the horizontal plane. Thus, a yearly increase of 16.5% is obtained by optimally tilting the solar thermal collector according to the latitude of the implementation location and the specific climatic parameters. In this example, the tilt angle was optimized considering a maximum yearly gain. If a reduced period of the year is of interest (e.g. the summer months), the tilt angle should be accordingly assessed.

Comparing the thermal energy demand of the case study building with the solar energy received by a solar thermal collector having a surface of 1 m^2 (E_{G_n}) (Fig. 4.7), it is obviously that there is a mismatch between the high thermal energy demand periods and the low availability of the solar energy during the cold season.

Step 3. **Selection of the solar thermal collector type**

Further on, starting from the functionalities to be provided by the solar thermal system, the solar thermal collectors' type will be selected. Thus, a solar thermal system designed to provide thermal energy for space heating can work at low temperatures (25–35 °C) when a low-temperature heating system is used (e.g. water-based radiative heating systems, air-based heating systems). For DHW preparation, the system works at higher temperatures (60 °C) imposed by the protection measures against the Legionella bacteria. For space cooling, the solar cooling systems are based usually on absorption/adsorption cycles that need the highest supply temperature (85–95 °C). These working temperatures along with the outdoor air temperature and the received global solar radiation will define the working area of the solar thermal collectors.

To select the best candidate among the market available solar thermal collectors, their efficiency has to be plotted depending on the difference between the mean temperature of the solar thermal collector and the outdoor air temperature. The solar thermal collector efficiency (η) is calculated using

Table 4.6 Main functional properties of the three selected solar thermal collectors (C1, C2 and C3)

Parameter	Symbol	C1	C2	C3
Nominal efficiency	η_o	85.1	77.1	64.4
First-order thermal losses coefficient (W/m^2 K)	a_1	3.681	1.894	1.749
Second-order thermal losses coefficient (W/m^2 K^2)	a_2	0.017	0.015	0.005

$$\eta = \eta_o - \frac{a_1 \cdot (T_m - T_a)}{G_n} - \frac{a_2 \cdot (T_m - T_a)^2}{G_n} \tag{4.2}$$

where

- η_o is the nominal efficiency of the solar thermal collector;
- a_1 is the first-order thermal losses coefficient;
- a_2 is the second-order thermal losses coefficient;
- G_n is the global solar irradiance received by the solar collector;
- T_a is the outdoor air temperature;
- T_m is the mean temperature of the solar thermal collector calculated as the average value of the inlet (T_i) and outlet (T_o) temperatures of the thermal fluid using

$$T_m = \frac{T_i + T_o}{2} \tag{4.3}$$

The main functional properties of a flat plate (C1) and two evacuated tube (C2 and C3) solar thermal collectors are presented in Table 4.6.

Using Eq. (4.2) and the parameters extracted from the manufacturers' files (Table 4.6), the efficiencies of the three selected solar thermal collectors are calculated for a range of temperature differences ($T_m - T_a$) between 0 and 150 °C. Based on the results plotted in Fig. 4.8, it can be observed that there is not a solar thermal collector best performing over the entire temperature difference range. The ranking at low values of the temperature difference is completely reversed at higher ones. Thus, for systems working at low-temperature differences (up to 45 °C) as in the case of DHW preparation mainly during the summer period when the average temperature of the solar thermal collector is about 60 °C and the average outdoor air temperature is 25 °C, the C1 solar thermal collector is the best candidate. If the solar thermal system is used only during the wintertime, when the outdoor temperature often decreases below 0 °C, the C2 solar thermal collector will perform better. The C3 solar thermal collector can be selected only for solar thermal systems intended to provide hot water (95 °C) when the outdoor air temperature is usually lower than −10 °C.

This case study only considered the technical properties of the solar thermal collectors. However, their initial costs and their operation and maintenance costs should also be considered, as discussed in Sect. 4.7.

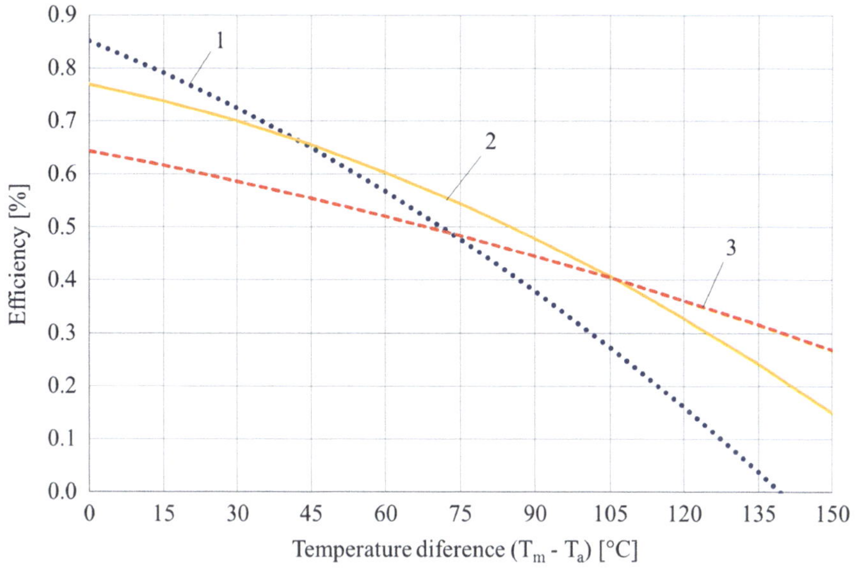

Fig. 4.8 Efficiency variation with the temperature difference ($T_m - T_a$) of the solar thermal collectors C1 (1), C2 (2) and C3 (3)

Step 4. **Estimation of the number of solar thermal collectors**

After selecting the solar thermal collector type according to the main functionality of the solar thermal system and the meteorological profile of the implementation location, the optimum number of solar thermal collectors has to be calculated. In this process, there are two main design requirements: the thermal energy demand must be covered, as much as possible, using the solar energy and the thermal energy overproduction that occurs during the summer season should be avoided, to limit the negative effects of high stagnation temperatures in the solar thermal collector.

The solar thermal collectors' efficiencies so far presented were recorded in standard testing conditions (constant solar irradiance of 1000 W/m^2, normal incidence of the solar radiation on the collector surface, constant wind speed, etc.). The real (outdoor) working conditions should be further considered to correctly assess the conversion efficiency and the thermal energy output of the solar thermal system. For large solar thermal systems, numerical simulations have to be performed using dedicated software (e.g. TRNSYS, Polysun, TSOL).

Based on the hourly average irradiance values of the received global solar radiation in the tilted plane of the solar thermal collector (G_n), on the outdoor air temperature (T_a) and on the mean temperature of the selected solar thermal collector (C2), the hourly efficiencies are obtained using numerical TRNSYS simulations and their monthly average values are included in Table 4.7. As expected, the highest efficiencies of the solar thermal collector are observed during the summer months due to the higher outdoor air temperature that limits the thermal losses. During winter, the

Table 4.7 Monthly average efficiency values of the C2 solar thermal collector

Month	January	February	March	April	May	June	July	August	September	October	November	December	Year
η	0.41	0.44	0.52	0.54	0.54	0.55	0.56	0.59	0.55	0.52	0.48	0.38	0.51

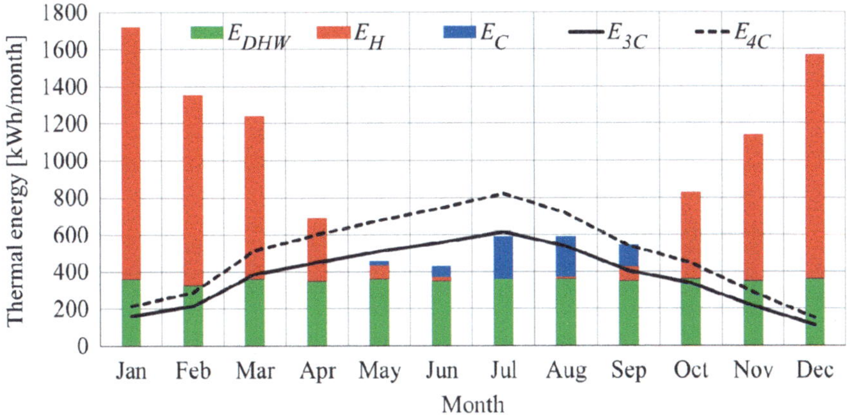

Fig. 4.9 Thermal energy demand for domestic hot water (E_{DHW}), space heating (E_{H}) and space cooling (E_{C}), and the thermal energy supplied by three ($E_{3\mathrm{C}}$) and four ($E_{4\mathrm{C}}$) solar thermal collectors

combined effect of the lower outdoor air temperature and lower received global solar radiation leads to efficiencies that are significantly decreased, the solar gain is mostly used to compensate for the thermal convection losses of the solar thermal collector to the colder outdoor air.

Based on these values, the optimum number of solar thermal collectors is further evaluated. This number can be obtained by incrementally increasing the surface of the solar thermal system until the thermal energy demand is fully covered during the summer months. Thus, overheating of the solar thermal collector is avoided. In the investigated case study, an optimum number of three solar thermal collectors results. The monthly values of the thermal energy provided by these three solar thermal collectors ($E_{3\mathrm{C}}$) are plotted along with the thermal energy demand of the building in Fig. 4.9. From May till July, the entire thermal energy demand of the building is covered with a slight (acceptable) excess in June. The thermal energy demand for DHW is entirely covered from March till October while the space heating energy demand is partly covered during the transition months such as April and September. During the months with uncovered thermal energy demand, an auxiliary source has to be used to back up the solar energy. During the entire year, the solar fraction obtained using the three solar thermal collectors results as high as 38%. Further increasing the number of solar thermal collectors to four increases the thermal energy produced ($E_{4\mathrm{C}}$), in Fig. 4.9, and the solar fraction to 52% but an overproduction of thermal energy occurs from May till August that can rise problems if solutions to use this excess energy are not found (e.g. higher storage capacity, recovering of the ground temperature).

Step 5. **Estimation of the storage capacity**

When the periods of thermal energy demand do not (entirely) overlap the period when the solar energy is available, a storage tank is required. Its storage capacity must be estimated based on the consumption profile of the users and on the solar energy

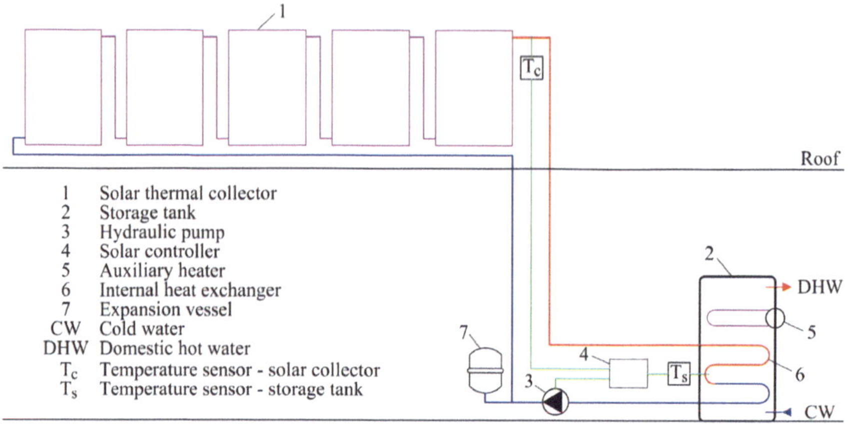

Fig. 4.10 Serial connection of the solar thermal collectors

potential. For small residential solar thermal systems implemented in continental temperate climate, a specific storage capacity of 50 L for each square metre of installed solar thermal collector is recommended [4]. For large systems, the optimum storage capacity must be estimated through numerical simulation.

Step 6. **Interconnection of the solar thermal collectors**

After deciding the optimum number of solar thermal collectors to be installed, the solar circuit must be designed by choosing the interconnection type. The solar thermal collectors can be serially or parallelly connected, depending on their number and on the functional conditions. The easiest way to connect the solar thermal collectors is serially (Fig. 4.10) because there is no need to balance the flow between them, as the flow rate through each solar thermal collector, pipe and pump is the same.

According to each manufacturer recommendations, the maximum number of serially connected solar thermal collectors is limited (usually to four or five) because of the successive temperature increase in the thermal fluid that worsens the conversion efficiency (due to increased thermal losses at the level of the warmer solar thermal collectors). When the optimum number of solar thermal collectors results higher than the maximum number recommended by the manufacturer to be serially connected, parallel connections must be also used. The parallel connection is recommended in locations with high solar energy potential, to avoid overheating the solar thermal collectors while the serial connection is preferred in cold areas where this overheating more seldom occurs. As an example, if ten solar thermal collectors are required (Fig. 4.11), two groups of five serially connected solar thermal collectors are recommended to be parallelly connected and measures to balance the thermal fluid flow rate between these must be considered. An advantage of the parallel connection is the decrease of the pressure losses while the flow rate of the thermal fluid increases. In this case, the flow rate through each serially connected solar thermal collectors will be the same while the pump must provide two times higher flow rate.

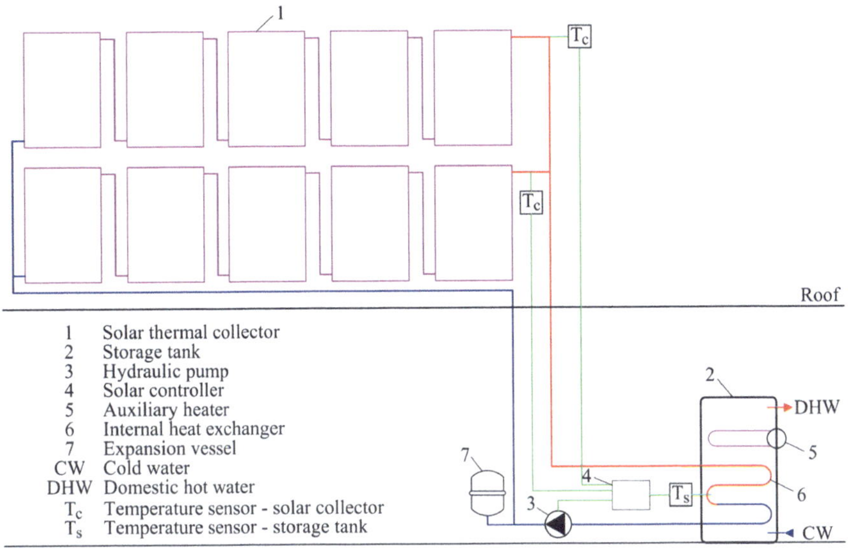

Fig. 4.11 Solar circuit with two parallelly connected rows of five serially connected solar thermal collectors

Step 7. **Design of the piping system**

Once selected the connection type of the solar thermal collectors in the solar circuit, the flow rate of the thermal fluid needs to be calculated for each solar thermal collector and pipe. If the collectors' manufacturer recommends a specific flow rate through the collector, this value must be used. If such a value is not indicated, a specific flow rate of 0.02 kg/(s m^2) can be considered as this is an optimal value for hydraulic pumps with a constant flow. Nowadays, variable flow, energy-efficient hydraulic pumps are often used: lower flow rates during periods with low solar irradiance will decrease the pumping energy consumption while higher flow rates are required when the solar irradiance is higher, and the thermal energy must be faster removed from the solar thermal collector to avoid overheating.

The diameter of the pipes is designed based on the thermal fluid flow rate in each section of the solar circuit. This is an optimization problem: for a given flow rate, a reduced diameter of the pipes reduces their costs and the amount and cost of the thermal fluid; however, the fluid velocity through the pipes increases and also the pressure losses, requiring a more powerful pump that will consume more electrical energy.

Thus, the minimum inside diameter of the pipes is calculated using

$$d_{\min} = \sqrt{\frac{4 \cdot Q}{\pi \cdot v_o}} \tag{4.4}$$

Table 4.8 Cost-optimal fluid velocity in pipes [77]

d_n (mm)	10	15	20	25	32	40	50
v_o (m/s)	0.75	0.80	0.90	1.00	1.10	1.20	1.30

where

- Q is the thermal fluid flow rate through the pipes;
- v_o is the cost-optimal fluid velocity in pipes (Table 4.8).

Based on the minimum inside diameter (d_{min}) established using Eq. (4.4), a standardized pipe is selected and the real fluid velocity is calculated based on the inside diameter (d_i) of the selected pipe using

$$v = \frac{4 \cdot Q}{\pi \cdot d_i^2} \tag{4.5}$$

Step 8. **Selection of the hydraulic pump**

The next step consists of selecting the hydraulic pump to circulate the thermal fluid through the solar circuit. Two parameters are required: the flow rate of the thermal fluid and the pressure losses in the solar circuit. The flow rate is that previously evaluated in the outlet pipe of the storage tank where the pump is usually installed.

The pressure losses through the solar circuit are evaluated according to the American Society of Heating, Refrigerating and Air-Conditioning Engineers [2] using

$$\Delta p = \Delta p_{lin} + \Delta p_{loc} \tag{4.6}$$

where

- Δp_{lin} are the linear pressure losses;
- Δp_{loc} are the local pressure losses.

The linear pressure losses occur in the linear parts of the pipes and have to be evaluated for each segment "*i*"

$$\Delta p_{lin} = \sum_{i=1}^{n} (\Delta p_i \cdot L_i) \tag{4.7}$$

where

- Δp_i is the specific linear pressure loss for the pipe having the inside diameter (d_i) and fluid velocity (v_i); for copper pipes several examples are given in Table 4.9;
- L_i is the length of the pipe with an inside diameter (d_i).

Table 4.9 Specific pressure losses of copper pipes [2]

Flow rate (L/s)	DN10 (12 × 1)		DN12 (15 × 1)		DN15 (18 × 1)		DN20 (22 × 1)		DN25 (28 × 1)	
	v_i (m/s)	Δp_i (mbar/m)	v_i (m/s)	Δp_i (mbar/m)	v_i (m/s)	Δp_i (mbar/m)	v_i (m/s)	Δp_i (mbar/m)	v_i (m/s)	Δp_i (mbar/m)
0.02	0.25	1.4	0.18	0.4	0.10		0.06		0.04	
0.06	0.76	9.0	0.53	2.5	0.30	0.9	0.19	0.5	0.12	
0.10	1.27	21.0	0.88	6.0	0.50	1.9	0.32	0.8	0.20	
0.20	2.55		1.77	20.0	1.00	7.1	0.64	4.0	0.41	1.0
0.40	5.10		3.54		1.99	25.0	1.27	14.0	0.82	3.4

The local pressure losses occur in those places of the solar circuit where the direction or the section of the flow is changed (e.g. elbows, tees, reductions, coil of the storage tank, external heat exchanger). These losses are proportional with the square of the flow velocity and are calculated using

$$\Delta p_{\mathrm{loc}} = 5 \cdot \sum_{i=1}^{n} \left(\xi_i \cdot v_{\mathrm{i}}^2\right) \tag{4.8}$$

where

- v_{i} is the velocity of the fluid through the *i*th element;
- ξ_i is the local pressure loss coefficient for the *i*th element (Table 4.10).

In Eq. (4.8), the coefficient 5 is used to calculate the pressure losses in mbar when the water velocity in pipes is expressed in m/s.

Based on the calculated values of the thermal fluid flow rate (Q) and of the total pressure losses (Δp), the hydraulic pump can be selected using the pressure/flow rate diagram (Fig. 4.12) provided by the manufacturer.

Thus, for an imposed flow rate through the solar circuit (e.g. 2 m^3/h) the hydraulic pump will be selected according to the pressure losses in the entire solar circuit as presented in Table 4.11. The characteristic points of the solar circuit (A–D) must be located below the pressure/flow rate characteristic of the hydraulic pump. If the characteristic point is not in the range of the selected pumps, other type of hydraulic pump must be selected. For an imposed type of hydraulic pump, a higher number of pumps can be installed. These pumps will be serially connected if higher pressure is required. To overcome a larger flow rate, the pumps will be parallelly installed; in this case, it is important to use identical pumps.

Step 9. **Estimation of the expansion vessel capacity**

The expansion vessel must compensate for the volumetric variation of the thermal fluid during its temperature increase/decrease. Thus, when selecting an expansion vessel, its volume (V_{E}) can be evaluated according to the Association of German Engineers [4] using

$$V_{\mathrm{E}} = (V_{\mathrm{C}} \cdot N_{\mathrm{C}} + \beta \cdot V_{\mathrm{T}} + V_{\mathrm{F}}) \cdot D_{\mathrm{F}} \tag{4.9}$$

Table 4.10 Local pressure loss coefficient through usual solar circuit screwed elements [2]

Nominal diameter	Elbow 90°	Elbow 45°	Teeline	Tee branch	Ball valve	Tank inlet	Tank outlet
10	2.5	0.38	0.9	2.7	20	0.5	1.0
15	2.1	0.37	0.9	2.4	14	0.5	1.0
20	1.7	0.35	0.9	2.1	10	0.5	1.0
25	1.5	0.34	0.9	1.8	9	0.5	1.0
32	1.3	0.33	0.9	1.7	8.5	0.5	1.0

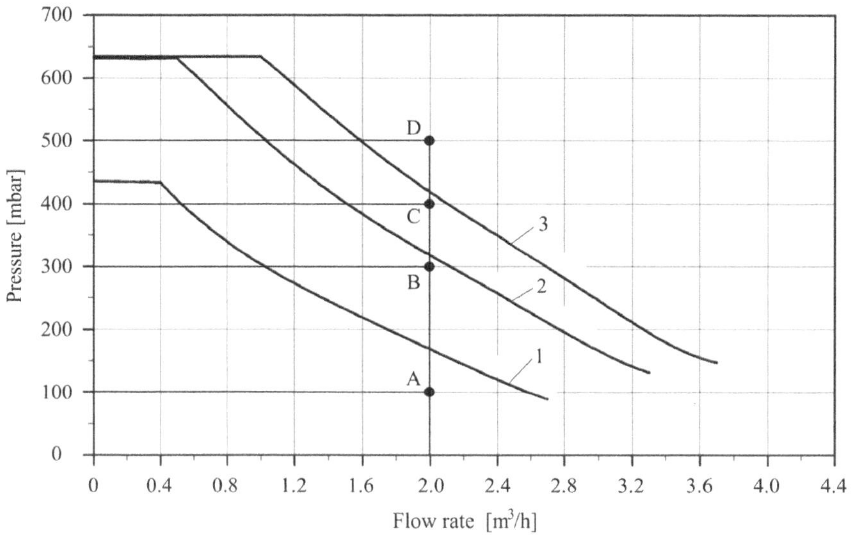

Fig. 4.12 Pressure/flow rate characteristic of three hydraulic pumps: Alpha 1L XX-40 (1), Alpha 1L XX-60 (2) and Alpha 1L XX-65 (3) [29]

Table 4.11 Selection of the hydraulic pump for an imposed flow (2 m^3/h) through the solar circuit

Pressure loss (mbar)	Characteristic point	Selected hydraulic pump
100	A	Alpha 1L XX-40
300	B	Alpha 1L XX-60
400	C	Alpha 1L XX-65
500	D	No solution, other type of hydraulic pump must be selected

where

- V_C is the volume of the thermal fluid in one solar thermal collector;
- N_C is the total number of solar thermal collectors;
- β is the thermal expansion coefficient of the thermal fluid;
- V_T is the total volume of the thermal fluid in the solar circuit;
- V_F is the volume of the thermal fluid in the expansion vessel;
- D_F is a pressure coefficient.

The total volume of the thermal fluid (V_T) is calculated using

$$V_T = V_C \cdot N_C + V_S + V_P \tag{4.10}$$

where

- V_S is the volume of the thermal fluid in the storage tank inferior coil;
- V_p is the volume of the thermal fluid in the pipes of the solar circuit.

The volume of the thermal fluid in the expansion vessel (V_F) should be at least 3 L and is calculated using

$$V_F = 4\% \cdot V_T \tag{4.11}$$

The pressure coefficient D_F can be calculated using

$$D_F = \frac{p_{max} + 1}{p_{max} - p_0} \tag{4.12}$$

where

- p_{max} is the maximum admissible pressure in the solar circuit established based on the safety valve pressure (p_{sv}) using

$$p_{max} = 0.9 \cdot p_{sv} \tag{4.13}$$

- p_0 is the preliminary pressure in the solar circuit calculated based on the static height (h_0) of the solar circuit using

$$p_0 = 1 + 0.1 \cdot h_0 \tag{4.14}$$

4.3 Optimizing the Thermal Output of Solar Thermal Systems by Using Solar Tracking

Flat plate and evacuated tube solar thermal collectors are typical solar energy converters implemented in the built environment for covering the domestic hot water (DHW) and, to some extent, the heating energy demand. The building-integrated flat plate solar thermal collectors (FPSTC) are usually fixed installed, at an optimized tilt angle, on the buildings' sloping roofs, on terraces and/or on facades or nearby the buildings. Although uncomplicated, this approach has the drawback of a lower thermal output, since optimally fixed and tilted convertors can collect only a limited amount (generally, less than 60–70%) of the available solar energy-requiring, thus a larger capacity of solar thermal systems to meet the thermal energy demand. Furthermore, the heat produced during the hot season commonly overcomes the demand and the solar thermal system enters in the stagnation regime that leads to the collector's overheating, thus limiting its durability, if not controlled.

The energy output and the durability of the solar thermal systems using flat plate collectors can be improved by *solar tracking*. *Forward tracking* is a well-known path for increasing the amount of solar energy received on the surface of the conversion equipment [41]. Moreover, *inverse tracking* proved to be a viable option for protecting the collector against overheating [64, 65]. Thus, *solar tracking* supports a better fit

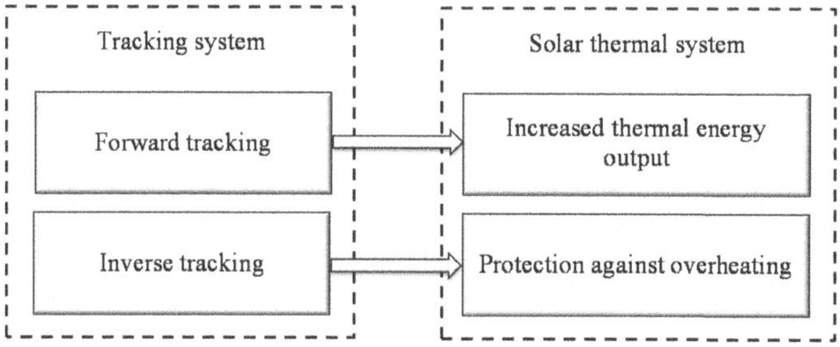

Fig. 4.13 Role of tracking systems for solar thermal collectors

of the thermal output with the thermal energy demand, in the concept "*when needed, in the required amount*".

Thus, a tracked solar thermal system can be operated in two tracking modes as outlined in Fig. 4.13:

- *Forward tracking*, aiming at increasing the thermal energy output, according to the thermal energy demand;
- *Inverse tracking* used to avoid the collector's overheating during *stagnation* and longer periods of *zero heat load*, e.g. during holidays.

Two application types of tracked solar thermal collectors are further considered:

- *Solar thermal collectors installed on the rooftop or near the buildings*, where *larger strokes of the tracking motions* can be exploited;
- *Façade mounted solar thermal collectors*, where only *limited stroke tracking* can be applied.

The analysis of the thermal energy output considering forward and inverse tracked versus the fixed tilted solar thermal collectors during days with different solar radiation profile (sunny, cloudy) is a recommended step for the optimal design and use of the solar thermal systems. A special attention should be paid to the *tracking algorithms* for tracked solar thermal systems that have to optimally combine *forward* and *inverse tracking* according to the thermal energy demand and the available solar radiation.

Forward tracking aims at increasing the amount of solar radiation received on the solar thermal surface by decreasing the incidence angle (ν) between the sunray and the normal line ($\vec{n}$) on the FPSTC surface (Fig. 4.14) while inverse tracking maintains the FPSTC temperature close to a preset maximum value or minimizes its temperature during holidays.

Such tracking algorithms require infield data of the solar irradiance, of the inlet and outlet temperatures of the thermal fluid, of the outdoor air temperature and of the storage tank temperature [64, 91]. The forward and inverse tracking can be employed using specific *tracking programs*:

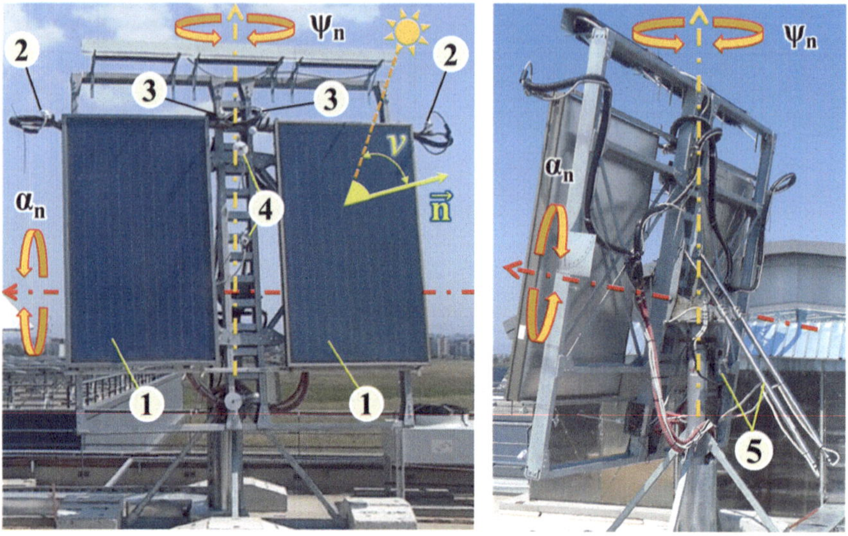

Fig. 4.14 Experimental outdoor testing rig (Brasov, Romania) of a dual-axis tracking system with two FPSTCs (1), inlet temperature sensors (2), outlet temperature sensors (3), pyranometers (4), actuators (5)

- The *forward tracking program* is adapted to normal operation periods, when the water temperature in the storage tank is lower than the maximum allowed (or required) value, and the direct solar radiation represents the largest share in the global solar radiation. This program increases the amount of solar radiation received on the FPSTC surface by using an optimized, stepwise tracking program [7]. During fully cloudy periods of the day, the solar thermal collector is brought to the horizontal position (if allowed) or in the closest position to horizontal;
- The *inverse tracking program* is used during *stagnation* periods, when the pumping system is turned off as a result of water temperature in the storage tank that reached the maximum allowed value; the program is applied to increase the incidence angle aiming at keeping constant the collector's temperature, close to its maximum allowed value. Consequently, the solar thermal system supports a faster heat delivery when the thermal energy demand increases. During holidays, when zero thermal energy load occurs over a long period, the collector is kept in the extreme West position from sunrise to noon, and in the extreme East position during the afternoon, thus protecting the collector against (over)heating.

4.3.1 *Increasing the Thermal Energy Output by Forward Tracking*

Various tracking solutions are reported in the literature, based on preset tracking algorithms mainly using single-axis tracking systems [19] or gravity-based tracking systems [20]. Yadav and Chandel [104] reported a thermal output increase of 35% for tracked solar thermal collectors as compared with the output of identical collectors installed fixed, facing South at an optimized tilt angle.

The effect of solar tracking on the collector's thermal output was also investigated using the outdoor testing rig installed in the Transilvania University of Brasov, Romania, as presented by Neagoe et al. [65]. This testing rig (Fig. 4.14) consists of two FPSTCs connected to a heat storage tank by independent hydraulic circuits, using a 40% ethylene glycol aqueous solution as thermal fluid, and a dual-axis tracking system.

The dual-axis tracking system of azimuth type is actuated by a rotary actuator for the diurnal motion (Ψ_n) and by linear actuators for the elevation motion (α_n). It has large angular strokes, with the diurnal angle (Ψ_n) ranging between $-100°$ and $+100°$ and the altitude angle $\alpha_n = 10°–90°$. The testing rig allows recording the collectors' inlet and outlet temperatures, the outdoor air and the storage tank temperatures, as well as the global solar irradiance measured in the collectors' plane.

Experimental testing of the tracked versus the fixed tilted FPSTC was done during several representative days (clear and mostly cloudy sky days) during September and October 2014, with the thermal load for DHW heating, stored in a 300 L tank. Each collector has the absorber area of 2.2 m^2 and the nominal efficiency of 80.5%. During the thermal load, the thermal fluid was continuously pumped at a constant flow of 0.2 m^3/h.

The experimental data recorded during the sunny (Fig. 4.15) and the cloudy (Fig. 4.16) days allow formulating the following conclusions:

- The hourly, stepwise, dual-axis tracking program (Fig. 4.15a) allows increasing the thermal energy production with 47.5% (Fig. 4.15b), compared to an optimally fixed and tilted system at 45° (Fig. 4.15c), as result of a larger solar energy amount received by the forward tracked collector [8.7 kWh/(m^2 day)] as compared with the fixed tilted collector [6.6 kWh/(m^2 day)]. The daily thermal energy produced by the tracked collector was about 5.9 kWh/(m^2 day), while the fixed collector delivered about 4 kWh/(m^2 day);
- During cloudy days (Fig. 4.16), the global solar energy received is higher for the FPSTC horizontally fixed as compared to the tracked FPSTC, i.e. 1.5 versus 0.77 kWh/(m^2 day). As a result, the thermal energy produced by the horizontal FPSTC is higher than that of the tracked one, i.e. 1.1 versus 0.55 kWh/(m^2 day). It is therefore recommended to horizontally position the collectors during cloudy days, when the solar thermal system increases the heat output and reduces the electrical energy amount required to power the actuators. Moreover, this position better protects the system against the wind that usually blows during cloudy weather;

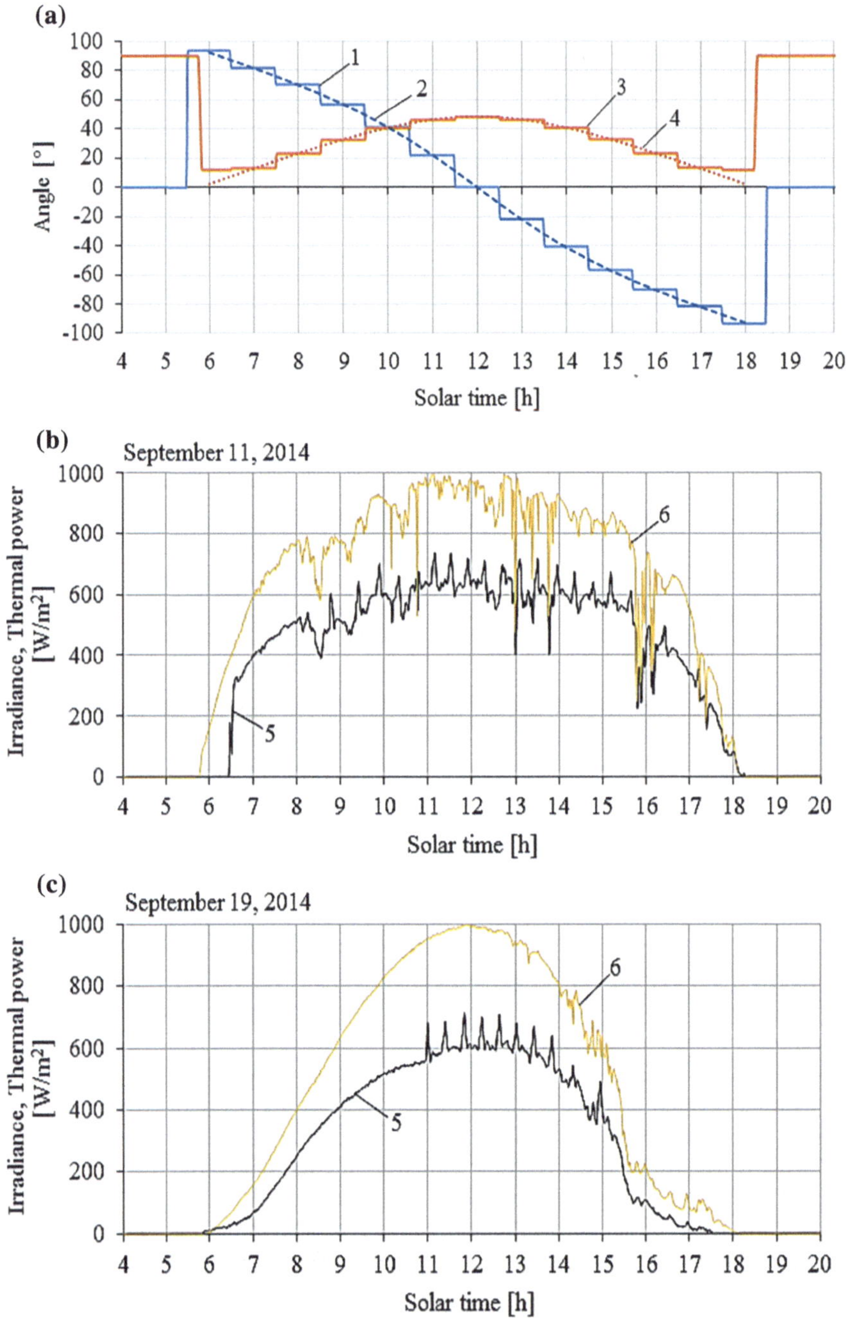

Fig. 4.15 Forward tracked versus fixed tilted FPSTC experimental results (Brasov, Romania): **a** tracking (1) and solar (2) azimuth angles, tracking (3) and solar (4) altitude angles, thermal power (5) and received global solar irradiance (6) for: **b** tracked FPSTC; **c** fixed tilted FPSTC at 45°

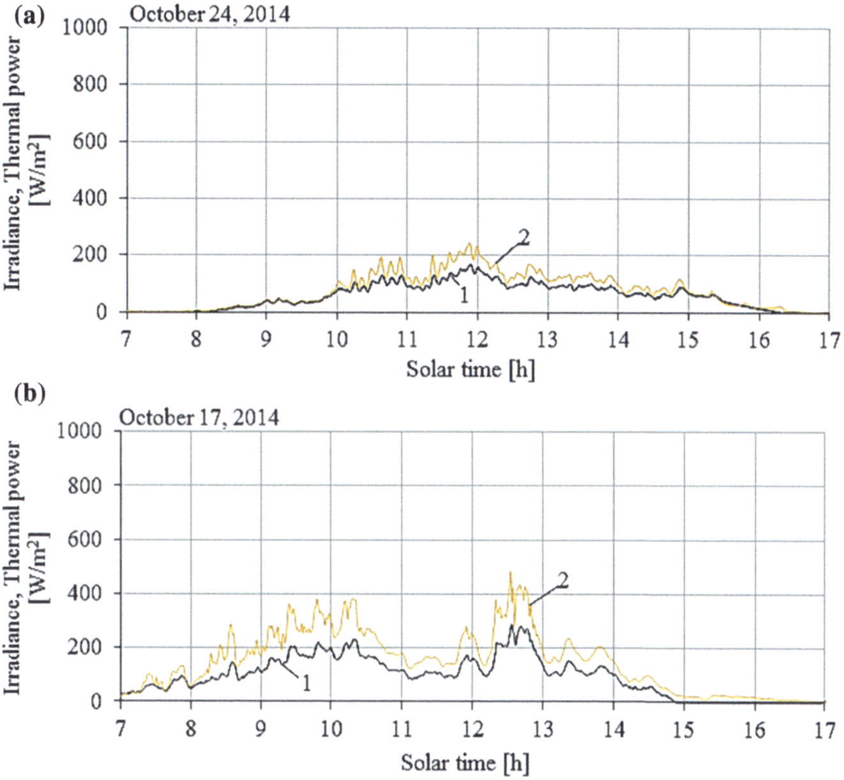

Fig. 4.16 Thermal power (1) and global solar irradiance received by the FPSTC (2) during cloudy sky days using: **a** dual-axis forward tracked FPSTC; **b** horizontally fixed FPSTC (Brasov, Romania)

- The average daily conversion efficiency of the global solar energy into thermal energy is about 60–68% during sunny days and about 71–73% on cloudy days.

Thus, forward tracking with high diurnal and elevation angular strokes allows collecting higher amounts of the available solar energy, and consequently, a significant heat gain can be delivered, on demand, during sunny or even during mixed (sunny–cloudy) days. Moreover, the tracking systems support the optimal collection of the diffuse solar radiation during cloudy days by (close to) horizontally positioning the FPSTCs.

4.3.2 Protection Against Overheating by Inverse Tracking

Overheating the solar thermal collectors may occur during stagnation periods and lead to increased temperatures in and on the collector, higher than the designed critical value.

The flat plate solar thermal collectors enter stagnation when the heat carrier no longer removes the output thermal energy, i.e. no thermal fluid flow is carried out from the collector [34, 98].

The solar circuit pump may be out of work for various reasons, in the normal operation mode when the maximum allowed temperature of the heat storage tank is reached, as well as in abnormal situations as pump failure, power supply blackout, failure of other equipment components, etc. [49, 84]. The collector is overheated if the temperatures on the absorber plate and of the thermal fluid exceed the threshold values, thus their degradation may occur [25]. The main negative impacts of overheating are:

- Permanent deformation of the thin absorber plate under thermal expansion [32] that results in micro-cracks in the ceramic coating [64] on the absorber plate, thus decreasing the collector's efficiency and lifetime;
- Chemical decomposition of the antifreeze thermal agent into corrosive substances (acids) that may damage the solar circuit metallic components [33]. Typically, the common antifreeze agent degradation starts at 120 °C and is complete and fast at 220 °C [32, 84];
- Vapour bubbles and water hammer effect, caused by the evaporated and overheated thermal fluid, can damage the system's components [32, 49];
- Environmental pollution by the thermal fluid vapours released into the atmosphere [83].

Technical solutions for avoiding overheating or at least limiting its effects are reported in the literature and are commonly used in practice:

- Reducing the solar radiation received on the absorber plate by:
  - *Covering or shading the collector*. Manual coverage with a protective casing or automatic systems to partially or totally cover the collector can be used [32]. The total coverage can block about 90% of the incident solar radiation [73];
  - Covering the collector glazing with *thermotropic materials*. The thermotropic materials can significantly modify their transmittance and reflectance at high temperature [27, 61], e.g. from 86% at 30 °C down to 52% at 110 °C as stated by Resch and Wallner [76];
  - Introducing *prismatic polycarbonate structures* between the glazing and the absorber plate that can reflect, back into the atmosphere, over 90% of the received solar radiation. During the normal operation mode, a switching fluid with a suitable refraction index flows into and around the prismatic structure redirecting the light towards the absorber plate [80];
- *Solar collector ventilation*, with additional cost for cooling the absorber plate during stagnation [32];
- *Removing the thermal fluid to a storage tank* using a drain-back solar system when the pump is turned off [25]. In the normal operation mode, the pump pushes the fluid back to the collector in a non-pressurized closed circuit;

- *Limiting the thermal fluid temperature* at a designed value using self-limiting aluminium heat pipes [39];
- *Heat transfer to the secondary thermal load* by using a heat dump package to remove the excess heat stocked in the thermal fluid during stagnation and to transfer it to a secondary thermal load (e.g. a pool) or even to release it in the atmosphere [25].

Most of these solutions are applied only after the heat was generated, and thus, the thermal stress is still active on the collector. Neagoe et al. [64] proposed a new approach of the *inverse tracking* that allows decreasing the solar radiation received by the collector during stagnation, thus limiting the heat production and the increase of the collector's temperature.

The collector's temperature can be kept at or close to a specific value by using an *adaptive inverse tracking program*, or can be minimized by a *maximum inverse tracking program*, as outlined in Fig. 4.17.

The *adaptive inverse tracking program* can be successfully applied to maintain an almost constant temperature in the collector by appropriately changing the incidence angle, depending on the variation in the ambient air temperature, solar angles and/or solar irradiance. Considering these, experimental testing was done during a sunny day, with the collector in stagnation, aiming at keeping the collector temperature close to 85 °C, by changing the diurnal solar radiation angle (ψ_n) while the elevation angle remains constant ($\alpha_n = 45°$), using the testing rig in Fig. 4.14.

According to Fig. 4.17a, the collector was exposed to the Sun during the morning by using accurate dual-axis forward tracking up to 9:45, followed by adaptive inverse tracking up to about 12:00. The collector's temperature can be maintained, with slight fluctuations, close to the preset stagnation temperature, by imposing an azimuth angle with negative values, i.e. not following the sunray (an incidence angle of 60°–70°). As a result, by adaptive inverse tracking the temperature of the collector in stagnation can be well controlled and kept lower than the admissible temperature, at a constant preset value.

The experimental data on *maximum inverse tracking program* are presented in Fig. 4.17b. The tracking system was maintained in the far West stationary position during the morning ($\Psi_n = -100°$), followed at solar noon by a short transition (about 2 min) to the far East stationary position ($\Psi_n = +100°$). The thermal fluid temperature remains close to the ambient temperature (up to 13 °C difference), except the transition time at noon when the difference increases to about 20–23 °C.

These experimental results on inverse tracking allow formulating the following conclusions:

- *Inverse tracking* is able to significantly reduce the amount of solar energy collected by the FPSTC—the main factor that causes fast variations in the thermal fluid temperature during stagnation;
- *Adaptive inverse tracking* represents a good option to keep the collector's temperature close to a preset value during (partial) heat load, thus being ready to deliver thermal energy when a new thermal energy demand occurs;

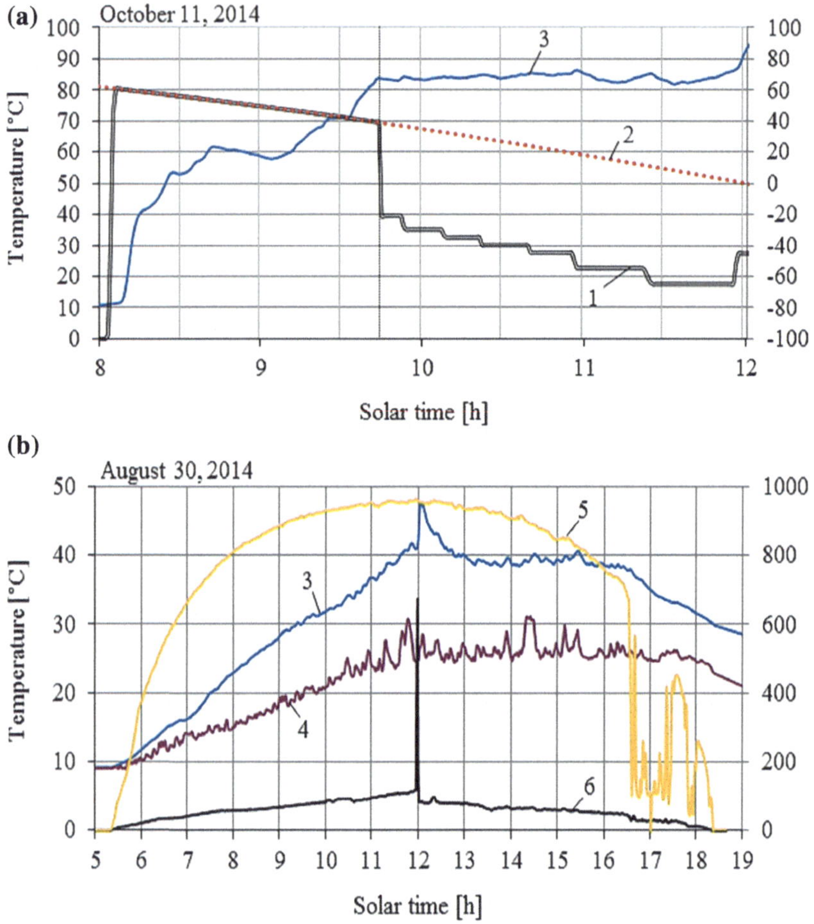

Fig. 4.17 Experimental results on inverse tracking during stagnation in clear sky days (case study: Brasov, Romania): **a** adaptive inverse tracking with variable stepwise azimuth angle Ψ_n (1), along with the solar azimuth angle Ψ (2) and the collector temperature (3); **b** maximum inverse tracking: the collector temperature (3) and the outdoor air temperature (4), the available (5) and the received (6) global solar irradiance

- *Maximum inverse tracking* is able to reduce, close to zero, the direct solar radiation and to significantly decrease the diffuse solar radiation that reaches the collector. It allows the collector's temperature to closely follow the ambient air temperature (with differences of about +10 to +15 °C); thus, it proves to be an effective path for reducing the overheating risk and to minimize the thermal stress on the collector during long stagnation periods (e.g. during holidays, when the owners leave their residence).

The solar tracking algorithm also supports a successive combination of forward and inverse tracking during one day, in direct relation with the time evolution of the thermal energy demand and of the available solar radiation. In a common use, forward tracking is applied during the morning hours to comply as much as possible with the required thermal energy demand, followed by adaptive inverse tracking when the collector reaches its maximum temperature. This combination of forward and inverse tracking can be repeated during the day, whenever required. The maximum inverse tracking can be also used if no more thermal energy is required, to protect the collector against overheating.

4.3.3 Tracking Solar Thermal Collectors Applied on the Buildings' Facades

The global quest for nearly Zero Energy Buildings or for energy-positive buildings requires an efficient use of the limited available space in the built environment to install renewable energy systems. Thus, installing FPSTCs on suitably oriented façades (South-facing or with small deviations from the South, in the northern hemisphere) represents an affordable option to fully or at least partially cover the thermal energy demand (e.g. for DHW) by converting the solar energy.

Installing FPSTCs on facades requires to carefully consider during the design process the significant reduction of the solar energy received on vertical surfaces, especially during summer and, consequently, the reduced amount of thermal energy that can be produced (as compared to the solar thermal systems with optimal fixed tilt angle). One way, still not enough explored, to mitigate this drawback refers to the *limited stroke tracking* for collectors mounted on the buildings facades. Theoretical studies and experimental investigations are recommended, considering the following assumptions:

- The use of *single-axis tracking systems* with diurnal (*vertical axis*) or elevation (*horizontal axis*) motion, with *limited strokes* (max. 30°, i.e. ±15°) imposed by the narrow space (usually less than 200–300 mm) between the collectors and the facade. Dual-axis trackers can be also used, but the difficulties in implementing them on the facade and in operating complex two-axis systems have to be considered in the design process;
- A single string of FPSTCs, vertically or horizontally tracked, is mounted on the façade. When several parallel strings are considered, mutual shading may occur and optimized tracking algorithms have to be implemented and/or the distance between the strings has to be optimized;
- The most favourable implementation option, on South-facing facades for buildings located in the northern hemisphere, is considered. For the southern hemisphere, the problems can be similarly approached, but for North-facing facades;
- Only clear sky days are considered, when the tracking effect is expected to be significant, and the diffuse component of the global solar radiation can be neglected. The maximum value of the diffuse solar radiation is observed in the horizontal plane; thus, diffuse radiation becomes even lower (about half) on vertical surfaces;

therefore, the received solar radiation on a façade can be well approximated only considering the direct solar radiation;
- The summer season is the most significant time period over the year considering the solar energy input and the need for protecting the collector during stagnation.

The forward (further denoted by Max) and the inverse (Min) tracking investigations, using modelled versus experimental results, are addressed considering the Brasov location, Romania. The estimation of the available direct solar radiation is based on the model proposed by Meliss [52], as described in Chap. 2, where the turbidity factor (T_R) has monthly values extracted from the global solar database SoDa [81].

Even if only reduced angular strokes are considered in this case, a dual-axis tracked system mounted on the vertical facade can bring a significant variation in the amount of the received solar energy, compared to a vertical fixed surface (symbolized by V), as presented in Fig. 4.18:

- In June, characterized by the maximum available solar radiation and maximum solar angular ranges, the solar energy collection time period increases or decreases with about 4–5 h during the day through forward, respectively, inverse tracking, as compared to the time period of the sunray falling on the fixed vertical surface, Fig. 4.18a;
- When using *forward tracking,* a significant increase of the received solar energy during the warm season leads to a rather uniform amount of solar energy received over the entire year (between ~4400 and 6400 Wh/(m^2 day), Fig. 4.18b; a maximum increase of ~145% is reached in June, i.e. 4830 versus 1970 Wh/(m^2 day), and over 110% increase is obtained during all the summer months (June, July and August);
- The *inverse tracking* has a significant impact on the solar energy amount collected by the FPSTC: a decrease below 200 Wh/(m^2 day) in June and less than 800 Wh/(m^2 day) in the other summer months (Fig. 4.18b) could be observed. Thus, inverse tracking proves highly efficient during the warmer period of May–August.

Despite these advantages, dual-axis tracking on the facades has the drawback of a complex tracking system that includes a dual-axis movable mechanical structure and a more complex driving and control system. Therefore, a more rational solution is the implementation of single-axis tracking systems that perform controlled motions on:

- *The horizontal axis* (H-axis), by changing only the altitude angle ($\alpha_n = -15°$ to $+15°$ while $\psi_n = 0°$);

  or on:
- *The vertical axis* (V-axis), by changing only the azimuth angle ($\psi_n = -15$ to $+15°$ while $\alpha_n = 0°$).

The results in Fig. 4.19 show that the H-axis system has a more important contribution than the V-axis system during the warmer season (April to August) during both, forward and inverse, tracking:

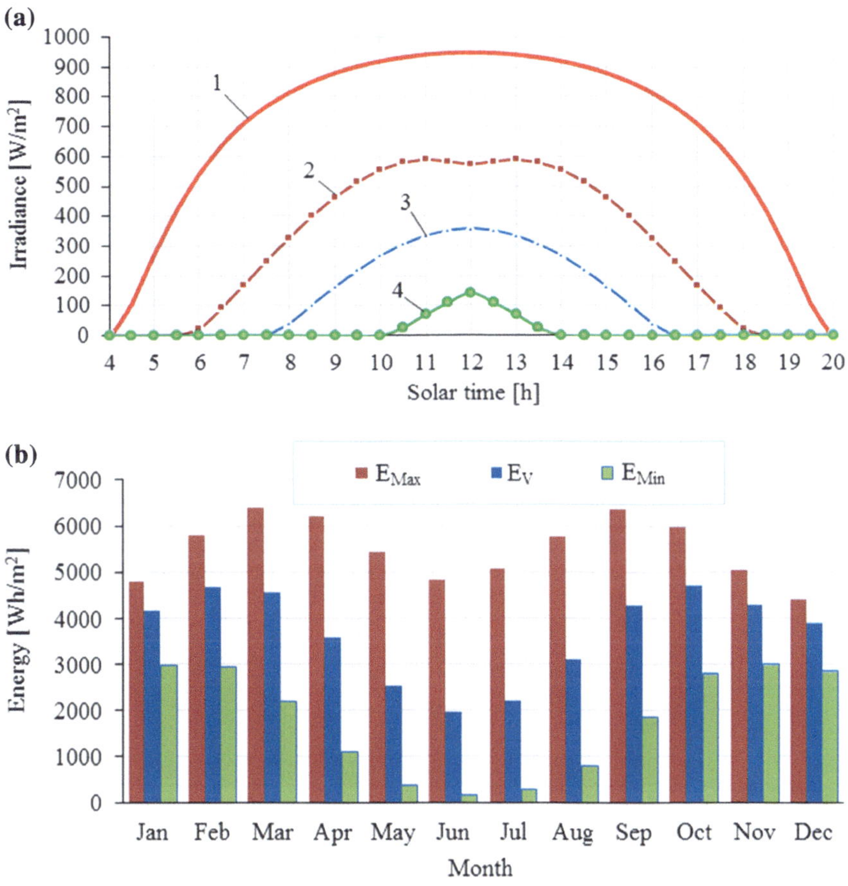

Fig. 4.18 Modelled direct solar radiation collected by the dual-axis tracked versus fixed vertical FPSTCs in Brasov, Romania: **a** the available (1) and the received direct solar radiation in June during forward tracking (2) and maximum inverse tracking (4) versus a fixed, vertical, South-facing façade (3); **b** the daily direct solar energy received by forward (E_{Max}) and maximum inverse (E_{Min}) tracked versus vertical (E_{V}) FPSTCs

- The highest increase in the received solar energy by the tracked FPSTC (compared to the vertically fixed collector) through forward tracking is recorded in June, when $E_{\mathrm{V}} = 1970$ Wh/(m^2 day), and reaches 83% for the *H*-axis ($E_{\max} = 3600$ Wh/(m^2 day), Fig. 4.19a) and 55% for the *V*-axis system ($E_{\max} = 3060$ Wh/(m^2 day), Fig. 4.19b). Significant amounts of the solar energy collected by the *H*-axis versus *V*-axis systems are also observed in May and in July, while during the rest of the year both systems lead to almost similar results;
- The best performance of the maximum inverse tracking is reached in June (450 Wh/(m^2 day) for *H*-axis versus 1200 Wh/(m^2 day) for *V*-axis). The superiority of the *H*-axis tracking is also remarkable in April, May, July and August.

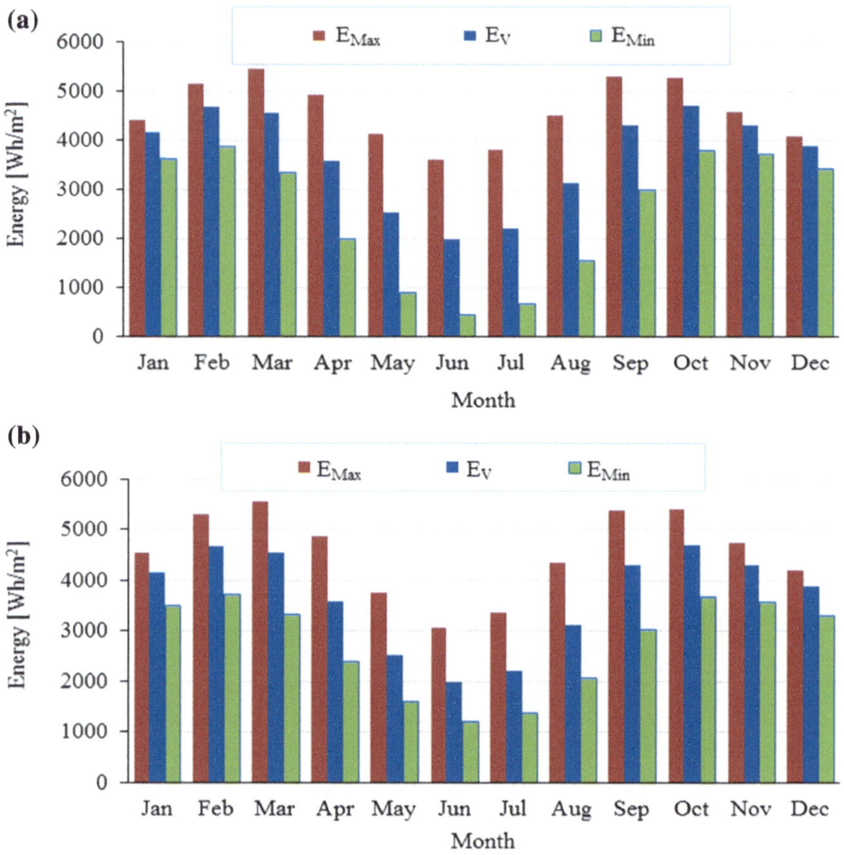

Fig. 4.19 Single-axis tracking for façade integrated solar thermal collectors installed in Brasov, Romania: monthly variation of the direct solar energy received by forward (E_{Max}) and by the maximum inverse (E_{Min}) tracked versus fixed vertical (E_{V}) FPSTCs using: **a** *H*-axis tracking; **b** *V*-axis tracking

As a result, the *H*-axis system leads to better results over the entire year compared to the *V*-axis system in both forward and inverse tracking. Thus, it is recommended to organize the FPSTCs in horizontal strings, tracked using *H*-axis systems with limited stroke. Avoiding reciprocal shading among the strings has to be well considered when designing the distance between multiple strings. Moreover, tracking during the winter season has little impact on the increase in the received solar energy, and thus, the FPSTCs can be kept in a fixed vertical position during this season.

The validation of the modelling results was recorded during experimentally testing the effects of limited tracking on the thermal energy production. The experimental testing rig (Fig. 4.20) is outdoor installed on the terrace of a laboratory building in the R&D Institute of the Transilvania University of Brasov. The *H*-axis tracking was used in testing the forward and inverse tracking programs, implemented using a

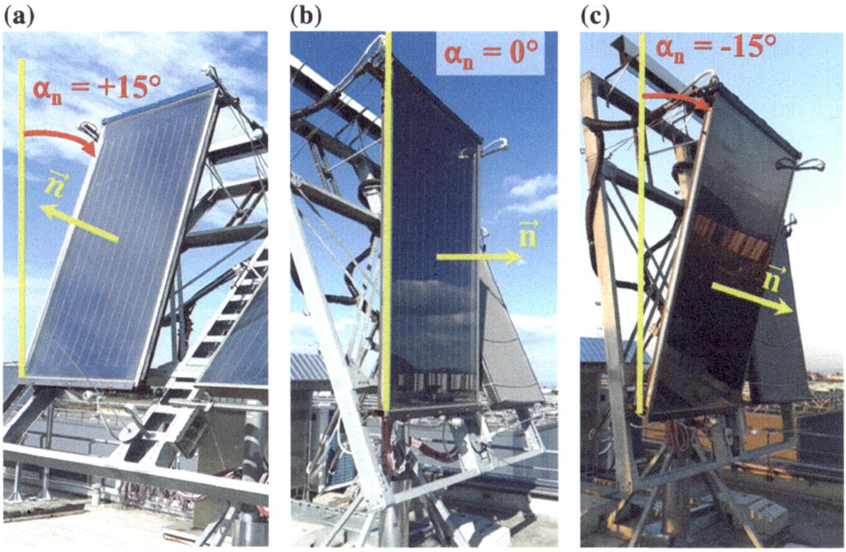

Fig. 4.20 Outdoor dual-axis tracking rig for testing FPSTCs by: **a** forward tracking; **b** fixed vertical; **c** inverse tracking

single-step approach for both forward ($\alpha_n = 0° \rightarrow +15°$) and inverse ($\alpha_n = 0° \rightarrow -15°$) tracking.

The experimental tests were developed during clear sky days in September 2016, with almost similar values of the available solar energy and outdoor air temperature. The vertically fixed (Fig. 4.20b) and the tracked (Fig. 4.20a, c) FPSTCs were considered as follows:

- *Forward tracking with limited stroke* compared to the vertical fixed position (Fig. 4.21a) allows producing about 9% more thermal energy (heat) than the vertically mounted collector, in Fig. 4.21b (i.e. 2660 versus 2440 Wh/(m^2 day), while about 27% more solar energy [5120 versus 4030 Wh/(m^2 day)] is received;
- *Maximum inverse tracking* during stagnation throughout the day, Fig. 4.21c: the collector temperature is maintained in an acceptable range (<100 °C), below the critical temperature of the thermal fluid.

In conclusion, the limited stroke *H*-axis tracking of facade integrated FPSTCs allows, according to the modelled and experimental results, both the increase of the amount of heat generated by forward tracking and the overheating protection in stagnation by inverse tracking.

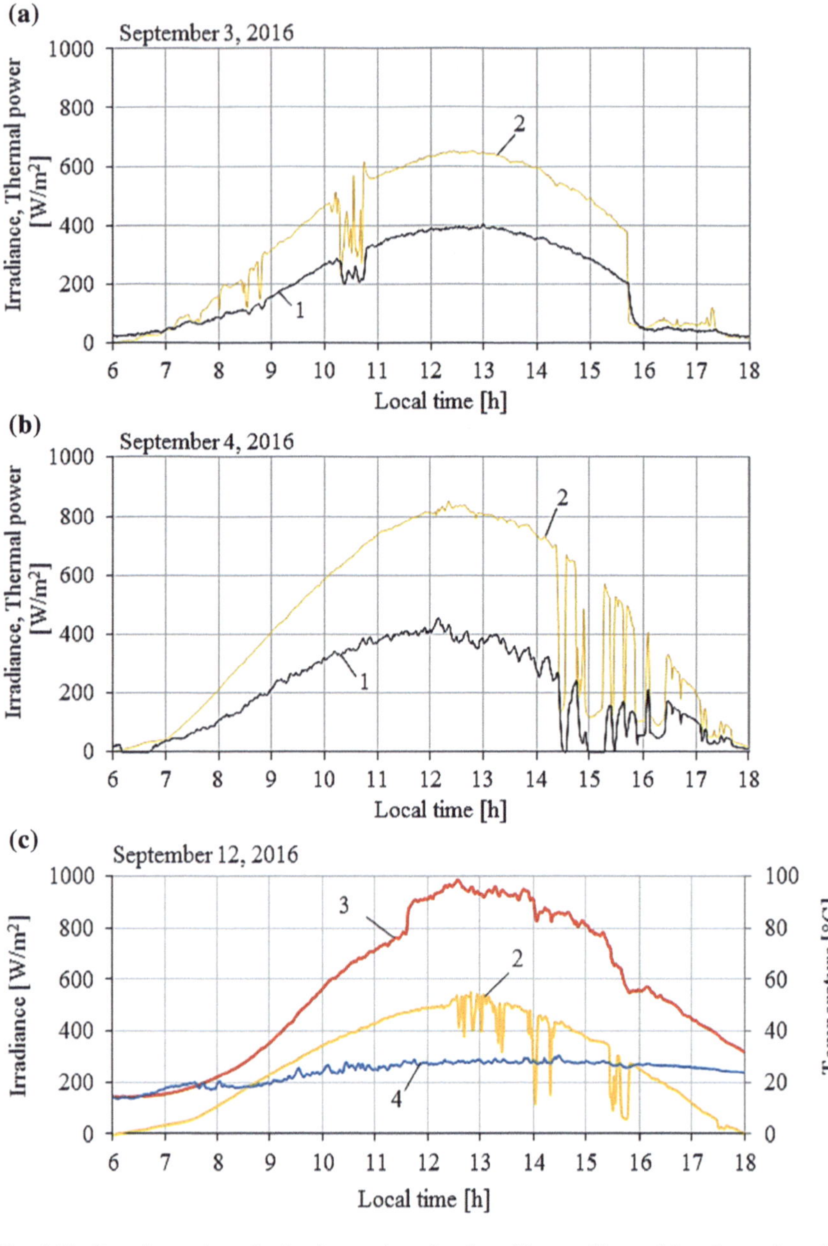

Fig. 4.21 Experimental results in almost clear sky days (Brasov, Romania): **a** forward tracked FPSTC; **b** fixed vertical FPSTC; **c** maximum inverse tracked FPSTC during stagnation: the thermal power (1), the received global solar irradiance (2), the collector temperature (3) and the outdoor air temperature (4)

4.4 Increasing the Solar Energy Share in Meeting the Thermal Energy Demand of a Building Through Solar Thermal Facades

The surface of a single-family building's roof is large enough to install the solar thermal collectors required to feasibly meet the thermal energy demand for DHW preparation. This is also valid for multifamily buildings up to a certain number of floors [56]. However, for very tall multifamily buildings (with a small ratio of the roof area reported to the number of families) or for the buildings with a rooftop used for other different purposes (e.g. for installing communication equipment, green roofs) or for buildings where the solar energy is also considered for space heating and cooling (thus requiring a larger number of solar thermal collectors), the building's facades represent good candidates to install solar thermal collectors in arrays known as *solar thermal facades*. The main barriers in implementing solar thermal facades are represented by the size and the shape of the available surfaces on the suitably oriented facades and by the architectural acceptance [63]. The size of the suitable surfaces on the building's facades could be smaller than the dimensions of the common solar thermal collectors (2 m × 1 m); thus, only a reduced number of solar thermal collectors may be installed [95]. An example of a three floors multifamily building facade is presented in Fig. 4.22. Due to the narrow space available between the windows, only 15 rectangular solar thermal collectors can be installed with an overall solar thermal surface of 30 m^2.

Moreover, the shape of the vertical surface could be non-rectangular, further reducing the number of commercial solar thermal collectors that can be implemented [94]. An example of such a facade is presented in Fig. 4.23, where only 3 rectangular solar

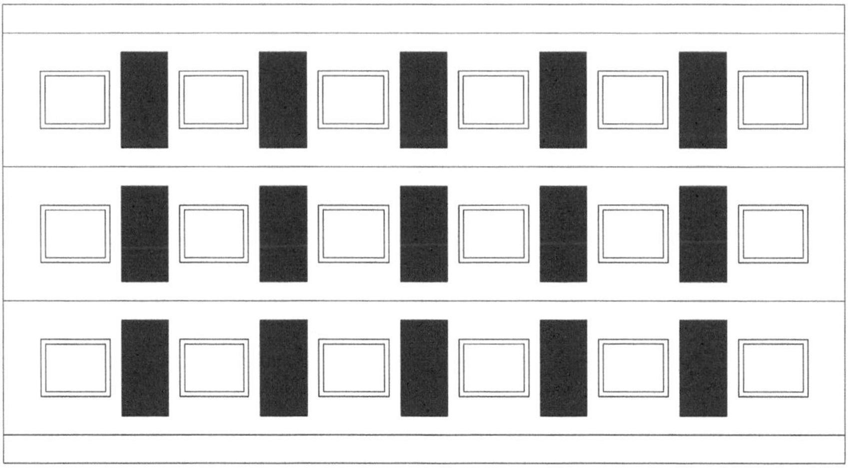

Fig. 4.22 Rectangular commercial collectors implemented on a multifamily building facade

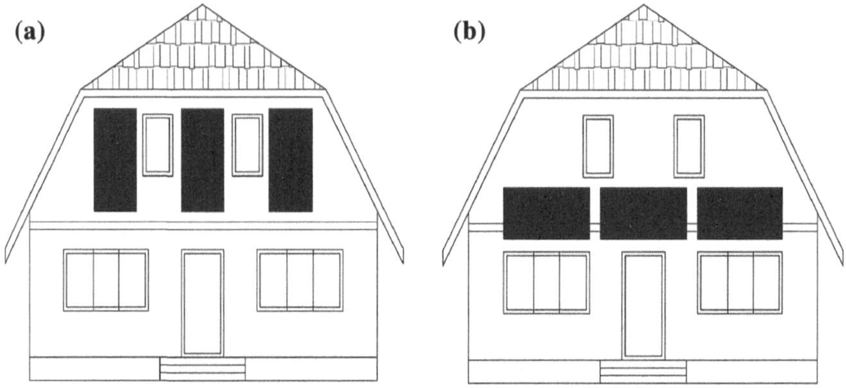

Fig. 4.23 Possibilities to implement rectangular collectors on a single-family building facade: **a** vertical position; **b** horizontal position

thermal collectors (6 m^2) may be vertically (Fig. 4.23a) or horizontally (Fig. 4.23b) installed.

To increase the coverage factor of the solar thermal facade (defined as the ratio between the surface of the installed solar thermal collectors and the available surface of the facade), and therefore its thermal energy output, the size and the shape of the solar thermal collectors have to be adapted to the available surfaces. One alternative consists of tailored made solar thermal collectors, according to each building's facade; however, this is not a cost-effective solution. Another alternative relies on considerably smaller solar thermal collectors with non-rectangular shapes allowing to increase the coverage factor of the facades with smaller sized available surfaces.

The concept of novel solar thermal collectors with non-rectangular shapes (e.g. isosceles trapeze or equilateral triangle) and small dimensions (0.67 m^2 and 0.22 m^2 respectively) was formulated by Visa et al. [87]. Various architectural integration possibilities offered by the triangle and trapeze shapes [92] promote the concept of novel, Lego-type solar thermal facades that support a better use of the available implementation surface, with increased coverage factor. Thus, for the multifamily buildings, the surface of the solar thermal facade covered with solar thermal collectors may increase up to 84.42 m^2 by installing 126 trapeze solar thermal collectors (Fig. 4.24), increasing the coverage factor with 181.4% while allowing the designers to enhance the architectural acceptance.

A higher increase of the coverage factor can be reached when considering a single-family house by using 22 trapeze solar thermal collectors (Fig. 4.25a) or a combination of 16 trapeze and 18 triangle solar thermal collectors (Fig. 4.25b) with a total surface of 14.74 m^2 and 14.68 m^2, respectively corresponding to coverage factors increased with 145.7% and 144.7%, respectively.

Further on, the architectural acceptance has to be addressed. Several criteria were formulated for the architectural acceptance of building-integrated solar thermal systems [62]. The most influential drawback is the dark colour (black or dark blue) of

Fig. 4.24 Novel trapeze solar thermal collectors implemented on a multifamily building facade

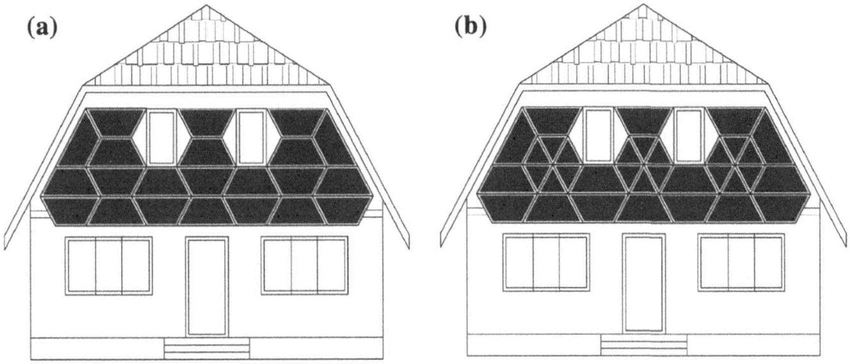

Fig. 4.25 Single-family building facade with: **a** novel trapeze solar thermal collectors; **b** combination of trapeze and triangle solar thermal collectors

the commercial solar thermal collectors as the dark colours are not usual choices for the building's facades and lead to a hardly accepted monotony.

A survey of urban facades showed that there is a need for several different colours to satisfy the architects' and designers' demands, with the preferred chromatic mainly including blue and red, followed by yellow and green. This is why several studies were developed on coloured solar thermal coatings for absorber plates and/or glazing, and promising results were reported [102].

Wu et al. [103] designed a multilayered structure based on metallic titanium (Ti) and dielectric aluminium nitride (AlN) deposited by magnetron sputtering to obtain five coatings with black, purple, yellowish-green, red and yellowish-orange colours.

Table 4.12 Solar absorptance and thermal emittance of selective coatings applied on the absorber plates of the solar thermal collectors

Colour	Absorptance/emittance				
	Wu et al. [103]	Dan et al. [11]	Zhu and Zhao [106]	Yang et al. [105]	Duta et al. [14]
Black	0.92/0.05		0.91/0.06		
Purple	0.94/0.05			0.956/0.085	
Blue			0.86/0.08	0.963/0.081	
Blue-green				0.957/0.086	
Blue sky		0.948/0.08			
Red	0.82/0.27		0.86/0.04		0.62/0.05
Dark red			0.95/0.09		
Yellowish-orange	0.92/0.06				
Orange-yellow				0.95/0.089	
Yellowish		0.902/0.08			
Yellow			0.81/0.06		
Yellow-green				0.951/0.078	0.55/0.23
Yellowish-green	0.88/0.26				
Green			0.92/0.08		
Dark green					0.95/0.36
Grey		0.753/0.08			

The results show that the solar absorptance ranges between 0.82 and 0.94 and the thermal emittance between 0.05 and 0.27 (Table 4.12).

A similar technique was applied by Dan et al. [11] to obtain grey, yellowish and sky blue selective coatings on a stainless steel substrate (Table 4.12) using tungsten (W), aluminium nitride (AlN), aluminium oxynitride (ALON) and aluminium oxide (Al_2O_3). The absorptance/emittance properties were improved starting from 0.387/0.12, corresponding to the stainless steel substrate, to 0.948/0.08 obtained for the blue sky colour resulted through the successive deposition of the layers.

A variety of colours (yellow, blue and green) were obtained by Zhu and Zhao [106] using DC magnetron sputtering, by controlling the film thickness of titanium aluminium nitride and the number of layers. Solar absorptance values higher than 0.80 and thermal emittance values lower than 0.08 were obtained (Table 4.12) with the best results reached for green selective coatings (absorptance: 0.92 and thermal emittance: 0.08).

Yang et al. [105] obtained purple, blue, blue-green, yellow-green and orange-yellow selective coatings by depositing $Cu/Ti/SiO_2/Ti/TiO_2/SiO_2$ through magnetron sputtering. These coatings exhibit solar absorptance higher than 0.95 and thermal emittance lower than 0.09 (Table 4.12).

Sol–gel synthesis followed by cold spray deposition was used by Duta et al. [14] to obtain red, yellow-green and dark green selective coatings using iron and vanadium oxides, copper monosulfide and gold nanoparticles. The highest solar absorptance was obtained for dark green, Cu_xS-based thin films but their emittance was as high as 0.36. The best performing coating considering the optical parameters (absorptance, emittance and their ratio—the spectral selectivity S) corresponds to the cermet structures of Fe_2O_3 and Au nanoparticles embedded in an aluminium oxide porous matrix ($S = 12.4$) or to a similar solar thermal coating additionally containing an exterior TiO_2 protection layer ($S = 9.29$).

Colouring the glazing less affects the properties of the absorber plate but this solution limits the access of the solar radiation into the solar thermal collector. Usually, a deposition technique of multilayered interference filters on the collector glazing is employed to produce a coloured reflection, hiding the black (ideal) absorber plate, while transmitting the non-reflected radiation entirely to the absorber. Mertin et al. [53] energetically optimized coatings for the solar thermal collector's glazing, produced, by reactive magnetron sputtering, in a variety of colours: green, green-blue, blue and orange filters. The coloured collector glasses exhibit a solar energy loss ranging between 0.4 and 4.5% at normal incidence with a VIS reflectance higher than 10.8% and a solar transmittance ranging between 88.9 and 91.4% (in comparison with 91.8% of the uncoated glass). Pana et al. [67] obtained a yellow-green reflecting coating for the solar thermal collector glazing through the SiO_2–TiO_2 multilayer deposition using RF magnetron sputtering. The coloured glazing showed an integrated solar transmittance of 80.3% and a VIS reflectance with a peak of 54% centred at 574 nm corresponding to the yellow-green colour. Schuler et al. [78] investigated the potential of quarter wave stacks in tuning the layers thicknesses, the peak position and thus the colour of the reflection. Using this type of coloured glazing, the absorber coating on the heating plate can have dark colours, deposited using up scalable techniques. Following this idea, Dudita et al. [12] reported absorber coatings with very high spectral selectivity ($S > 30$) obtained by the spray pyrolysis deposition of Cu_xO and NiO dark layers from an aqueous—alcoholic precursor systems.

Further on, the full architectural integration must be addressed considering the mounting of the solar thermal collectors in various positions. The newly proposed shapes (triangle and trapeze) for the solar thermal collector offer a good flexibility in obtaining planar or spatial configurations as presented by Comsit et al. [9]. This is possible due to the modularity of the units (e.g. equilateral triangle) allowing to develop various other polygonal shapes such as diamond, trapeze and hexagon (Fig. 4.26), useful for covering small- and medium-sized facades.

Fig. 4.26 Polygonal shapes obtained based on the equilateral triangle solar thermal collector

Fig. 4.27 Possible combinations between triangle and trapeze solar thermal collectors

When larger surfaces are available on the building's facades, isosceles trapeze derived from equilateral triangle is recommended for cost-effective arrays. Thus, several combinations of the triangle and trapeze solar thermal collectors, further considered as units, can result (Fig. 4.27).

The number of possible combinations increases with the number of the considered units as selectively presented in Fig. 4.28 for a number of two units.

Several possible configurations of three triangle and trapeze solar thermal collectors are presented in Fig. 4.29.

Thus, architects and designers can use various shapes and colours to develop novel arrays of solar thermal collectors to be integrated into the building's visible parts, as the facades are. To reduce the installation complexity, the manufacturer can deliver groups of collectors in several typologies. Each collector has plug-and-play connections that allow the serial or parallel interconnection, targeting an increased efficiency. Several examples are given in Fig. 4.30 for a single-family house. The building in Fig. 4.30a has groups of six triangles forming hexagons that can be serially preinstalled as well as groups of two trapezes. These two types of groups, having similar active surfaces, can be further parallelly connected to the solar thermal system

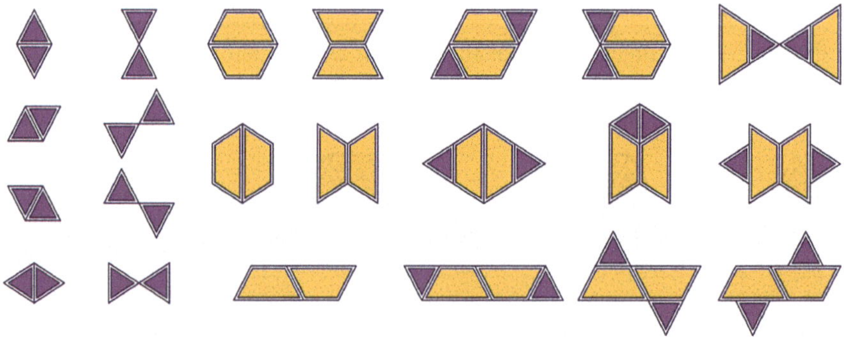

Fig. 4.28 Possible combinations of two units of triangle and trapeze solar thermal collectors

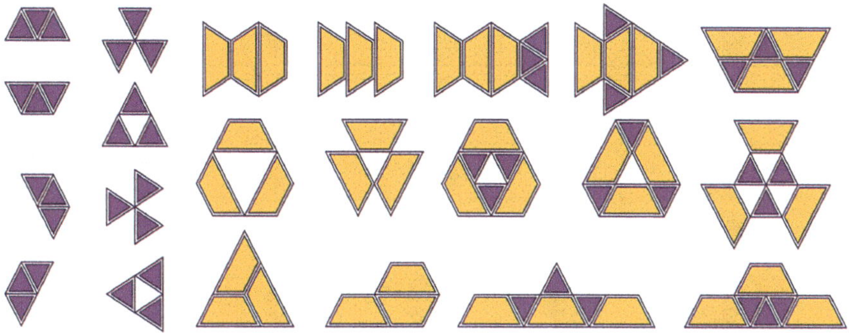

Fig. 4.29 Possible combinations of three units of triangle and trapeze solar thermal collectors

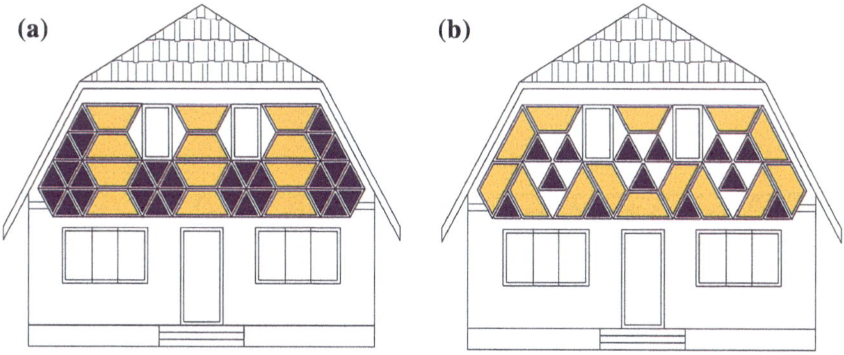

Fig. 4.30 Coloured, non-conventional configurations of triangle and trapeze solar thermal collectors on the facade of a single-family house

installed on the building. Similar connections could be designed for the homogenous groups of collectors (e.g. the groups of three triangle collectors), in Fig. 4.30b.

Another example is presented in Fig. 4.31 for a multifamily building. Here, two possibilities may occur: a centralized solar thermal system with large capacity storage tanks that share the thermal energy to each dwelling in the building or distributed solar thermal systems, one for each dwelling. In the first case, the solar thermal collectors are serially grouped to form similar active surfaces and further parallelly connected to a central distribution system; the thermal energy storage system requires a common space to install large storage tanks as well as the other components of the solar thermal system. In the second case, the collectors are grouped for each dwelling where individual storage tank, pumping group, solar controller, backup source, etc., are installed.

In office buildings, the thermal energy demand for DHW is lower than for residential buildings, and thus, solar thermal facades are usually considered to contribute also to the space heating (Fig. 4.32).

Fig. 4.31 Coloured, non-conventional configurations of triangle and trapeze solar thermal collectors on the facade of a multifamily building

Fig. 4.32 Coloured configurations of triangle and trapeze solar thermal collectors on the facade of an office building

An additional functionality of the solar thermal collectors that can be exploited when implemented in the built environment is their use as shading devices. An example is given in Fig. 4.33 outlining that trapeze solar thermal collectors can be horizontally installed or optimally tilted as overhangs above windows on the southern facade. On eastern and western facades, these trapeze solar thermal collectors can be installed as vertical wing walls for each window.

Heritage buildings, where several constraints are imposed to maintain the original aspect of the building's facade or roof, represent a particular implementation case. Architectural acceptance is highly restrictive in this case, and the solutions should be evaluated considering the collector shape and size along with aesthetic and heritage

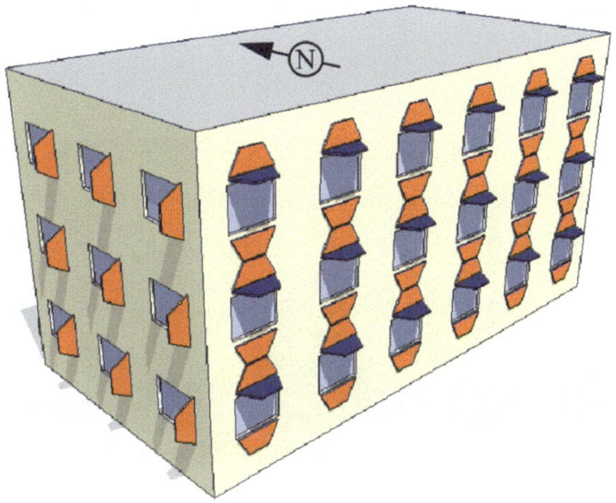

Fig. 4.33 Solar thermal facade with trapeze solar thermal collectors also used as shading devices: overhangs for the southern windows and vertical wing walls for the eastern and western windows

conservation prerequisites. An obvious solution consists of installing the solar thermal collectors on the less visible parts of the building (e.g. on the roof). However, traditional roofs have non-rectangular shapes (Fig. 4.34a) and cannot accommodate a large number of commercial, rectangular solar thermal collectors. The triangle and trapeze shapes of the novel solar thermal collectors, along with tailored colours, traditional (dark blue) or non-traditional (dark brown) could be a viable option for this type of buildings (Fig. 4.34b).

Based on the requirements imposed to increase the coverage factor and the architectural acceptance, a novel, small-sized (0.63 m^2) flat plate solar thermal collector with trapeze shape and black absorber plate was firstly developed as demonstrator and tested on an indoor testing rig (Fig. 4.35).

This demonstrator was successively tested and constructively optimized [90] through eight variants, and, for the best-performing ones, (one with aluminium and one with copper absorber plate) the experimental conversion efficiencies are plotted in Fig. 4.36. A nominal efficiency of 60.59% was reached for the demonstrator with the aluminium absorber plate while for the copper absorber plate the nominal efficiency was 57.90%.

The black demonstrator with aluminium absorber plate was installed on the southern facade of an outdoor testing rig along with two commercial solar thermal collectors (one of flat plate type and the other one with evacuated tubes) for comparative testing (Fig. 4.37). Each solar thermal collector was connected to identical DHW storage tanks, pumping groups, expansion vessels and solar controllers installed inside the outdoor testing rig [96]. Thus, similar testing conditions were insured considering both the energy source and the energy demand. A pyranometer was installed in the vertical plane of the southern facade to measure each minute the global solar irradiance. Each solar thermal collector has temperature sensors installed at the inlet and outlet connections, and ultrasonic flowmeters to evaluate the thermal power/energy output.

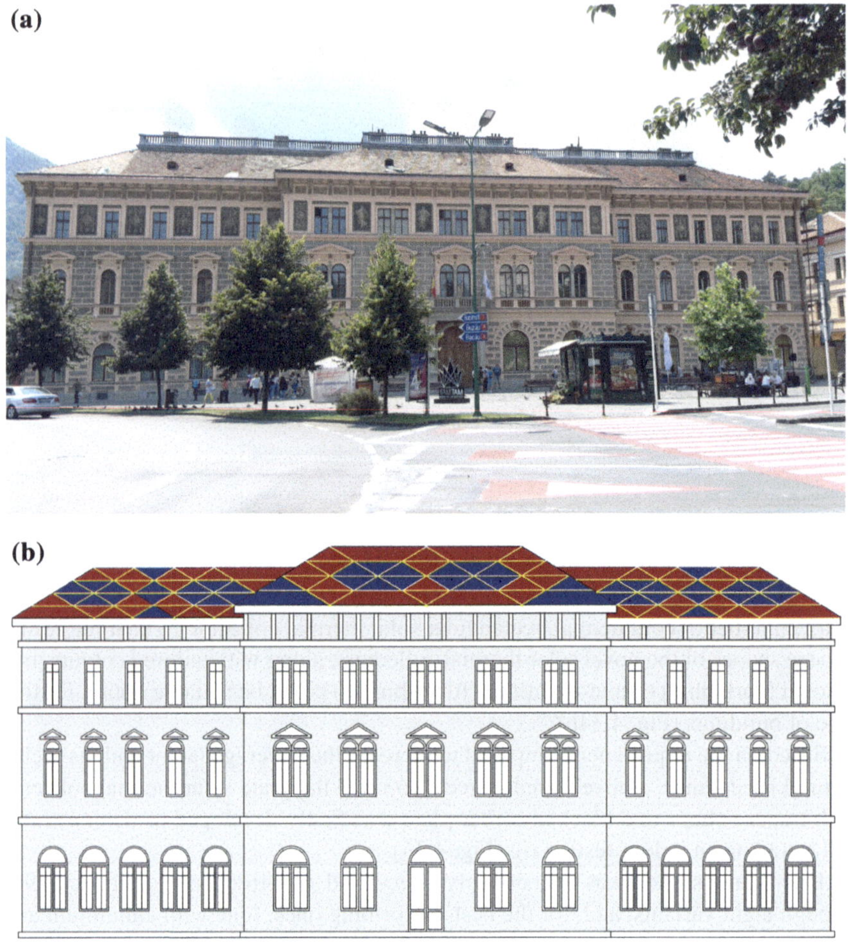

Fig. 4.34 Rectorate of the Transilvania University of Brasov, Romania: **a** current situation; **b** proposed implementation of trapeze and triangle solar thermal collectors on the roof

Based on the measured values of the solar irradiance and of the thermal power output, the efficiencies of the flat plate solar thermal collectors are calculated and the results are plotted in Fig. 4.38 for a sunny winter day (14 December 2014). As expected, the efficiency of the commercial solar thermal collector (1) has better values due to its better nominal efficiency as compared to the efficiency of the novel trapeze solar thermal collector (2).

The next step consisted of developing other trapeze solar thermal collector prototypes having aluminium absorber plate and differently coloured absorber plates (black, green, blue and red). After indoor testing, these were mounted on the southern facade of the outdoor testing rig to form an array of trapeze solar thermal collectors

Fig. 4.35 Black trapeze solar thermal collector installed on the indoor testing rig

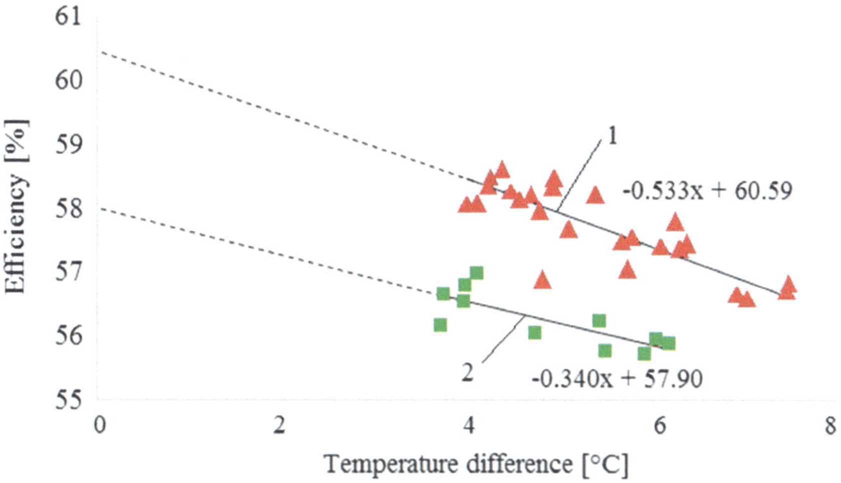

Fig. 4.36 Experimental efficiencies of two trapeze solar thermal collectors, one with an aluminium absorber plate (1) and the second one with copper absorber plate (2)

(Fig. 4.39). This array consists of three parallelly connected black trapeze collectors (the upper row) and six serially connected, coloured trapeze collectors [94].

The highest conversion efficiency (Fig. 4.40) was recorded for the tilted (1) and for the vertical (2) commercial solar thermal collectors. The serial connection of the trapeze solar thermal collectors aimed at assessing the influence of the temperature increase on the collector efficiency. Considering as reference the efficiency of the first trapeze solar collector (3), there is a visible efficiency drop for the second (4) and the third (5) serially connected trapeze collectors, especially during the morning

Fig. 4.37 Outdoor testing rig with solar thermal collectors installed on the southern facade

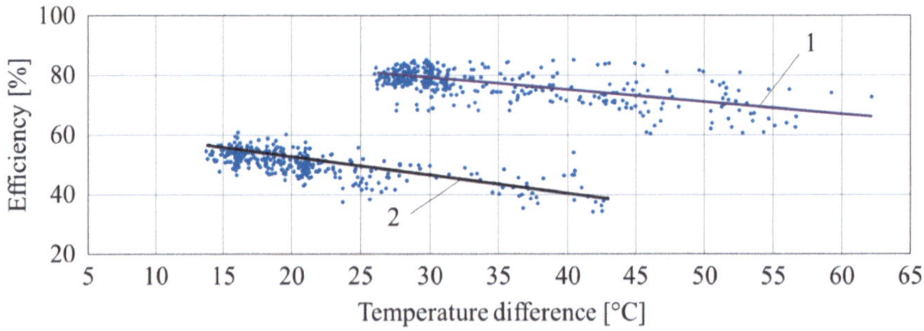

Fig. 4.38 Efficiency of the commercial (1) and of the trapeze (2) flat plate solar thermal collectors installed on the outdoor testing rig

Fig. 4.39 Outdoor testing rig with tilted (1) and vertical (2) commercial solar thermal collectors and trapeze solar thermal collectors (3, 4, 5)

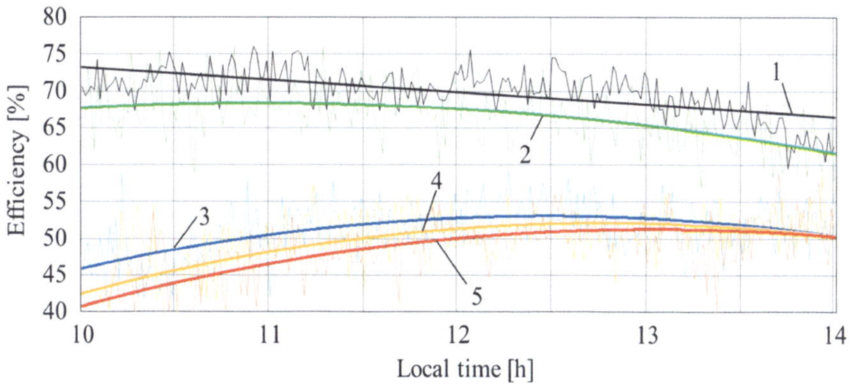

Fig. 4.40 Efficiency of the flat plate solar thermal collectors installed on the outdoor testing rig: commercial tilted (1), commercial vertical (2), first (3) second (4) and third (5) trapeze FPSTC

hours when the outdoor air temperature was significantly lower than the collector's temperatures.

To further increase the coverage factor and the architectural acceptance, a triangle solar thermal collector with a smaller size (0.083 m^2) was developed [97]. Because of its reduced dimensions (the side length of the absorber plate is 350 mm), the internal pipes were replaced by a cavity below the entire absorber plate to transfer the thermal energy from the absorber plate to the thermal fluid (water or an antifreeze solution) flowing through the collector. This concept raises two problems: a significant deformation (25 mm) of the absorber plate (Fig. 4.41a) and preferential flow zones below

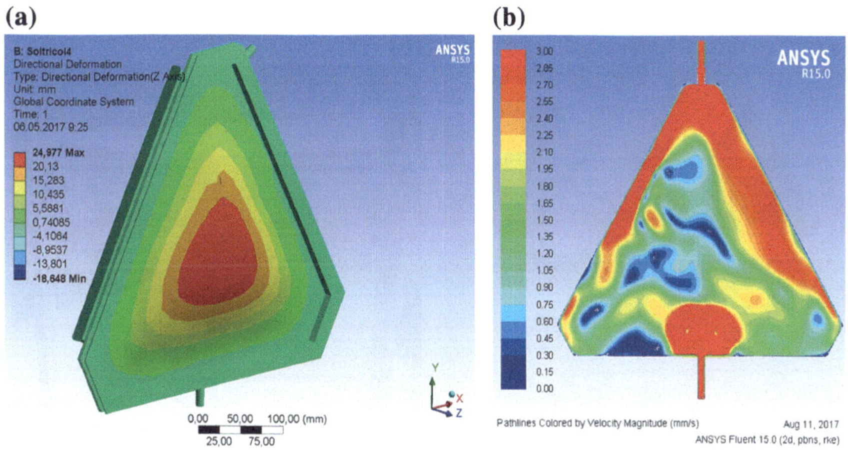

Fig. 4.41 ANSYS simulations: **a** the deformation of the absorber plate; **b** fluid velocity into the central body cavity of the novel triangle solar thermal collector

the absorber plate of the novel triangle solar thermal collector (Fig. 4.41b) that was evidenced through ANSYS simulations.

To mitigate these effects, several solutions were investigated. A solution based on seven internal fixing elements was found optimum for both problems: the deformation decreased down to 1.8 mm (Fig. 4.42a) and an increased uniformity of the fluid flow was reached (Fig. 4.42b).

Based on the results of the ANSYS simulations, four triangle solar thermal collectors were developed at demonstrator level (Fig. 4.43a). These demonstrators have

(a) (b)

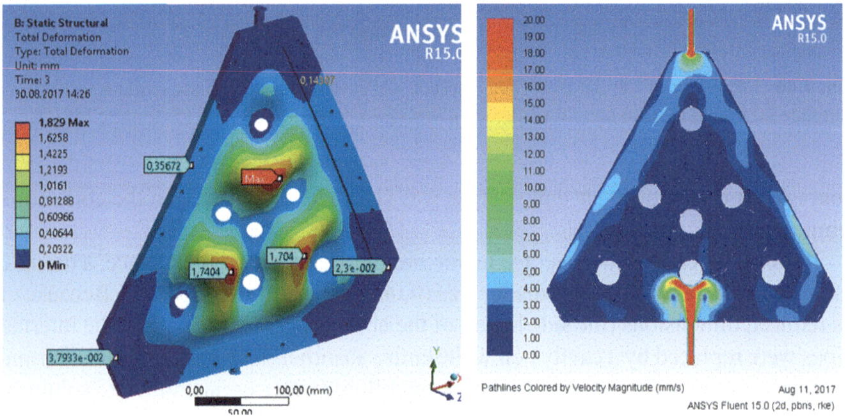

Fig. 4.42 ANSYS simulations: **a** the central body with seven internal fixing elements; **b** fluid velocity below the absorber plate

(a) (b)

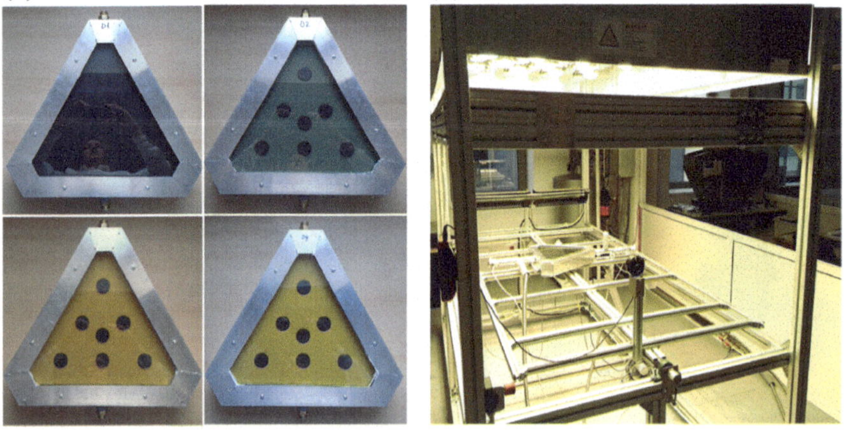

Fig. 4.43 **a** Novel triangle solar thermal demonstrator collectors; **b** testing the collector on the indoor testing rig

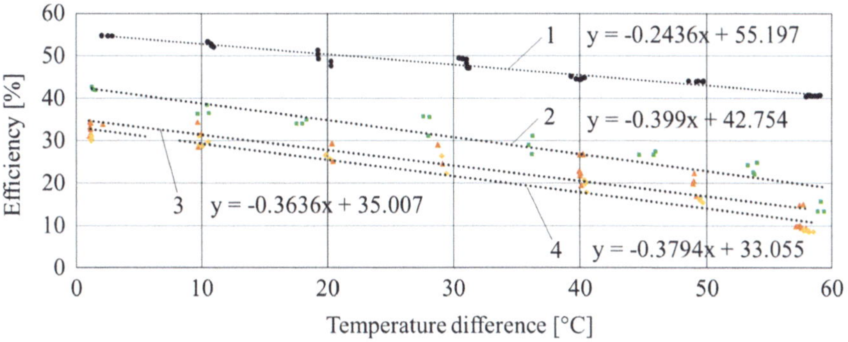

Fig. 4.44 Efficiency of the black (1), green (2) and orange (3, 4) triangle demonstrators recorded on the horizontal indoor testing rig

identical construction (design and materials) but different colours of the spectrally selective absorber coatings: one black, one green and another two with shades of orange.

Each of the four triangle demonstrators was individually tested on the indoor horizontal testing rig (Fig. 4.43b) to evaluate the conversion efficiency, and the results are comparatively plotted in Fig. 4.44 for the black, green and orange demonstrators. The black demonstrator has the best conversion efficiency (55%), as expected, while for the demonstrators with non-traditionally coloured absorber plates, a decreased nominal conversion efficiency was recorded (42% for the green, 35 and 33% for the orange absorber plates) due to the lower absorptance coefficients under 900 W/m^2 simulated solar irradiance.

After individually testing each demonstrator, their serial and parallel connection was further investigated through ANSYS flow simulations, followed by indoor testing on a vertical solar simulator and finally on an outdoor testing rig. The ANSYS flow simulations showed that stagnation areas are highly limited (at the corners) inside the solar thermal collectors regardless of the connection type (serial or parallel) or the triangles position (Fig. 4.45).

The conversion efficiency of the four demonstrators was further evaluated on an indoor testing rig for vertically installed solar thermal collectors [57] mounted in a serial (Fig. 4.46a) and in a parallel (Fig. 4.46b) connection.

The indoor testing with vertically installed demonstrators allowed recording the experimental results presented in Fig. 4.47, comparable to those obtained on the horizontal indoor testing rig (Fig. 4.44). The nominal efficiency of the solar collectors was slightly lower: with 0.08% for the black collector (1), with 1.64% for the green collector (2) and with 1.58 and 1.21% for the orange collectors (3 and 4).

After indoor testing on the horizontal and on the vertical testing rigs, the four triangle solar thermal collectors (three non-traditionally coloured and one black) were installed on the southern facade of the outdoor testing rig for a long-term testing procedure (Fig. 4.48). The collectors have a parallel connection structure. As in the case of the trapeze solar thermal collectors, each triangle collector is equipped

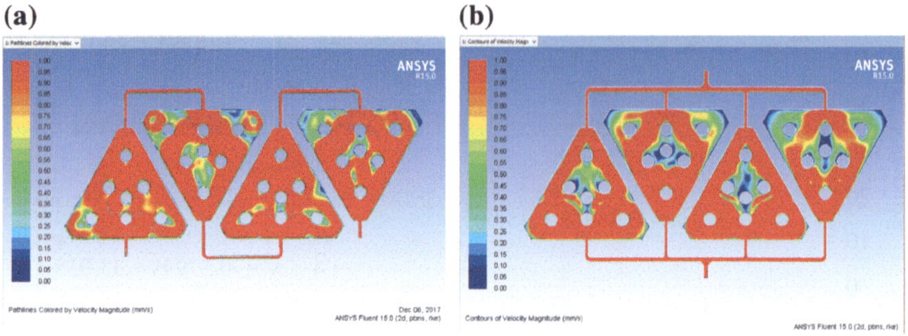

Fig. 4.45 ANSYS flow simulations through the triangle solar thermal collectors: **a** serially connected; **b** parallelly connected

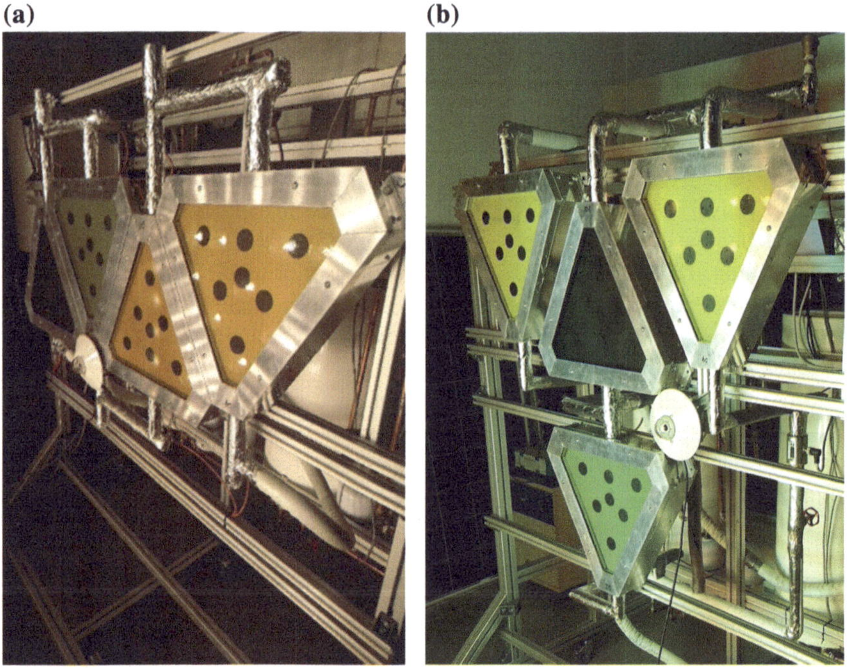

Fig. 4.46 Vertical indoor testing of solar thermal collectors: **a** serially connected; **b** parallelly connected

with temperature sensors at the inlet and outlet and with digital flowmeters that allow evaluating the specific thermal output power for each demonstrator.

The conversion efficiencies obtained by reporting the thermal output power of each solar thermal collector to the solar irradiance received in the vertical plane are plotted in Fig. 4.49. These efficiencies well match the values registered on the

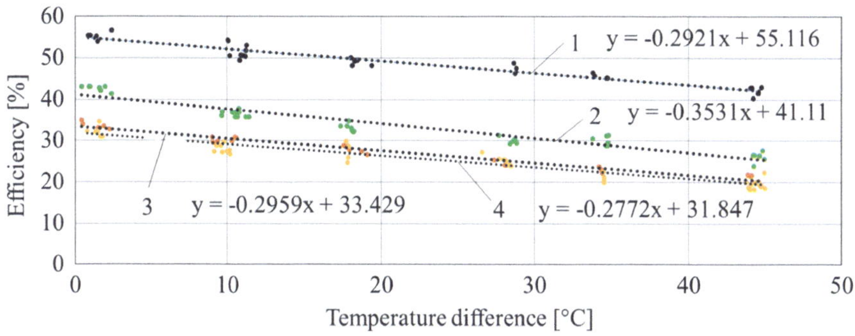

Fig. 4.47 Efficiency of each solar thermal collector in serial connection evaluated on the indoor testing rig: black (1), green (2) and orange (3 and 4) solar thermal collectors

Fig. 4.48 Outdoor testing rig with coloured trapeze and triangle solar thermal collectors

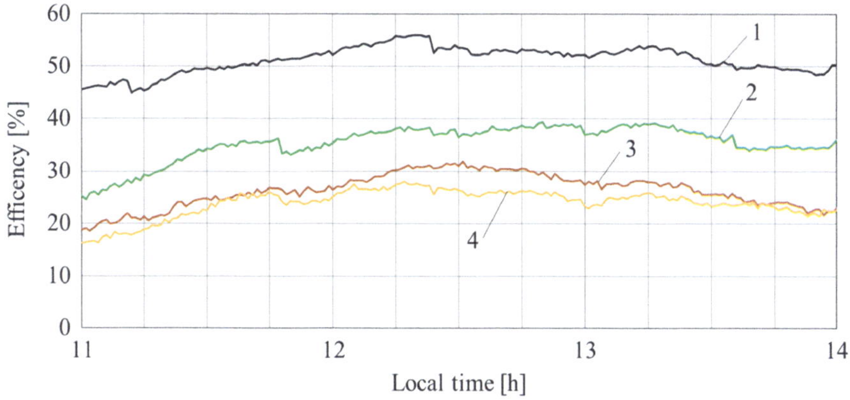

Fig. 4.49 Conversion efficiencies of the triangle solar thermal collectors recorded on the outdoor testing rig in a sunny summer day (20 August 2018): black (1), green (2) and orange (3 and 4) solar thermal collectors

indoor vertical testing rig, and thus, the testing procedure was extended for a longer period of time to investigate the behaviour of the novel solar collectors in various meteorological conditions.

4.5 Exploitation and Maintenance of the Solar Thermal Systems Implemented in the Built Environment

After the solar thermal system is installed, its exploitation and maintenance become the responsibility of the owner who, according to the complexity of the system, can perform itself these activities or must hire a specialized company for specific tasks to reach the expected efficiency and lifetime.

The exploitation of the solar thermal system aims at maximizing the thermal energy output while protecting the user against a too high temperature of the DHW and the system against overheating. Usually, a solar thermal system includes a solar controller that changes the state of the components according to the monitored temperatures. This solar controller allows the user to select the operation mode (standard, holiday, special schedule, manual, protection, etc.) and to adjust the functional parameters (the temperature differences between the solar thermal collector and storage tank when the pump is turned on and off, the flow rate of the thermal fluid, the maximum allowed temperature in the storage tank, etc.).

The solar thermal systems, as any other energy conversion systems, require a regular maintenance of their components to ensure a safe and efficient functioning, to reduce the operation costs and to extend their lifetime. The maintenance should be periodically done (daily, monthly, seasonally, yearly, etc.), according to the complexity of the system, to the specific features of the implementation location and to the instructions received from the components/system manufacturer(s).

Usual maintenance tasks should focus on checking the following:

- The aspect of the thermal fluid and its freezing point, followed by its replacement if required;
- The filters, followed by cleaning or replacing them if required;
- The safety relief valve, the discharge pipe, the valves;
- The hydraulic pump for blockage or corrosion;
- The air vents;
- The pressure drop on the heat exchangers;
- Leakages at all joints and pipe connections;
- The temperature sensors, the electrical connections;
- The display panel for errors or service codes;
- The system's operating settings;
- The safety controls;
- The state of the glazing and the joints of the solar collectors, followed by cleaning if necessary;

- The fixing elements of the components, especially of the solar thermal collectors and of the connecting pipes.

Finally yet importantly, maintenance should also involve checking and restoring the following parameters:

- The gas pressure in the expansion vessel;
- The pressure of the thermal fluid;
- The flow rate of the thermal fluid.

Maintenance also implies checking and repairing (if necessary) the thermal insulation of the pipes and checking if any shading effect occurs on the collector. After the maintenance operations are finalized, the solar thermal system is ready to be used and should be set for an optimal operation.

Some of the maintenance tasks such as the visual inspection or the exterior cleaning of the glazing can be done by the system's user but most of the tasks require qualified personnel.

The main categories of the maintenance tasks are further presented, grouped according to their occurrence and their influence on the system performance.

Maintenance of the thermal fluid is very important because its properties have a significant influence on the heat transfer at the level of the solar collector and in the internal/external heat exchanger of the storage tank.

The aspect of the thermal fluid is one of the first indicator of its state. It can be checked directly on the glass of the flowmeter, usually included in the pumping group, or by taking a sample of thermal fluid. It has to be clear, colourless (if water), light green or pink (if a mixture of water and glycol is used), and without any suspended particles. If one of these features is not respected, the source of the issue must be firstly identified. The thermal fluid usually degrades because of the lack of filters installed in the solar circuit to prevent the accumulation of the suspended particles in sensitive points (e.g. thinner pipes of the solar thermal collectors, hydraulic pumps). These accumulations can decrease the flow rate through the entire system and thus through the solar thermal collectors. The reduced flow rate increases the temperature of the solar thermal collectors and of the thermal fluid, accelerating its degradation. Moreover, in oversized solar thermal systems, where stagnation more often occurs, the degradation of the thermal fluid can cause scaling that ends up by clogging the solar circuit.

The thermal fluid must be drained out; the filters must be cleaned or replaced; the circuit must be cleaned and filled with a new, clean, thermal fluid using a flushing/cleaning/filling pump for solar thermal systems (Fig. 4.50a) connected to the solar circuit as presented in Fig. 4.50b. After the entire air is vented from the solar circuit, the pressure is set to the required value for an optimum operation of the system and the filling pump is stopped.

When the solar thermal system is installed in a location where the outdoor air temperature drops below 0 °C, the freezing point of the thermal fluid must be lower than the lowest possible outdoor air temperature and this parameter can be checked using a refractometer (Fig. 4.51a) or a glycol meter (Fig. 4.51b).

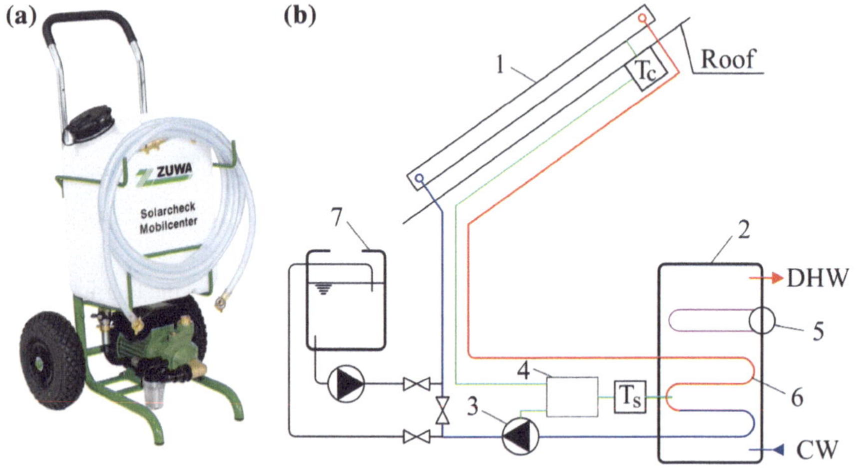

Fig. 4.50 **a** Flushing/cleaning/filling pump for solar thermal systems; **b** connection scheme: solar thermal collector (1), storage tank (2), hydraulic pump (3), solar controller (4), auxiliary heater (5), internal heat exchanger (6), flushing/cleaning/filling pump (7)

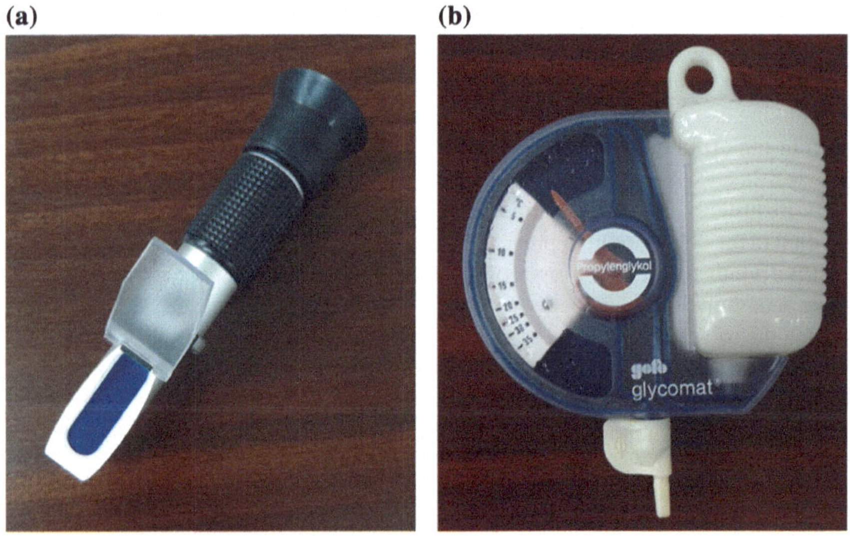

Fig. 4.51 Instruments for checking the freezing point: **a** refractometer; **b** glycol meter

The flow rate of the thermal fluid in the solar circuit can be checked at the level of the pumping group of the solar thermal system (Fig. 4.52a), on the scale of the glass flowmeter (Fig. 4.52b) or by other type of flowmeters included in the circuit and adjusted by a control valve.

(a) (b)

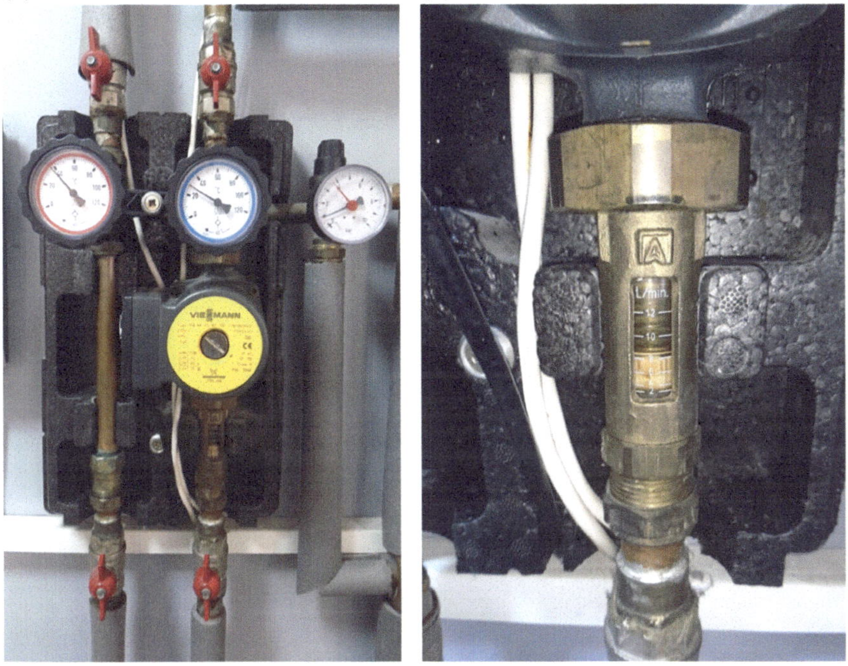

Fig. 4.52 **a** Solar thermal system pumping group; **b** flowmeter

The maintenance of the solar thermal collectors must be more often performed in dusty climates, with low amounts of precipitations and high-temperature differences between day and night, where the water vapours contained in the ambient air condense during the night on the glazing, increasing the adherence of the dust on the glazed surface (Fig. 4.53).

The soiling mechanism can be divided into three phases: generation, deposition and adhesion. The dust generation sources could be natural (in sandy or deserted areas, vegetation–pollen, etc.) or anthropogenic (agriculture, fossil fuel-based power plants, traffic, etc.). When the wind speed is high enough, the dust particles are lifted and transported. Due to gravity and turbulence, these particles are deposited according to their size: firstly, particles with a diameter higher than 0.5 mm will settle on the surfaces, followed by the particles with smaller sizes (down to 60–70 μm). Once deposited, several adhesion mechanisms occur governed by van der Waals forces, electrostatic forces or capillarity. The adhesion is supported by the interaction between the air humidity and the water-soluble components in the particles [44].

To avoid soiling, the glazing of the solar thermal collectors is covered with a hydrophobic (self-cleaning) coating allowing the deposited dust to be naturally removed by wind and rain. However, in dry and/or in heavy polluted areas the solar thermal collectors must be periodically cleaned as soiled or dusty collectors poorly perform as a result of the solar radiation partly blocked at the glazing level.

Fig. 4.53 Soiled solar thermal collectors

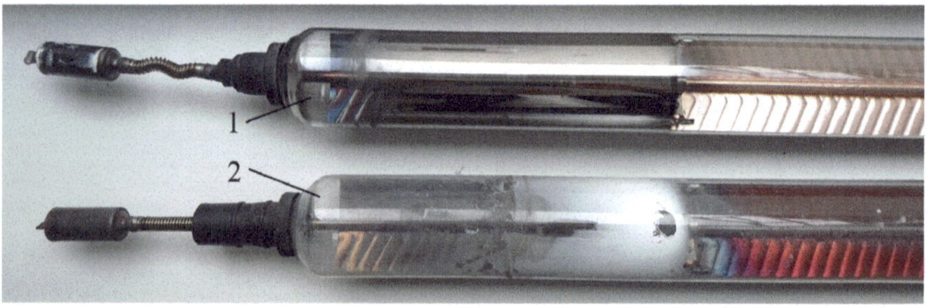

Fig. 4.54 Solar thermal collector's evacuated tubes: well-sealed tube (1), tube with air inside (2)

In the case of the evacuated tubes solar thermal collectors, the integrity of the sealing must be periodically checked. As shown in Fig. 4.54, in a well-sealed tube (1) a silver coating can be observed inside the glass while for tube (2) with air inside, this layer turns white. The tube (2) is still functioning but at a lower efficiency because of increased thermal convection losses between the absorber/air/glass components and should be replaced with a new one.

Periodical visual checking of the collector shading by the vegetation grown over time or by new constructions in the neighbouring properties is required because shading largely affects the thermal energy output.

For safety reasons, the fixing elements of the solar thermal collectors must be yearly checked for integrity, corrosion, etc., to avoid the fall of the solar thermal collector that can cause serious damages to the building and people. This task must also be performed after each period with high wind speed or large amounts of snowfall.

The maintenance of the hydraulic pump: it is important to keep the pump clean and well-maintained for its best operation and longevity. A filter must be installed in the circuit before the pump to keep the foreign materials away from the rotor area, thus avoiding its surface damage. This filter must be periodically checked and cleaned/replaced. The seals of the pump connections must be yearly replaced, using the replacement kits according to the service instructions.

Maintenance of the safety components refers to the safety relief valve, its discharge pipe and the expansion vessel. The safety relief valve must be checked for proper functioning by increasing the pressure in the solar circuit up to the maximum allowed limit where the safety relief valve must discharge the extra pressure. If this does not happen, the safety relief valve must be replaced with a new one that must be checked too. The discharge pipe connected to the safety relief valve will be checked against clogging. The nitrogen pressure cushion of the expansion vessel must be further verified and its value should be set according to the static height of the system.

4.6 Renewable Energy Mixes Based on Solar Energy in nZEB

In Europe, the Energy Performance of Buildings Directive 31/2010 imposes that all newly designed or major refurbished public buildings must comply with the nearly Zero Energy Building (nZEB) standard starting with 1 January 2019; this standard will be mandatory for all types of buildings after 1 January 2021 [18]. To reach this standard, cost-effective solutions must be firstly applied to reduce the building's energy demand. Both, thermal energy (for DHW preparation, space heating and cooling) and electrical energy (for lightning and ventilation) are subject of energy-saving measures aiming to reach the low energy building standard. This reduced energy demand must be supplied to a very large extent by renewable energy produced on-site or nearby the building.

Visa et al. [89] developed a design method for nZEB following an algorithm that consists of the following main steps: assessing the current energy status of the building, reducing the building's energy demand and increasing the energy supplied by renewable-based energy mixes.

The current energy status is defined based on input data such as the building's and the implementation location-specific features. According to these input data, the building energy demand for space heating and cooling, for DHW preparation and for lighting is calculated. Further on, for buildings where different renewable energy systems have already been installed, as results of various policies supporting their implementation, the renewable energy share is evaluated, and if the nZEB standard is reached, the algorithm stops. There are pilot projects aiming to fully cover the energy demand of the buildings using renewable energy sources, transforming these in Zero Energy Building (ZEB) or Plus Energy Buildings (PEB) when extra renewable energy is provided for consumers not directly related to the building (e.g. electric vehicles).

Unfortunately, most of the buildings still have a high energy demand, seldom covered by renewable energy systems (to a large extent); in this case, the next two steps have to be followed to reach the nZEB standard.

If the energy demand of the building is higher than the values imposed by the standardized energy efficiency indicators (e.g. the yearly specific energy consumption), energy-saving solutions and energy efficiency measures must be applied starting with the largest consumer(s). Usually, the space heating and/or cooling energy demand represents the highest share in the energy demand of a building, followed by the energy demand for hot water preparation and for lighting. Thus, to reduce the space heating/cooling energy demand, the building's envelope must be thoroughly refurbished using energy-efficient materials and specific solutions for its opaque and transparent elements along with the application of passive solar design methods. After reducing the space heating/cooling energy demand, the share of the thermal energy demand for DHW becomes higher or even the largest (e.g. for Passive Houses). Decreasing this energy demand can be obtained by eliminating all the leakages, by thermally insulating the distribution pipes and the storage tank, by using energy-efficient faucets and showerheads (with optimized flow, automatic mixing valves and presence sensors), etc.

Finally yet importantly, the energy demand for lighting must be reduced too, and this can be easily done by maximizing the share of natural lighting and using efficient light bulbs (e.g. LED) and presence and crepuscular sensors. Special attention has to be paid to the visual comfort according to each room functionality. Increased energy consumption for lighting is reported for commercial buildings, e.g. supermarkets with large floor surface and low windows to floor ratio where artificial lighting is required during day and night.

After reducing the building's energy demand, the newly obtained value has to be compared again with the renewable energy produced by the existing renewable energy systems, if any. There might be situations, when the renewable energy share is increased enough by reducing the energy demand, and the nZEB standard is reached. If not, the procedure further runs to the next step, increasing the energy supplied by renewable energy systems.

Further on, the renewable energy potential in the implementation location has to be evaluated to implement an optimal, renewable-based energy mix. As stated in Chap. 2, the solar energy conversion systems have to face the problem of the solar energy variability. This variability is more obvious in temperate and cold climates where considerable differences occur between the solar energy available during the summer and during the wintertime. As a result of this variability, a solar thermal system designed to fully cover the thermal energy demand for DHW during the summer months will only partly cover the energy demand in the transition months and even less during the winter season. In this case, a backup source is required to supply the difference. If the nZEB standard is targeted, this backup energy source must be also provided by a renewable energy system. For meeting the thermal energy demand, the backup system is usually represented by a ground-coupled heat pump system or by a bioenergy system. When electrical energy should also be covered to a large extent using renewable energy sources, wind turbines or micro-hydro systems can back up the photovoltaic systems or even biofuel-based combined heat and power systems can represent a feasible option. Therefore, renewable-based energy mixes should be designed to obtain efficient, cost-optimal solutions supporting the building to reach the nZEB standard.

4.6.1 Solar Thermal–Geothermal Energy Mixes in Buildings

Solar thermal–geothermal energy mixes are used to cover the thermal energy demand of the buildings (for DHW and for space heating and cooling). This type of renewable-based energy mixes was studied for decades [3, 6, 10, 26, 85] and the interest increased when the technological maturity was reached by both technologies. Thus, following the development of the heat pumps mainly for single-family houses [17], most of the research on solar thermal–geothermal energy mixes was developed focusing on the residential sector [42, 55]. Several researchers tried to identify an optimal design strategy as Buker and Riffat [6] and Poppi et al. [70] and concluded that this is a difficult task due to the broad variety of building types and characteristics, climates and performance indicators.

Many successful design configurations are described in the literature and were also marketed in countries as Switzerland, Austria and Germany as reviewed by the International Energy Agency [36] in its program on solar heat pumps.

Among these configurations, Haller et al. [31] and Hadorn [30] concluded that the parallel configuration, with the solar heat directly used in the building, is recommended to obtain a maximum seasonal performance factor. Besides the advantages of improved performance, a reduction in the drilling depth of the boreholes is possible [8] followed by lower investment costs. Another advantage is represented by the possibility to inject into the ground the thermal energy supplied in excess by the solar thermal collectors during summer, thus protecting the solar collectors from overheating and recovering the temperature in the ground [16], thus solving the freezing problems in the deep borehole heat exchangers [66].

Moreover, in large systems (e.g. at community level), the focus was set on thermal energy storage into the ground, primarily seeking to increase the solar fraction in buildings by seasonally storing the solar thermal energy. It was suggested that cost-effective systems can be developed only for thermal energy loads larger than 500 MWh per year [10]. Most of the ground storage systems for solar thermal energy were designed considering low mean temperatures (10–40 °C) and were used in combination with heat pumps [72]. When a higher solar fraction is targeted, higher temperatures (50–70 °C) must be used [79].

Besides the higher temperature, a careful design is required for the ground storage field, with dense spacing and small depths, in order to create a compact volume and minimize heat losses, as suggested by Eslami-Nejad et al. [16] based on a TRNSYS simulation of the thermal energy supplied by 200 m^2 of unglazed solar collectors injected in the central nine boreholes of a 5 $\times$ 5 boreholes field with a depth of 151.3 m. Without this heat injection, the depth of each borehole should be increased to 236.7 m to maintain the heating capacity of the heat pump.

Similar results are reported by Reda [74] and Reda and Laitinen [75] when comparing various configurations, controls and borehole depths for a Finnish multifamily house, showing that ground recharging benefits on the decreased depth of the boreholes.

Emmi et al. [15] simulated a multifamily house in six European cities with cold climates, finding that 50% of the boreholes could be removed when the thermal energy supplied by 176 m^2 of glazed solar collectors is connected to the borehole circuit via a buffer tank.

Moldovan et al. [58] evaluated several functional scenarios to outline the optimal solution for a solar thermal–geothermal energy mix used in a low energy office building located in Brasov, Romania (mountain area with a temperate continental climate) in the Renewable Energy Systems and Recycling Research Centre of the Transilvania University of Brasov. The outdoor air design temperatures for this location were −21 °C and +27 °C for the heating and cooling periods, respectively, while the indoor temperatures were set at +20 °C during winter and +24 °C during summer. The building is developed on three floors, each with a surface of 450 m^2 and a useful height of 3.8 m, one of them being partly (2.20 m) under the ground level (Fig. 4.55). The rooftop is flat and accommodates various renewable energy systems supplying both thermal energy and electrical energy for the building.

The building was designed to meet the low energy building standards. Thus, it has a well-insulated envelope with the thermal resistance of 6 m^2 K/W for all the buildings' elements in direct contact with the ground (basement floor and exterior walls), 4 m^2 K/W for the exterior walls, 7 m^2 K/W for the ceiling and 1.6 m^2 K/W for the glazed surfaces. Additionally, the building has an optimal orientation, with the long sides facing South and North, with large curtain walls on the southern and western facades to increase the passive use of the solar energy during winter (Fig. 4.56). Natural ventilation is a major contributor in decreasing the cooling energy demand during summer [59]. Thermally activated building elements (the slabs between the floors

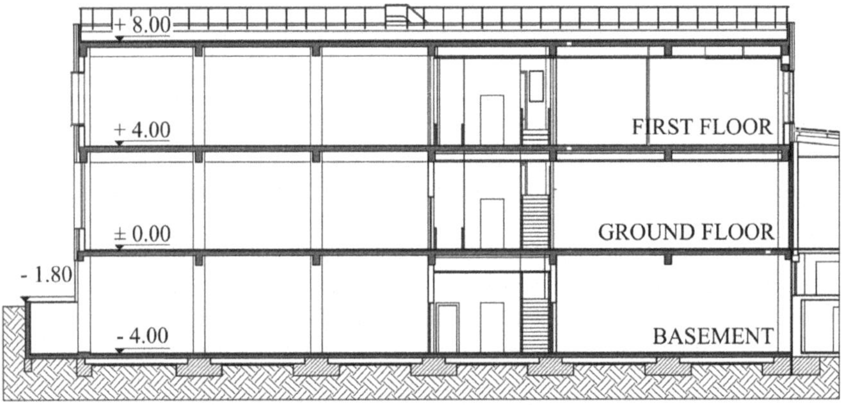

Fig. 4.55 Vertical section of the office building

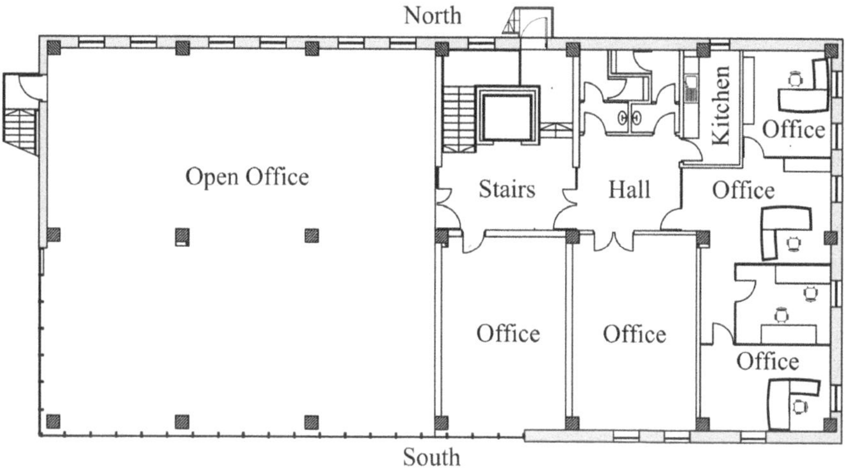

Fig. 4.56 Horizontal section through the upper floor of the office building

and the slab above the first floor) are used for both spaces, heating and cooling, efficiently distributing the thermal energy provided by a ground-coupled heat pump system. A thermal zoning system allows the optimal use of the thermal fluid only in the areas requiring heating or cooling; thus, the energy consumption is reduced while the thermal comfort is reached. The large curtain wall facades, along with the Sun pipes inserted in the ceiling on the upper floor, provide natural light consistently reducing the energy demand for lighting.

The thermal energy demand of the building was evaluated based on its characteristics and on the on-site measured meteorological parameters with a Delta-T automatic Weather Station (outdoor air temperature and relative humidity, wind speed and direction, precipitations level) and a Kipp & Zonen SOLYS2 Sun Tracker (for the direct and for the horizontal diffuse and global solar radiation).

The monthly thermal energy demand for DHW and for space heating and cooling is presented in Table 4.13. The total thermal energy demand of the investigated office building is 90,030 kWh/year, that divided with the entire surface of the building (1350 m^2) which corresponds to an yearly specific energy demand of 66.69 kWh/(m^2 year) thus including the building in the low energy building standard. The largest energy consumption corresponds to space heating with 78,027 kWh/year, i.e. 57.8 kWh/(m^2 year), followed at a considerable difference by the space cooling energy demand of 7759 kWh/year thus, a specific cooling energy demand of 5.75 kWh/(m^2 year) and DHW production with 4244 kWh/year, corresponding to 3.14 kWh/(m^2 year).

A solar thermal–geothermal energy mix is considered to cover the entire thermal energy demand of the building, its design relying on iterative simulations for functional scenarios with various ratios of the thermal energy demand to be covered by the solar thermal system and the heat pump system, respectively. The results obtained for a 50% share covered by the solar thermal system and 50% share covered by the

Table 4.13 Monthly thermal energy demand of the analysed office building

Month	Energy demand for DHW (kWh)	Energy demand for space heating (kWh)	Energy demand for space cooling (kWh)
January	360	14,132	0
February	326	11,534	0
March	360	9875	0
April	349	6316	0
May	360	3803	868
June	349	0	1681
July	360	0	2260
August	360	0	2082
September	349	3625	868
October	360	6867	0
November	349	9447	0
December	360	12,429	0
Year	4244	78,027	7759

heat pump system are presented in Fig. 4.57a. To reach this high share using the solar thermal system(s), the surface of the solar thermal collectors has to be of 167.3 m^2, resulting a high thermal energy excess (37.6 MWh/year) during summer. A solution should be found to use this thermal energy, such as other buildings requiring thermal energy during the summer or high capacity seasonal storage systems using water or phase-change materials or ground thermal energy storage system.

When only a 20% share is considered to be covered by the solar thermal system (Fig. 4.57b), the surface area of the solar thermal collectors decreases to 49.2 m^2 and also does the thermal energy excess, to 1.5 MWh/year, mostly concentrated in June. In both scenarios, the thermal energy supplied by the solar thermal system is firstly used and only the uncovered energy demand is supplied by the geothermal system.

4.6.2 *Solar Thermal–Geothermal–Photovoltaic Energy Mixes*

After a solar thermal–geothermal energy mix is designed to partly or fully cover the thermal energy demand of a building, the associated electrical energy consumption should be also addressed when targeting the nZEB standard. Thus, a photovoltaic system can be included in the energy mix to cover, on a yearly basis, the electrical energy demand required to drive the compressor of the heat pump, the hydraulic pumps and control systems, along with the usual consumption of the lighting system in the building.

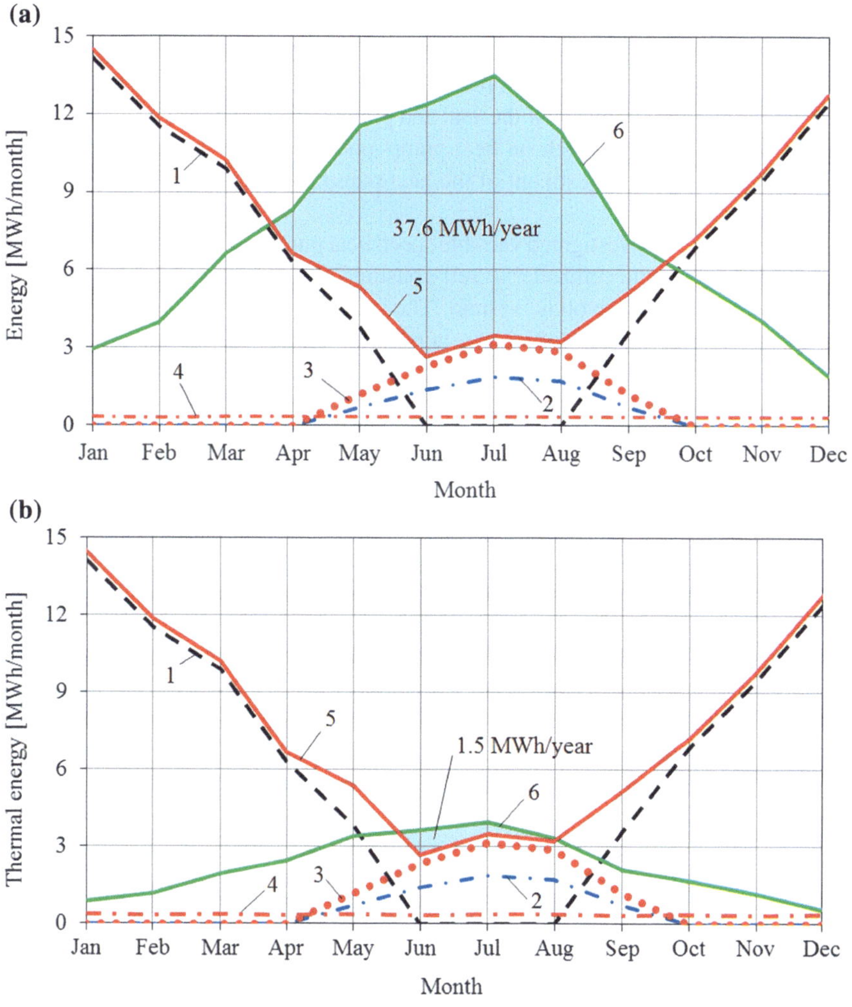

Fig. 4.57 Thermal energy demand of the building for space heating (1), for space cooling (2), for the thermally driven chiller (3), for DHW (4), total energy demand for heating, cooling and DHW (5) and thermal energy supplied by the solar thermal system (6) for the energy mix with: (a) 50% share of solar thermal and 50% of heat pump systems; (b) 20% share of solar thermal and 80% of heat pump systems

Another approach aims at increasing the self-consumption of the electrical energy supplied by a photovoltaic system. This is possible when the energy demand is well matched by the energy supply as in the case of the heat pumps used in reversed mode for cooling, as the peak cooling loads usually occur during periods with large amounts of solar energy.

The continuous decrease of the photovoltaic system's cost and the governmental incentives had as result a substantial increase in the number and capacity of photovoltaic systems installed during the recent years [37], that has subsequently levelled the installed capacity of the solar thermal and photovoltaic systems [100]. This trend also led to an intense research on heat pump–photovoltaic systems energy mixes, typically involving the adjustment of the heat pump controls according to the solar energy availability.

Fischer et al. [23] investigated variable speed heat pumps in a German multifamily house and modified the controller to self-consume as much as possible the electricity supplied by the photovoltaic system. The results show that economic gains are expected even if the system efficiency is reduced due to higher storage tank temperatures that reduce the heat pump's coefficients of performance and increases the tank losses. When also including a solar thermal collector in this energy mix [22], the electrical energy demand of the heat pump is reduced and the electricity supplied by the photovoltaic system can be injected in the grid.

There are many studies on photovoltaic–heat pump systems, and a more comprehensive analysis is developed in the review of Fischer and Madani [21] concluding that heat pumps are considered a major technology able to provide flexibility to a smart grid while providing efficient heating and cooling solutions when implemented in the built environment.

Hirvonen and Sirén [35] developed a photovoltaic-assisted heat pump system for a district heating system of low energy residential buildings in a Finnish community with ground storage of the thermal energy. Based on a Pareto front simulation of cost-optimized solutions, they showed that a photovoltaic heat pump energy mix has lower life cycle costs than a solar thermal heat pump system.

Visa et al. [89] developed an algorithm to improve the renewable energy mix to reach the nearly zero energy status for a building and applied this algorithm to a Solar House built in 2009 according to the solar passive design principles, with large windows to take full advantage of the solar gain during the heating season (Fig. 4.58).

The Solar House is located along with various renewable energy systems on the Colina Hill Campus of the Transilvania University of Brasov, Romania (Fig. 4.59). All these systems are fully monitored and are used as laboratories for the Renewable Energy Systems Engineering—B.Sc. program and for the Sustainable Product Design and Environment Protection—M.Sc. program.

A share of 83.83% from the total energy demand (thermal and electrical) of the Solar House is covered using renewable sources (solar and geothermal). This share can be further increased by improving the thermal resistance of the transparent envelope's elements responsible for the largest thermal energy loses.

The thermal energy demand of the Solar House is covered in a large extent (95.8%) by a 10 kW heat pump extracting the thermal energy through a horizontal geothermal heat exchanger of 600 m^2, and a solar thermal system consisting of six flat plates and three evacuated tube solar thermal collectors.

The solar thermal collectors are installed on the rooftop (Fig. 4.60) fixed at a tilt angle optimized considering the entire year. The six flat plate solar thermal collectors are parallelly connected to avoid overheating during summer. Two of the evacuated

Fig. 4.58 Lecture hall in the Solar House

Fig. 4.59 Renewable energy mix of the Solar House located in the Transilvania University of Brasov, in the Colina Hill Campus

tube solar thermal collectors are serially connected to increase the temperature of the thermal fluid during the wintertime while the third one is separately included in the system and used as a reference to compare the effect of the serial connexion on the conversion efficiency [60].

The thermal energy supplied by the solar thermal collectors is stored in a DHW tank (1) with a 300 L capacity, and in other two larger storage tanks (3), each with a capacity of 1000 L installed in a solar thermal laboratory (Fig. 4.61a). The solar thermal system provides DHW for the Solar House and for the locker rooms of a sport hall in the close neighbourhood.

The heat pump system is installed in a heat pump laboratory (Fig. 4.61b). The heat pump (4) supplies the thermal energy to a storage tank (5) with a capacity of

Fig. 4.60 Flat plate and evacuated tube solar thermal collectors installed on the rooftop

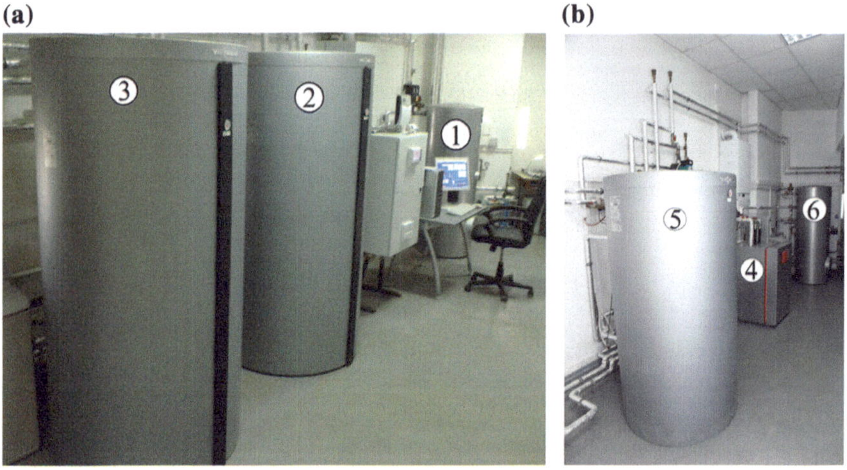

Fig. 4.61 Renewable energy systems for thermal energy: **a** solar thermal laboratory; **b** heat pump laboratory: DHW storage tank (1), warm water storage tanks (2 and 3), heat pump (4), warm water storage tank (5) and DHW storage tank (6)

120 L connected to the heating system of the Solar House and to a DHW storage tank (6) with a capacity of 300 L. The warm water from the storage tank (5) is distributed in the underfloor heating system of the Solar

House allowing to decrease the temperature of the warm water and to increase the coefficient of performance of the heat pump.

The electrical energy demand for the heat pump system and for the lighting system of the Solar House is partly covered (40%) by fixed and by tracked photovoltaic platforms with an installed capacity of 10 kW and 1.28 kW, respectively (Fig. 4.62). Both photovoltaic platforms are grid connected. Thus, the excess electrical energy produced during summer is supplied to the grid to compensate for the consumption

Fig. 4.62 Fixed (1) and tracked (2) photovoltaic platforms of 10 kW and 1.28 kW, respectively

during winter. The fixed photovoltaic platform (1), consisting of 48 polycrystalline silicon photovoltaic modules each with a power of 210 W, is optimally tilted to produce the maximum amount of electrical energy over the entire year. The tracked photovoltaic platform (2) supports four types of photovoltaic modules: monocrystalline, polycrystalline and amorphous silicon and CIS thin film. It was designed to comparatively assess the output response of these modules for different types of tracking algorithms [88]. An identical platform is installed at the Zwickau University, in Germany.

Moldovan and Visa [56] developed TRNSYS numerical simulations to evaluate the optimal renewable-based energy mix for a multifamily building community targeting the Nearly Zero Energy Community standard [93]. Five building types were analysed (Fig. 4.63): a single-family building B1 (100 m^2), further considered as a reference unit; a multifamily building B2 with two units on one single floor; a multifamily building B3 with four units on one single floor and other two multifamily buildings B4 and B5 with two and four floors, respectively, each of these with four units per floor. When grouping the units, the shape factor of each building (the ratio between the surface of the building's envelope and the volume of the building) is reduced as presented in Table 4.14.

Firstly, the influence of the building configuration on its thermal energy demand for space heating was evaluated and the results are presented in Table 4.15. For the single-family building B1, 62.51 kWh/(m^2 year) resulted in the yearly specific thermal energy demand for space heating. This indicator decreases with the increase of the units number per multifamily buildings, reaching the lowest value of 32.98 kWh/(m^2 year) for the building B5.

The results show that significant reductions in the thermal energy demand for heating, up to 47%, can be reached by simply merging single-family buildings into multifamily buildings. The energy savings obtained for the four multifamily buildings (B2–B5) using as reference the single-family building B1 are plotted in Fig. 4.64 as function of the shape factor of each building (the single-family building has a shape

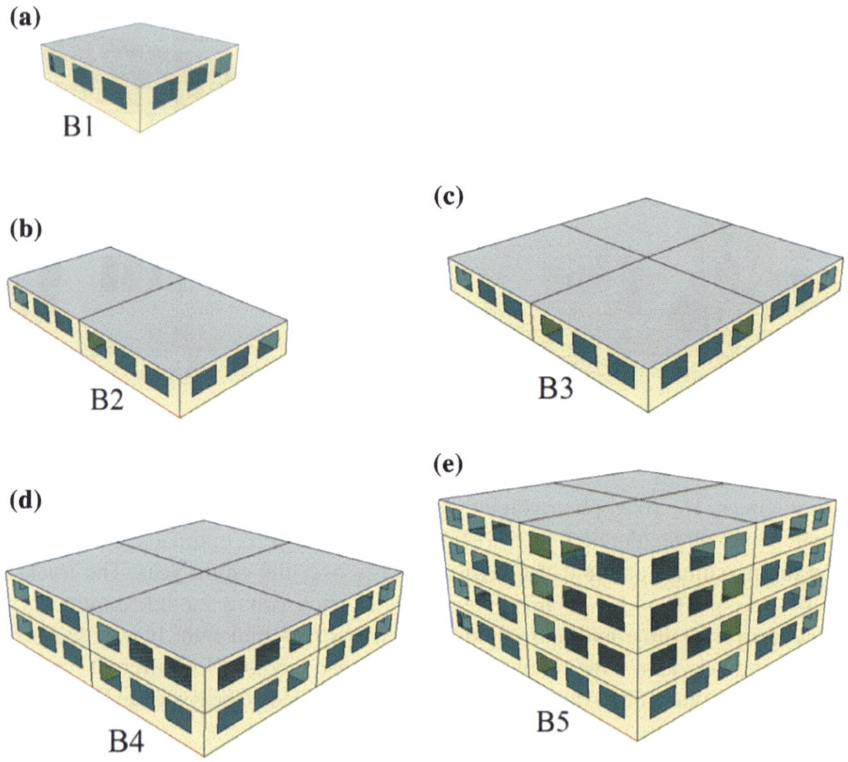

Fig. 4.63 Volumetric configuration of the: **a** single-family building (B1); **b–e** multifamily buildings (B2–B5)

Table 4.14 Specific features of the single-family (B1) and multifamily (B4–B5) buildings

Feature	B1	B2	B3	B4	B5
Number of units/floors	1/1	2/1	4/1	8/2	16/4
Length/Width (m)	10/10	20/10	20/20	20/20	20/20
Height (m)	2.5	2.5	2.5	5	10
Total floor surface area (m^2)	100	200	400	800	1600
Envelope surface area (m^2)	300	550	1000	1200	1600
Volume (m^3)	250	500	1000	2000	4000
Shape factor (m^2/m^3)	1.2	1.1	1.00	0.60	0.40

factor of 1.2). It has to be noticed that these energy savings, once implemented, will not generate operating costs during the entire lifetime of the buildings.

Further on, the thermal energy demand for DHW and the electrical energy demand for lighting were assessed. The results are included in Table 4.16 along with the total energy demand of each building. As the results show, all the buildings reach the

Table 4.15 Heating energy demand of single-family (B1) and multifamily (B4–B5) buildings

Features	B1	B2	B3	B4	B5
Number of units in the building	1	2	4	8	16
Total floor area of the building (m^2)	100	200	400	800	1600
Yearly thermal energy demand for space heating (kWh/year)	6251	11,629	22,203	36,597	52,766
Specific total yearly energy demand [kWh/(m^2 year)]	62.51	58.15	55.51	45.75	32.98

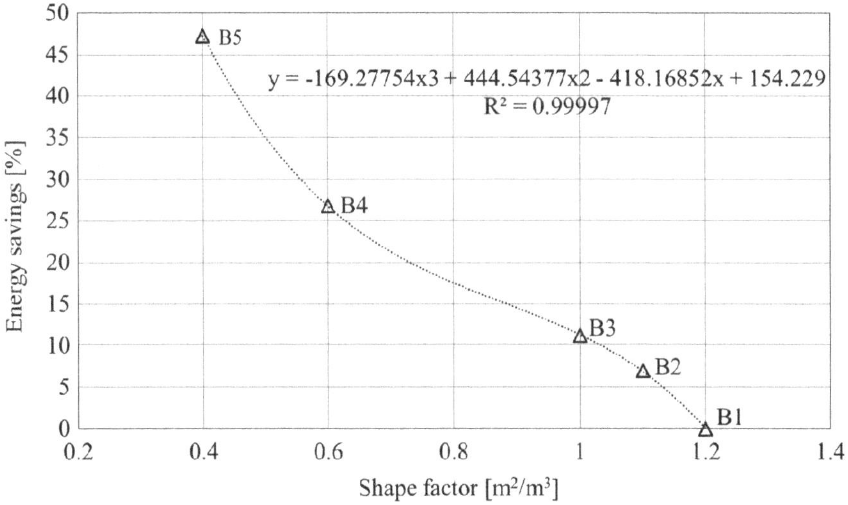

Fig. 4.64 Variation of the energy savings with the shape factor of the single-family (B1) and multifamily (B2–B5) buildings

Table 4.16 Energy demand of single-family (B1) and multifamily (B2–B5) buildings

Building	B1	B2	B3	B4	B5
Number of units in the building	1	2	4	8	16
Yearly thermal energy demand for DHW (kWh/year)	5305	10,610	21,221	42,441	84,883
Yearly electrical energy demand for lighting (kWh/year)	1460	2920	5840	11,680	23,360
Total yearly energy demand (kWh/year)	13,017	25,159	49,263	90,719	161,009
Specific total yearly energy demand [kWh/(m^2 year)]	130.17	125.80	123.16	113.40	100.63

low energy building standard ranging in EU between 50 and 150 kWh/(m^2 year) according to Kurnitski [45]; thus, no further measures are required to decrease the energy demand.

Among the components of the buildings' energy demand, the energy demand for DHW preparation is close to that for space heating for the first three building types and becomes the highest component starting with building B4. Thus, the solar thermal system is firstly considered in the design process of the renewable-based energy mix. The number of solar thermal collectors to be installed results from the requirement of maximizing the thermal energy output of the solar thermal system while avoiding the system overheating, by keeping below 10% the excess thermal energy. To maximize the received solar energy and thus the thermal energy output, the solar thermal collectors will be installed fixed on the building's rooftop at an optimal tilt angle of 40° and an azimuth angle of 0° (South oriented). This position maximizes the yearly amount of received solar energy, with the highest share during summertime when more sunny days with longer daylight time occur. The total surface of the required solar thermal collectors, the total thermal energy yearly supplied and its useful part, as well as the solar fraction, are included, for each building, in Table 4.17.

The monthly energy demand for DHW and the thermal energy that can be supplied by the solar thermal system are plotted in Fig. 4.65 for the B3 building. Overproduction occurs, as expected, during the summer months, with the highest values in July and August when measures to avoid overheating the solar thermal collectors have to be considered such as storing the excess thermal energy into large tanks or directing it to the ground heat exchanger. The first solution requires large space for installation while the second solution is beneficial for enhancing the ground temperature, thus increasing the coefficient of performance of a heat pump system that will use this thermal energy during the winter season.

The next step is to design the ground-coupled heat pump system to fully cover the thermal energy demand for DHW in the transition and winter months along with the thermal energy demand for space heating. A parallel system is considered, with the heat pump supplying the thermal output either to a storage tank for the warm water used in the low-temperature radiant system or to the upper coil of the DHW storage tank. The heat pump thermal power capacity is estimated through TRNSYS simulations, and the results obtained for the five buildings being inserted in Table 4.18

Table 4.17 Features of the solar thermal system

Building	B1	B2	B3	B4	B5
Total surface area of the solar thermal collectors (m^2)	6.3	12.6	25.2	50.4	100.8
Yearly thermal energy output (kWh/year)	4701	9362	18,501	35,736	64,310
Useful thermal energy output (kWh/year)	4234	8451	16,807	3302	62,168
Thermal energy demand for DHW (kWh/year)	5305	10,610	21,221	42,441	84,883
Solar fraction (%)	79.81	79.65	79.20	77.76	73.24

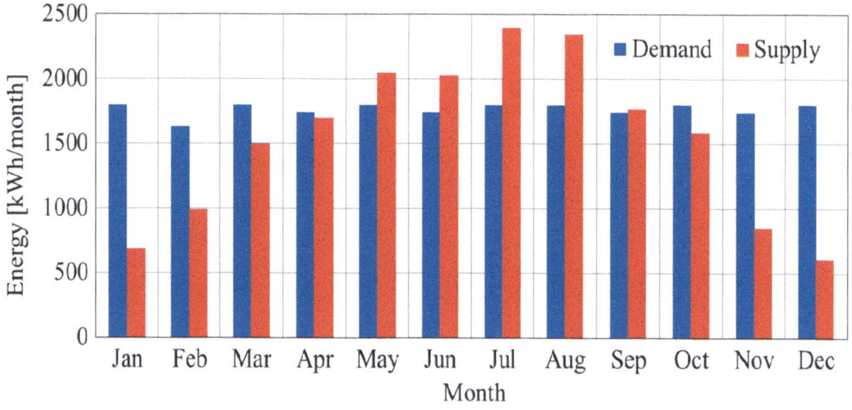

Fig. 4.65 Thermal energy demand and supply for domestic hot water preparation in the multifamily building B3

Table 4.18 Features of the ground-coupled heat pump system

Building	B1	B2	B3	B4	B5
Thermal capacity of the heat pump (kW)	3.94	7.55	14.54	25.67	47.82
Thermal energy supplied by the heat pump (kWh/year)	7322	13,788	26,616	46,037	75,481
Yearly heat pump coefficient of performance (–)	2.98	3.01	3.02	3.04	3.10

along with the thermal energy supplied and the seasonal coefficient of performance of the heat pump system.

The thermal energy demand (for domestic hot water—DHW and for space heating—SH) and the energy supplied (by the solar thermal system—STS and by the heat pump system—HP) all over the year are plotted in Fig. 4.66 for the B3 building. The heat pump system is designed to supply each month the thermal energy demand not covered by the solar thermal system. The largest shares of the thermal energy supplied by the heat pump occur during the winter months, between November and March.

In the last step, a grid-connected photovoltaic system is designed to yearly balance the electrical energy demand of the building. Because a heat pump system was considered for meeting the thermal energy demand for space heating and for DHW preparation that is not covered by the solar thermal system, the electricity required to drive the heat pump must also be included into the electrical energy demand of the building, along with the energy demand for lighting. The entire electrical energy demand of the building is detailed in Table 4.19 for the five assessed buildings.

Due to the limited space available on the buildings' rooftops, small-sized (1400 mm/635 mm) photovoltaic modules were selected to be mounted on the suitable surfaces on the building facade. The photovoltaic system was designed to fully

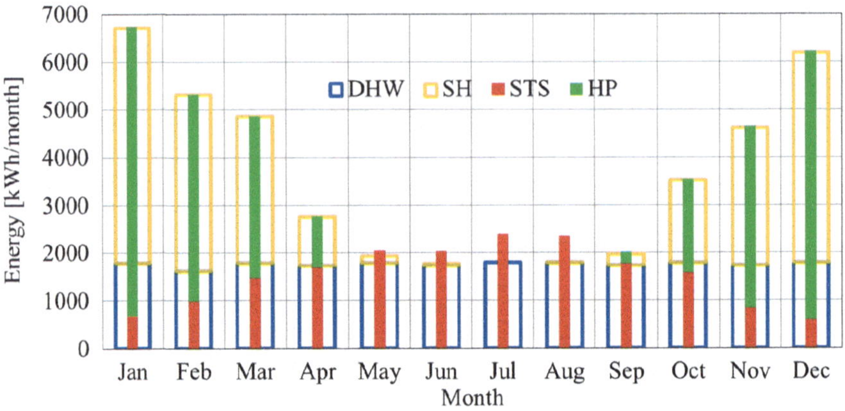

Fig. 4.66 Thermal energy demand for domestic hot water (DHW) and space heating (SH), and the energy supplied by the solar thermal system (STS) and by the heat pump system (HP) for the B3 building

Table 4.19 Electrical energy demand of the single-family (B1) and multifamily (B2–B5) buildings

Building	B1	B2	B3	B4	B5
Electrical energy demand for lighting (kWh/year)	1460	2920	5840	11,680	23,360
Electrical energy demand for the heat pump (kWh/year)	2457	4581	8813	15,144	24,349
Total electrical energy demand (kWh/year)	3917	7501	14,653	26,824	47,709

cover the electrical energy demand, firstly considering the available space on the rooftop and afterwards the building facades if the electrical energy demand cannot be covered by the rooftop installed photovoltaic modules. A special attention was paid to the rooftop installed photovoltaic system where shadowing between consecutive rows of photovoltaic modules was avoided by optimizing the distance between them. The photovoltaic modules installed on the facades were vertically positioned to avoid shading and to prevent various aesthetical issues. The installed capacities of the photovoltaic systems, with optimally tilted (at 40°) photovoltaic modules, that fully cover the electrical energy demand of each building are comparatively presented in Table 4.20.

After designing the components in the energy mix, the available surfaces for installing these components have to be assessed.

The rooftop of each building is large enough to accommodate the solar thermal collectors, even for an additional floor, with another four units. However, if the building has more than five floors, the rooftop is no longer enough to install all the required solar thermal collectors and the southern facade has also to be considered.

For the heat pump system, the geothermal heat exchanger can be developed in a horizontal plane, if enough surface area is available close to the building or in

Table 4.20 Features of the photovoltaic system with optimally tilted modules for the single-family (B1) and multifamily (B2–B5) buildings

Building	B1	B2	B3	B4	B5
Installed capacity of the PV system (kW)	4.20	8.00	15.60	28.60	50.80
Yearly supplied electrical energy (kWh/year)	3946	7516	14,655	26,868	47,724
Solar fraction (%)	100.73	100.20	100.01	100.17	100.03

Table 4.21 Ground surface required for the horizontal geothermal heat exchanger for the single-family (B1) and multifamily (B2–B5) buildings

Buildings	B1	B2	B3	B4	B5
Ground surface area (m^2)	98.5	188.7	363.4	642.0	1196.0

a vertical position (with vertical boreholes) if the available surface is low. In the analysed case study, only the surface of a horizontal geothermal heat exchanger is assessed, as the solution with larger requirement in terms of available surface for installation. The required ground surfaces are included in Table 4.21 for the analysed buildings.

The PV systems cannot be entirely installed on the rooftop of any of the five buildings. For the single-family building B1 and for the multifamily building B2, the rooftop and the southern facade are required. For the B3 building, the eastern or the western facades will be additionally required. In the case of the B4 and B5 buildings, besides the rooftops, southern, eastern and western facades, additional 282 m^2 (for B4) and 744 m^2 (for B5) of ground surfaces are required. In conclusion, the renewable energy that can be obtained "on building", by installing solar thermal collectors and PV modules, can cover the building energy demand with shares ranging between 63% for the B1 building and 50.7% for the B5 building. These shares are mainly relying on the rooftop area, the contributions of the southern facade being in the range 6.6–3.5% while the eastern or western facades can add only 3.1–2.4% each. The renewable energy share can be further increased by a ground-coupled heat pump system, requiring additional ground surface nearby the building (Table 4.21). Thus, a well-designed renewable energy mix can support the nZEB standard for an energy-efficient building, based only on the available surfaces on and near the building.

When the building rooftop has a limited available surface for installing solar energy convertors, hybrid photovoltaic thermal (PVT) modules, combining both solar technologies, may represent a solution to reach a higher solar share and energy efficiency than the individual photovoltaic modules and solar thermal collectors [28]. A low energy building system was designed in Switzerland including unglazed PVT modules and a ground-coupled heat pump with a novel dual-depth borehole field with solar recharging [5, 51] through which an average heating seasonal coefficient of performance COP = 6 was obtained. Lee et al. [47] found a reduction of 11% of the net energy demand (heating, cooling and electricity) in a community-scale PVT—ground source heat pump system in Korea as compared to a ground source heat pump system. Putrayudha et al. [71] later reached an additional 15% of energy savings using a fuzzy logic controller in the same system. Pardo García et al. [69] studied an energy mix based on glazed PVT and a heat pump system supplying a low-temperature district heating system for a multifamily house in Prague in multiple configurations. They obtained a maximum 7% reduction in the energy costs by connecting the PVT to the low-temperature district heating network; this was found as the most effective solution since there are no storage or load matching limitations, thus gaining a maximum production from the PVT modules. As in the case of solar thermal—heat pump energy mixes, PVT modules—ground source heat pump systems have the potential to reduce the borehole's depth [82] , thus decreasing the initial investment cost of the ground heat exchanger.

4.7 Economic and Financial Aspects of Solar Thermal Systems Implemented in the Built Environment

The feasibility of a solar thermal system is directly related to:

- The initial investment cost;
- The operating and maintenance costs;
- The costs of dismantling the system at the end of its lifetime;
- The cost of conventional fuels.

The initial investment cost of the solar thermal system is mainly associated with the solar thermal collectors, storage tanks, pumping group, solar controller, piping and accessories. If the capital for the initial investment has to be borrowed, the associated interests, taxes and insurances must also be considered in assessing the overall costs.

The solar thermal systems are characterized by low operating costs, as the energy source (the solar radiation) is free of charge. Thus, the only costs involved are related to the conventional fuel used as backup source and, with a lower influence, to the electrical energy required for powering the hydraulic pumps and the solar controller. Except the solar thermal collectors, the maintenance costs are similar to those related to any other heating system.

At the end of its lifetime, a solar thermal system has to be dismantled, the components recycled and these operations involve certain costs. Special attention has to be given to harmful or toxic materials contained in the components of the solar thermal system.

The costs of the conventional fuels have a significant impact on the feasibility of a solar thermal system, as low values will decrease its feasibility and this can be improved only by government incentives or supporting programs.

Thus, the decision of installing a solar thermal system can be considered an economic problem of comparing the costs generated by an initially known investment and by the future estimated operating and maintenance costs with the costs associated with a conventional system installed for the same functionality. The objective is to identify the optimal size of the solar thermal system that ensures the lowest cost of the supplied thermal energy. Many studies outlined that, due to the variability of the solar energy, a solar thermal system designed to fully cover the thermal energy demand it is generally not feasible. Such a solar thermal system must be designed considering the worst operating conditions (e.g. during the winter season with low solar radiation and very low outdoor air temperatures), resulting an oversized system that provides thermal energy in excess during the other (warmer) seasons; if this excess energy cannot be used the investment cannot be fully justified. Considering this, the best option is to design the solar thermal system in summertime conditions, thus avoiding overheating. During the rest of the year, a backup system is required to fully cover the thermal load.

The economic analysis of a solar thermal system can be developed using various methods, one of the simplest choice being based on the evaluation of the simple payback time.

The *simple payback time* (SPT) is the most common and the easiest method to evaluate the economic value of a project. The simple payback time is defined as the number of years required for the positive cash flows to surpass the initial investment, with no use of time value of money [43]. The simple payback time is the ratio of the initial investment cost of the system (C_s) to the annual cost savings (S)

$$\mathrm{SPT} = \frac{C_s}{S} \tag{4.15}$$

where the annual cost savings (S) are evaluated using

$$S = F \cdot E_{th} \cdot C_F \tag{4.16}$$

where

- F is the solar fraction defined as the ratio of the useful thermal energy supplied by the solar thermal system to the thermal energy demand of the building;
- E_{th} is the yearly thermal energy demand of the building;
- C_F is the cost of the saved conventional fuel by installing the solar thermal system.

If the simple payback time of a solar thermal system computed using Eq. (4.15) is shorter than or equal to the lifetime of the system, the project is accepted but otherwise it is rejected.

Considering a single-family building with an yearly thermal energy demand for DHW of 5305 kWh for which a solar thermal system was designed to cover a solar fraction of 79.81%, an initial investment cost of 2677.5 € for the solar thermal system and a cost of 0.1 €/kWh for the saved conventional fuel, the simple payback time of the solar thermal system

$$\mathrm{SPT} = \frac{2677.5}{0.7981 \cdot 5305 \cdot 0.1} = 6.32\,\text{years}$$

The usual lifetime of a solar thermal system ranges between 15 and 20 years; thus, the considered solar thermal system is a feasible one.

For projects developed over several years, the inflation rate for the conventional fuels and fuel-saving discounting rate has to be additionally considered for a more accurate economic analysis.

The payback time is defined as the time (usually expressed in years) required to get back the initial investment cost in the solar thermal system from the conventional fuel savings resulted when using it.

The payback time (PT) can be calculated, without discounting fuel savings, using the method presented by Duffie and Beckman [13]

$$\mathrm{PT} = \frac{\ln\left(\frac{C_s \cdot i_F}{F \cdot E_{th} \cdot C_{F1}} + 1\right)}{\ln(1 + i_F)} \tag{4.17}$$

where

- i_F is the fuel inflation rate;
- C_{F1} is the cost of the energy delivered from a unit of conventional fuel in the first year calculated based on the fuel higher heating value (HHV) and conventional system efficiency (η) using

$$C_{F1} = \mathrm{HHV} \cdot \eta \tag{4.18}$$

Considering a single-family building with an yearly thermal energy demand for DHW of 5305 kWh for which a solar thermal system was designed to cover a solar fraction of 79.81%, an initial investment costs of 2677.5 € for the solar thermal system, a cost of 0.1 €/kWh for the energy delivered using a unit of conventional fuel in the first year and a fuel inflation rate of 10%, the payback time of the solar thermal system

$$\mathrm{PT} = \frac{\ln\left(\frac{2677.5 \cdot 0.1}{0.7981 \cdot 5305 \cdot 0.1} + 1\right)}{\ln(1 + 0.1)} = 5.14\,\text{years}$$

The payback time can also be calculated with discounting fuel savings using Eq. (4.17) developed for the case of a fuel-saving discounting rate (d) different than the fuel inflation rate (i_F)

$$\mathrm{PT} = \frac{\ln\left(\frac{C_s \cdot (i_F - d)}{F \cdot E_{th} \cdot C_{F1}} + 1\right)}{\ln\left(\frac{1 + i_F}{1 + d}\right)} \quad \text{if}\, d \neq i_F \tag{4.19}$$

and for a fuel-saving rate (d) equal to the fuel inflation rate (i_F)

$$\mathrm{PT} = \frac{C_s \cdot (1 + i_F)}{F \cdot E_{th} \cdot C_{F1}} \quad \text{if}\, d = i_F \tag{4.20}$$

For the previous example, considering a fuel-saving rate of 7%, the payback time is

$$\mathrm{PT} = \frac{\ln\left(\frac{2677.5\cdot(0.1-0.07)}{0.7981\cdot5305\cdot0.1}+1\right)}{\ln\left(\frac{1+0.1}{1+0.07}\right)} = 6.28\,\text{years}$$

while in the case of a fuel-saving rate of 10% the payback time is

$$\mathrm{PT} = \frac{2677.5\cdot(1+0.1)}{0.7981\cdot5305\cdot0.1} = 6.96\,\text{years}$$

The simple payback time and the payback time are indicators easy to calculate but have the drawback of not showing the benefits of the investment during the lifetime of the solar thermal system. To this end, another economic indicator is used, net present worth (NPW), defined as the difference between the life cycle costs of a conventional fuel system and the life cycle cost of the solar thermal plus auxiliary energy system. It compares the present value of money with the value of money in the future, considering the inflation and returns. The net present worth is calculated using

$$\mathrm{NPW} = -C_\mathrm{S} + \sum_{j=1}^{N}\frac{S_j}{(1+i_\mathrm{F})^j} \tag{4.21}$$

where

- N is the lifetime of the system expressed in years;
- S_j represents the savings in the jth year.

The investment depends on the value of the net present worth. For a positive net present worth, the investment would add value to the investor and the project may be accepted. Negative net present worth means that the investor will lose money; thus, the project should be rejected. If the net present worth equals zero, the investment would neither gain nor lose value for the investor and the decision should be based on other criteria. For the previous example of the solar thermal system providing savings of 423.4 €/year, having an initial investment cost of 2677 €, a fuel inflation rate of 10%, and an expected lifetime of 20 years, the net present worth of the investment is

$$\mathrm{NPW} = -2677 + \sum_{j=1}^{20}\frac{423.4}{(1+0.1)^j} = 2599\,€$$

Thus, this investment is worth doing as the thermal energy provided by the solar thermal system will cover the investment costs and will generate a profit of 2599 €.

References

1. Amaral AR, Rodrigues E, Gaspar A Rodrigues, Gomes A (2018) Review on performance aspects of nearly zero-energy districts. Sustain Cities Soc 43:406–420
2. American Society of Heating, Refrigerating and Air-Conditioning Engineers (2017) ASHRAE handbook. Fundamentals. Atlanta, GA
3. Andrews JW (1981) Ground coupled solar heat pumps: analysis of four options. In: American Society of Mechanical Engineers (ed) 4th ASME solar energy division conference. American Society of Mechanical Engineers, Reno, Nevada
4. Association of German Engineers (2014) VDI 6002-1 Solar heating for potable water; Basic principles; System technology and application in residential buildings. Publishing House of German Engineers Association
5. Baetschmann M, Leibundgut H (2012) LowEx solar building system: integration of PV/T collectors into low exergy building systems. Energy Proc 30:1052–1059
6. Buker MS, Riffat SB (2016) Solar assisted heat pump systems for low temperature water heating applications: a systematic review. Renew Sustain Energy Rev 55:399–413
7. Burduhos BG, Visa I, Neagoe M, Badea M (2015) Modeling and optimization of the global solar irradiance collecting efficiency. Int J Green Energy 12(7):743–755
8. Cimmino M, Eslami-Nejad P (2017) A simulation model for solar assisted shallow ground heat exchangers in series arrangement. Energy Build 157:227–246
9. Comsit M, Visa I, Moldovan M, Isac L (2014) Architecturally integrated multifunctional solar thermal façades. In: Sustainable energy in the built environment-steps towards nZEB. Springer International Publishing, Berlin, pp 47–65
10. Dalenbäck J-O (1990) Central solar heating plants with seasonal storage—status report (D14:1990). IEA Solar heating and cooling programme—task VII. Swedish Council for Building Research, Stockholm, Sweden
11. Dan A, Chattopadhyay K, Barshilia HC, Basu B (2016) Colored selective absorber coating with excellent durability. Thin Solid Films 620:17–22
12. Dudita M, Isac L, Duta A (2012) Influence of solvents on properties of solar selective coatings obtained by spray pyrolysis. B Mater Sci 35(6):997–1002
13. Duffie J, Beckmann W (2013) Solar engineering of thermal processes. Wiley Interscience, New York
14. Duta A, Isac L, Milea A, Ienei E, Perniu D (2014) Coloured solar-thermal absorbers—a comparative analysis of cermet structures. Energy Proc 48:543–553
15. Emmi G, Zarrella A, De Carli M, Galgaro A (2015) An analysis of solar assisted ground source heat pumps in cold climates. Energy Convers Manage 106:660–675
16. Eslami-Nejad P, Langlois A, Chapuis S, Bernier M, Faraj W (2009) Solar Heat Injection into Boreholes. In: 4th Canadian solar buildings conference, Toronto, pp 237–246
17. European Heat Pump Association (2018) European Heat Pump Market and Statistics Report 2018, Brussels. https://www.ehpa.org/market-data/market-report/. Accessed 01 May 2019
18. European Parliament (2010) The Directive 2010/31/EU of the European Parliament and of the Council of 19 May 2010 on the energy performance of buildings. Official Journal of the European Union 53
19. Fahl P, Ganapathisubbu S (2011) Tracking benefits for solar collectors installed in Bangalore. J Renew Sustain Energy 3:1–12
20. Farooqui SZ (2013) A gravity based tracking system for box type solar cookers. Sol Energy 92:62–68
21. Fischer D, Madani H (2017) On heat pumps in smart grids: a review. Renew Sust Energy Rev 70:342–357
22. Fischer D, Toral TR, Lindberg KB, Wille-Haussmann B, Madani H (2014) Investigation of thermal storage operation strategies with heat pumps in German multi family houses. Energy Proc 58:137–144

23. Fischer D, Rautenberg F, Wirtz T, Wille-Haussmann B, Madani H (2015) Smart meter enabled control for variable speed heat pumps to increase PV self-consumption. In: 24th IIR International Congress of Refrigeration. https://doi.org/10.13140/RG.2.1.2566.3762
24. Florio P, Munari Probst MC, Sculler A, Roecker C (2018) Assessing visibility in multi-scale urban planning: A contribution to a method enhancing social acceptability of solar energy in cities. Sol Energy 173:97–109
25. Frank E, Mauthner F, Fischer S (2015) Overheating prevention and stagnation handling in solar process heat applications, Technical Report A.1.2, IEA SHC Task 49. http://task49.iea-shc.org. Accessed 5 Jun 2017
26. Freeman T, Mitchell J, Audit T (1979) Performance of combined solar-heat pump systems. Sol Energy 22:125–135
27. Gladen A, Davidson J, Mantell SC (2014) Selection of thermotropic materials for overheat protection of polymer absorbers. Sol Energy 104:42–51
28. Good C, Chen J, Dai Y, Grete A (2015) Hybrid photovoltaic-thermal systems in buildings—a review. Energy Proc 70:683–690
29. Grundfos (2019) One Alpha to fit them all, Denmark. http://pdf.directindustry.com/pdf/grundfos/alpha1-l/5420-782049.html. Accessed 01 May 2019
30. Hadorn JC (2015) Solar and heat pump systems for residential buildings, 1st edn. Ernst & Sohn GmbH & Co., Berlin
31. Haller MY, Carbonell D, Mojic I, Winteler C, Bertram E, Bunea M, Lerch W, Ochs F (2014) Solar and heat pump systems—summary of simulation results of the IEA SHC task 44/HPP annex 38. In: 11th IEA heat pump conference, Montreal, pp 1–12
32. Harrison S, Cruickshank CA (2012) A review of strategies for the control of high temperature stagnation in solar collectors and systems. Energy Proc 30:793–804
33. Hausner R, Fink C (2000) Stagnation behaviour of thermal solar systems. In: EUROSUN, Copenhagen, 19–22 June 2000
34. Hausner R, Fink C (2002) Stagnation behaviour of thermal solar systems. A report of IEA SHC—task 26, Solar Combysystems. http://task26.iea-shc.org. Accessed 7 Jun 2017
35. Hirvonen J, Sirén K (2018) A novel fully electrified solar heating system with a high renewable fraction—optimal designs for a high latitude community. Renew Energy 127:298–309
36. International Energy Agency (2013) solar heating and cooling task 44/HPP Annex 38 solar and heat pump systems. http://www.iea-shc.org/Data/Sites/1/publications/T44A38-Newsletter-2013-02.pdf. Accessed 02 Feb 2019
37. International Energy Agency (2017) Trends 2017 in photovoltaic applications, Task 1, Annex 32. http://www.iea-pvps.org/index.php?id=92&eID=dam_frontend_push&docID=4245. Accessed 01 Feb 2019
38. International Energy Agency (2018) Solar heat worldwide 2018, Solar Update, 67. https://www.iea-shc.org/solar-heat-worldwide. Accessed 29 May 2019
39. Institute for Solar Energy Research (2019) Heat pipes in solar collectors—thermodynamic fundamentals and evaluation plus new approaches of application. www.isfh.de/institut. Accessed 29 May 2019
40. Jani DE, Mishra M, Kumar Sahoo P (2018) A critical review on application of solar energy as renewable regeneration heat source in solid desiccant–vapour compression hybrid cooling system. J Build Eng 18:107–124
41. Kalogirou SA (2004) Solar thermal collectors and applications. Prog Energy Combust 30:231–295
42. Kamel RS, Fung AS, Dash PRH (2015) Solar systems and their integration with heat pumps: a review. Energy Build 87:395–412
43. Kaplan S (1983) Energy economics: quantitative methods for energy and environmental decisions. McGraw-Hill, New York
44. Kok JF, Parteli EJR, Michaels TI, Karam DB (2012) The physics of wind-blown sand and dust. Rep Prog Phys 75(10):106–901
45. Kurnitski J (2013) Cost optimal and nearly zero-energy buildings (nZEB). Definitions, calculation principles and case studies. Springer, London

46. Lai C-M, Hokoi S (2015) Solar façades: a review. Build Environ 91:152–165
47. Lee EJ, Entchev E, Soesanto AP, Kang EC (2014) Photovoltaic thermal (PVT) integrated ground source heat pump system annual energy performance simulation for a multi-load case. In: 11th international energy agency heat pump conference, Montreal, Canada
48. Lobaccaro G, Croce S, Lindkvist C, Munari Probst MC, Scognamiglio A, Dahlberg J, Lundgren M, Wall M (2019) A cross-country perspective on solar energy in urban planning: lessons learned from international case studies. Renew Sust Energy Rev 108:209–237
49. Marken C (2011) Overcoming overheating. Home Power Review 142. http://www.homepower.com/articles/solar-water-heating/domestic-hot-water/overcoming-overheating. Accessed 5 Jun 2017
50. Martinopoulos G (2018) Life cycle assessment of solar energy conversion systems in energetic retrofitted buildings. J Build Eng 20:256–263
51. Meggers F, Ritter V, Goffin P, Baetschmann M, Leibundgut H (2012) Low exergy building systems implementation. Energy 41:48–55
52. Meliss M (1997) Regenerative Energiequellen - Praktikum. Springer, Heidelberg
53. Mertin S, Caer VH-L, Joly M, Mack I, Oelhafen P, Scartezzini J-L, Schuler A (2014) Reactively sputtered coatings on architectural glazing for coloured active solar thermal facades. Energy Build 68:764–770
54. Ministry of Regional Development and Public Administration (2015) Government of Romania, Standard I9-2015 for the design, construction and operation of sanitary facilities for buildings. Official Journal of Romania, I, 830 bis
55. Mohanraj M, Belyayev Y, Jayaraj S, Kaltayev A (2018) Research and developments on solar assisted compression heat pump systems—a comprehensive review (Part-B: Applications). Renew Sustain Energy Rev 83:124–155
56. Moldovan M, Visa I (2017) Renewable energy systems for a multifamily building community. In: Nearly zero energy communities. Springer, Berlin, pp 129–147
57. Moldovan M, Visa I (2019) Development of an indoor testing rig for façade integrated solar thermal collectors. In: E3S Web Conference, vol 85, p 04005. https://doi.org/10.1051/e3sconf/20198504005
58. Moldovan M, Visa I, Neagoe M, Burduhos B (2014) Solar heating & cool-ing energy mixes to transform low energy buildings in nearly zero energy buildings. Energy Proc 48:924–937
59. Moldovan M, Visa I, Duta A (2017) Enhanced sustainable cooling for low energy office buildings in continental temperate climate. J Energy Eng 143(5):04017054-1-12
60. Moldovan M, Visa I, Burduhos B (2019) Experimental comparison of flat plate and evacuated tube solar thermal collectors for domestic hot water preparation in education facilities. J Sustain Develop Energy Water Environ Syst. http://dx.doi.org/10.13044/j.sdewes.d7.0285
61. Muehling O, Seeboth A, Ruhmann R, Eberhardt V, Byker H, Anderson C, De Jong S (2014) Solar collector cover with temperature-controlled solar light transmittance. Energy Proc 48:163–171
62. Munari Probst M, Roecker C (2007) Towards an improved architectural quality of building integrated solar thermal systems (BIST). Sol Energy 81:1104–1116
63. Munari Probst M, Roecker C (2012) Criteria for architectural integration of active solar systems IEA Task 41, Subtask A. Energy Proc 30:1195–1204
64. Neagoe M, Visa I, Burduhos BG, Moldovan M (2014) Thermal load based adaptive tracking for flat plate solar collectors. Energy Proc 48:1401–1411
65. Neagoe M, Visa I, Duta A, Cretescu N (2017) A new approach on the protection against overheating of flat plate solar-thermal collectors. In: Visa I, Duta A (eds) Nearly zero energy communities. Springer Proceedings in Energy, Heidelberg, pp 283–291
66. Nordell B, Ahlström A-K (2007) Freezing problems in borehole heat exchangers. In: Thermal energy storage for sustainable energy consumption, NATO Science Series (Mathematics, Physics and Chemistry). Springer, Dordrecht, pp 193–203
67. Pana I, Vitelaru C, Kiss A, Zoita NC, Dinu M, Braic M (2017) Design, fabrication and characterization of TiO_2-SiO_2 multilayer with tailored color glazing for thermal solar collectors. Mater Des 130:275–284

68. Parameshwaran R, Kalaiselvam S, Harikrishnan S, Elayaperumal A (2012) Sustainable thermal energy storage technologies for buildings: a review. Renew Sustain Energy Rev 16:2394–2433
69. Pardo García N, Zubi G, Pasaoglu G, Dufo-López R (2017) Photovoltaic thermal hybrid solar collector and district heating configurations for a Central European multi-family house. Energy Convers Manage 148:915–924
70. Poppi S, Sommerfeldt N, Bales C, Madani H, Lundqvist P (2018) Techno-economic review of solar heat pump systems for residential heating applications. Renew Sustain Energy Rev 81(1):22–32
71. Putrayudha AS, Kang EC, Evgueniy E, Libing Y, Lee EJ (2015) A study of photovoltaic/thermal (PVT)-ground source heat pump hybrid system by using fuzzy logic control. Appl Therm Eng 89:578–586
72. Rad FM, Fung AS (2016) Solar community heating and cooling system with borehole thermal energy storage—review of systems. Renew Sustain Energy Rev 60:1550–1561
73. Radiant Floor (2019) Overheating protection for solar collectors. www.radiantcompany.com/system/solar/heatdump/. Accessed May 2019
74. Reda F (2015) Long term performance of different SAGSHP solutions for residential energy supply in Finland. Appl Energy 144:31–50
75. Reda F, Laitinen A (2015) Different strategies for long term performance of SAGSHP to match residential energy requirements in a cold climate. Energy Build 86:557–572
76. Resch K, Wallner GM (2009) Thermotropic layers for flat-plate collectors—a review of various concepts for overheating protection with polymeric materials. Sol Energy Mat Sol C 93:119–128
77. Romanian Standards Association (1990) Sanitary installations. After supply for civil and industrial buildings. Main design specifications. Standardization Publishing House, Bucharest
78. Schuler A, Roecker C, Boudaden J, Oelhafen P, Scartezzini JL (2005) Potential of quarterwave interference stacks for colored thermal solar collectors. Sol Energy 79:122–130
79. Sibbitt B, McClenahan D, Djebbar R, Thornton J, Wong B, Carriere J, Kokko J (2012) The performance of a high solar fraction seasonal storage district heating system—five years of operation. Energy Proc 30:856–865
80. Slaman M, Griessen R (2009) Solar collector overheating protection. Sol Energy 83:982–987
81. SoDa (2019) Solar radiation data. www.soda-pro.com. Accessed 29 May 2019
82. Sommerfeldt N, Madani H (2016) Review of solar PV/thermal plus ground source heat pump systems for European multi-family houses. In: 11th ISES Eurosun Conference, Palma de Mallorca, Spain. https://doi.org/10.18086/eurosun.2016.08.15
83. Streicher W (2000) Minimising the risk of water hammer and other problems at the beginning of stagnation of solar thermal plants—a theoretical approach. In: EUROSUN, Copenhagen
84. Thiesen S (2009) Simple and safe solar heating: a whole systems approach. Pac J Sci Technol 10(1):117–122
85. Threlkeld JL (1953) Solar energy as a potential heat source for the heat pump (PhD Thesis). University of Minnesota
86. UrbanSol+ (2014) Solar thermal systems in multifamily house renovation. http://www.urbansolplus.eu/en/downloads. Accessed 31 May 2019
87. Visa I, Comsit M, Duta A (2014) Urban acceptance of facade integrated novel solar thermal collectors. Energy Proc 48:1429–1435
88. Visa I, Comsit M, Moldovan M, Duta A (2014) Outdoor simultaneous testing of four types of photovoltaic tracked modules. J Renew Sustain Energy 6:4871003
89. Visa I, Moldovan M, Comsit M, Duta A (2014) Improving the renewable energy mix in a building toward the nearly zero energy status. Energy Build 68:72–78
90. Visa I, Duta A, Comsit M, Moldovan M, Ciobanu D, Saulescu R, Burduhos B (2015) Design and experimental optimisation of a novel flat plate solar thermal collector with trapezoidal shape for facades integration. Appl Therm Eng 90:432–443
91. Visa I, Jaliu C, Duta A, Neagoe M, Comsit M, Moldovan M, Ciobanu D, Burduhos B, Saulescu R (2015) The role of mechanisms in sustainable energy systems. Transilvania University Pub. House, Brasov

92. Visa I, Comsit M, Duta A, Neagoe M, Moldovan M, Burduhos B, Perniu D, Enesca A, Isac L, Cosnita M, Totu I, Savvides I, Vassiliades C (2017) Novel solar thermal collectors/array with increased architectural acceptance for building integration. In: Building integration of solar thermal systems design and applications handbook, pp 373–391, COST Office
93. Visa I, Duta A, Moldovan M, Burduhos B (2017) Implementing renewable energy systems in nearly zero energy communities. In: Nearly zero energy communities. Springer, Berlin, pp 3–24
94. Visa I, Moldovan M, Comsit M, Duta A (2017) Infield output of a new solar-thermal Façade with increased architectural acceptance. Adv Mater - TechConnect Briefs 2:62–65
95. Visa I, Moldovan M, Comsit M, Neagoe M, Duta A (2017) Facades integrated solar-thermal collectors—challenges and solutions. Energy Proc 112:176–185
96. Visa I, Duta A, Moldovan M (2019) Outdoor performance of a trapeze solar thermal collector for facades integration. Renew Energy 137:37–44
97. Visa I, Moldovan M, Duta A (2019) Novel triangle flat plate solar thermal collector for facades integration. Renew Energy 143:252–262
98. Weiss W (2003) Solar heating systems for houses. a design handbook for solar combi-systems. James & James, London
99. Weiss W, Spörk-Dür M (2018) Solar heat worldwide, global market development and trends in 2017, Detailed market Figures in 2016. International Energy Agency Solar Heating & Cooling Programme. https://www.iea-shc.org/Data/Sites/1/publications/Solar-Heat-Worldwide-2018.pdf. Accessed 25 May 2019
100. Weiss W, Spörk-Dür M (2018) Solar heat worldwide. The Solar heating and Cooling Programme, International Energy Agency. https://www.iea-shc.org/solar-overview. Accessed May 2019
101. Weiss W, Spörk-Dür M (2019) Solar heat worldwide, global market development and trends in 2018, Detailed market figures 2017. International Energy Agency Solar Heating & Cooling Programme. https://www.iea-shc.org/Data/Sites/1/publications/Solar-Heat-Worldwide-2019.pdf. Accessed 27 Jul 2019
102. Wijewardane S, Goswami DY (2012) A review on surface control of thermal radiation by paints and coatings for new energy applications. Renew Sustain Energ Rev 16:1863–1873
103. Wu Y, Zheng W, Lin L, Qu Y, Lai F (2013) Colored solar selective absorbing coatings with metal Ti and dielectric AlN multilayer structure. Sol Energy Mat Sol C 115:145–150
104. Yadav AK, Chandel SS (2013) Tilt angle optimization to maximize incident solar radiation: a review. Renew Sustain Energy Rev 23:503–513
105. Yang R, Liu J, Lin L, Qu Y, Zheng W, Lai F (2016) Optical properties and thermal stability of colored solar selective absorbing coatings with double-layer antireflection coatings. Sol Energy 125:453–459
106. Zhu D, Zhao S (2010) Chromaticity and optical properties of colored and black solar–thermal absorbing coatings. Sol Energ Mat Sol C 94:1630–1635

Chapter 5
Increasing the Solar Share for Electrical and Thermal Energy Production in the Built Environment

5.1 PVT Systems

A photovoltaic thermal (PVT) system produces electrical and thermal energy by using a single solar energy conversion device, the PVT module, instead of a photovoltaic module and a solar thermal collector. This is why, besides the PVT module, a PVT system contains components specific to both solar thermal and photovoltaic systems.

A functional scheme of a PVT system that uses an antifreeze thermal fluid is presented in Fig. 5.1 outlining the electrical and thermal circuits:

- The PVT modules (1) are electrically connected through a direct current (DC) separator (2) to an inverter (3) connected through an alternating current (AC) separator (4) to the building's electrical distribution panel (5);
- The PVT modules (1) are thermally connected to the internal heat exchanger (6) in the storage tank (7) through a hydraulic pump (8) driven by the solar controller (9). The storage tank has a backup heater (10). The fluid volume variation is balanced using an expansion vessel (11).

The PVT modules (1) convert the solar energy to electrical energy similarly to the photovoltaic modules and to thermal energy as the solar thermal collectors do. Thus, in the electrical circuit, the modules are serially connected according to the maximum voltage accepted by the inverter (3). The DC separator (2) allows disconnecting the PVT modules (1) from the inverter (3) during the maintenance operations on the PVT system. The inverter (3) converts the DC supplied by the PVT modules (1) into AC required by the consumers. The AC separator (4) is used to disconnect the inverter (3) from the building's electricity distribution panel (5) if an intervention to the electrical grid is required. The building's electricity distribution panel (5) contains safety devices protecting both the equipment and the users and a bidirectional electric meter to measure the energy consumed from the grid and that one injected by the PVT system into the grid when not used in the building. Simultaneously with the

I. Visa et al., *Solar Energy Conversion Systems in the Built Environment*, Green Energy and Technology, https://doi.org/10.1007/978-3-030-34829-8_5

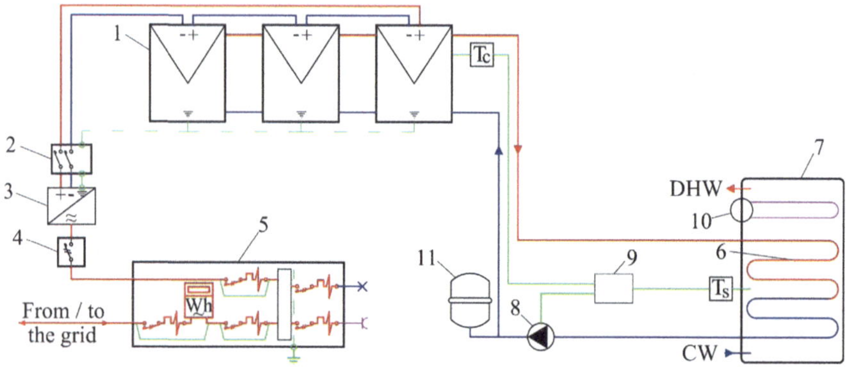

Fig. 5.1 Functional scheme of a PVT system: PVT modules (1), DC separator (2), inverter (3), AC separator (4), distribution panel (5), internal heat exchanger (6), storage tank (7), hydraulic pump (8), solar controller (9), electric heater (10), expansion vessel (11), cold water (CW), domestic hot water (DHW), temperature sensor—collector (T_c), temperature sensor—storage tank (T_s)

photovoltaic conversion, the PVT modules (1) are heated by the solar radiation, and the thermal energy is transferred to the cold water (CW) in the storage tank (7) through an internal heat exchanger (6) using a thermal fluid circulated by the hydraulic pump (8). The pump will be switched on by the solar controller (9) as long as the temperature measured by the sensor (T_c) installed on the PVT module outlet is higher than the temperature of the water in the storage tank measured by a sensor (T_s). When solar energy is scarce, a backup heater (10) is used to supply the thermal energy demand. The volume variations of the thermal fluid are taken over by the expansion vessel (11).

As in the case of individual photovoltaic and solar thermal systems, various types of PVT systems can be considered. For the electrical part, the PVT systems can be also designed as stand-alone systems, batteries, controllers and appropriate inverters thus being required. For the thermal part, the direct use of the heated water in the PVT modules through natural circulation represents an option in locations where the outdoor air temperatures are above the freezing point of water, thus simplifying the system and reducing the costs. Moreover, an external heat exchanger can be used when the thermal energy must be supplied to an existent domestic hot water storage tank with only one internal heat exchanger connected to a heating source (e.g. heat pump, electric heater, gas or biomass boiler). Various types of thermal fluids can be used to transfer the thermal energy from the PVT module to the user: liquids (water, antifreeze solution or nano-fluids), gases (air), refrigerants (when the PVT module plays the role of the vapourizer of a heat pump) or phase-change materials. For each type of thermal fluid, the components of the systems are accordingly designed.

PVT systems are good candidates for buildings with electrical energy demand larger than the thermal energy requirements, mainly in countries with warm climates where the large amounts of available solar energy can be used for solar cooling systems based on photovoltaic air-conditioning technologies and for domestic hot water preparation. Moreover, in these climates the solar thermal collectors already have reached a high penetration rate, and the acceptance of similarly looking PVT modules is increased [7]. Thus, the available surfaces off the buildings' roof or facades are extensively exploited for electrical energy production, additionally supplying thermal energy.

Usually, the thermal energy supplied by a PVT system represents about three quarters of the entire energy output while the electrical energy represents only one quarter. As an example, Solimpeks produces two PVT module types with the same surface: PowerVolt is designed for a higher electrical power (200 W) than PowerTherm producing only 180 W electrical power, while the thermal power is 630 W and 680 W, respectively [12]. Other PVT manufacturers as DualSun focus even more on the electrical energy production by designing modules with higher ratios (almost 50%) between nominal electrical (280 W) and thermal (570 W) powers DualSun [2].

The thermal energy produced by the PVT systems must be kept at relative low-temperature levels aiming at maintaining (or even increasing) the efficiency of the photovoltaic cells. These low temperatures can be used in various applications that require large amounts of water and rather low-temperature levels such as the low-temperature (radiant) building's heating systems or for domestic hot water preparation, as process heat for washing and cleaning the food or in the drinks, paper or textile industries. According to the Swiss Institute for Solar Technology, four options represent the most common uses in implementation locations similar to central Switzerland [3]:

- Warm water is produced for direct domestic use (35–40 °C), without the need to mix it with cold water; special attention has to be paid to avoid the bacterial contamination (e.g. with Legionella) in the large storage tanks that must be once each day heated up to 60 °C;
- Preheating the domestic hot water; further increasing its temperature requiring other sources (e.g. a gas or an electrical boiler);
- Heating the water in swimming pools;
- Regenerating the ground temperature where the geothermal energy is extracted using ground-coupled heat pumps.

The main advantages of a PVT system are as follows:

- A large share of the electrical energy demand of the building can be covered using all the available and suitable surfaces of the building's envelope while the thermal energy is supplied by using the same surfaces;
- Small available surfaces (e.g. 2–3 m^2) can be used to produce both electrical and thermal energy by converting solar energy;
- The combination of the two systems may lower the temperature of the photovoltaic cells. As described in Chap. 3, the conversion efficiency of photovoltaic modules decreases when the temperature increases (with roughly 0.2%/°C for thin film modules and up to 0.5%/°C for crystalline silicon-based modules). Thus, lowering the temperature of the photovoltaic cells in the PVT modules is welcome for stabilizing their conversion efficiency and the current–voltage characteristic curve and for increasing their effective lifespan [6];

- In cold regions, snow and ice can be (automatically) removed from the PVT modules surface by circulating the warm thermal fluid for a short time interval when the solar radiation is available.

The PVT systems also have several disadvantages:

- Reduced efficiency of the PV cells when thermal energy with high temperature is required;
- The need for a large storage capacity for the thermal energy supplied during the summer months, while only about 10% is supplied during the winter [10];
- Solutions to store these large amounts of low-temperature thermal fluid that have to be implemented are increasing the system's cost;
- Without these storage systems, the PVT systems should be designed similar to the solar thermal systems, to produce the required thermal energy only during the warm intervals, thus lower areas of PVT modules that will produce reduced amounts of electrical energy;
- If the PVT systems are designed to cover the electrical energy demand, the thermal component is likely to be oversized, generating overheating issues during the time intervals with large amounts of received solar energy; these issues affect the thermal system and the photovoltaic process;
- Their relatively high complexity that requires both electrical and thermal components; this might complicate even more the system when refurbishing older buildings. Moreover, professional designers and installers are required with an appropriate expertise in both fields [1].

5.1.1 PVT Modules

The most important component in a PVT system is the PVT module. It consists of a PV module and a system that remove the heat for a specific application (domestic hot water preparation, space heating, etc.). Thus, the temperature of the PV cells is kept at the lowest possible value, increasing the electrical energy output of the PVT module. This is usually done using air or water as heat extraction solutions [8]:

- *Air-cooled PVT modules*, with either natural or forced circulation, represent relatively simple and low-cost solutions to remove the heat from the PVT modules. The simplest type of PVT module (Fig. 5.2) consists of a glazing (1), a PV module (2), a back insulation (3) and a casing (4). The glazing allows solar radiation to reach the PV module while reducing the thermal losses through convection from this module to the outdoor air. The air gap between the glazing and the PV module is used to circulate an airflow that removes the heat from the PV module surface. Usually, the air ascendingly flows through the PVT module taking advantage of the natural buoyancy. The back insulation reduces the thermal losses of the PVT module to the ambient air. The casing provides mechanical resistance to all the PVT module's components and, being waterproof, also limits corrosion. The main

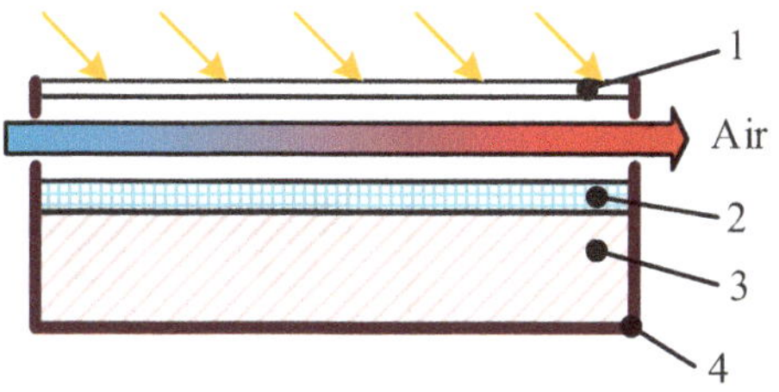

Fig. 5.2 Cross section through an air-cooled PVT module: glazing (1), PV module (2), back insulation (3), casing (4)

advantage of this PVT module type is the fact that it does not require any special electrical insulation of the PV part, as possible leakages are not dangerous and can be automatically compensated. Therefore, air-cooled PVT modules have a low manufacturing cost. The major disadvantage of this solution is represented by the low specific heat capacity of the air (about four times lower than that of water) imposing high flow rates that generate pressure losses and power consumption. This type of PVT modules is used mainly in air-based heating systems reaching high efficiency both in the electrical and thermal energy output at low air temperatures, at the air inlet and outlet. Thus, air-cooled PVT modules are used at higher latitudes or in altitude regions where the periods with higher outdoor air temperature are relatively short.
To further increase the heat transfer rate from the PV module to the thermal fluid (air), a second gap can be included between the PV module and the thermal insulation. The airflow through these gaps can be in the same direction (Fig. 5.3a) or in opposite directions (Fig. 5.3b). The first solution can be applied in natural circulation systems while the second one is implemented only in systems with forced circulation;

- *Liquid-cooled* PVT modules represent a possible option when water has to be (pre)heated. The liquid-cooled PVT modules (Fig. 5.4) consist of a glazing (1), a PV module (2), an absorber plate (3), pipes (4), a thermal insulation (5) and a

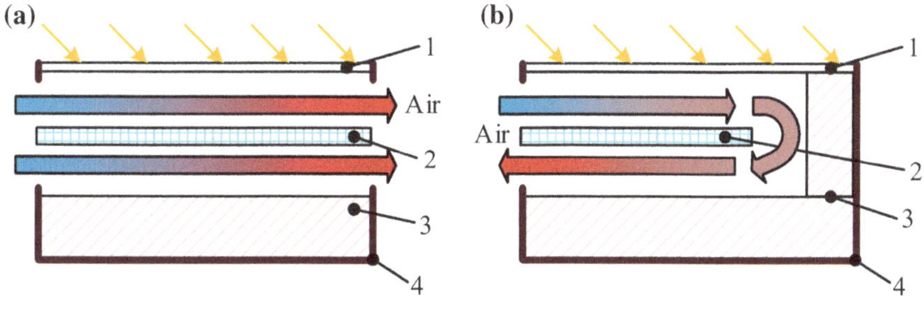

Fig. 5.3 Cross sections through air-cooled PVT module with: **a** two gaps with parallel flows; **b** two gaps with return flow: glazing (1), PV module (2), back insulation (3), casing (4)

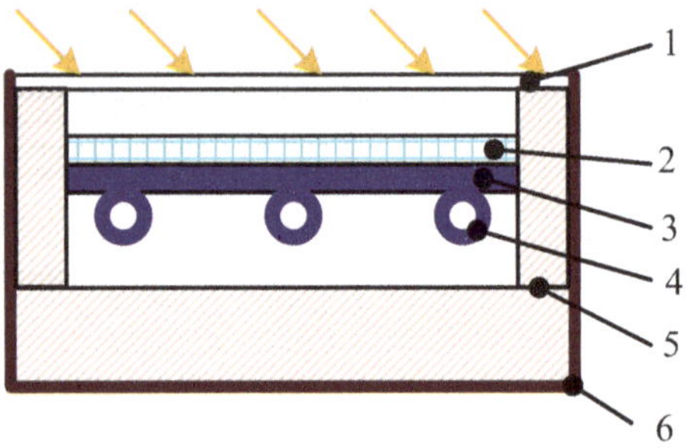

Fig. 5.4 Cross section through a liquid-cooled PVT module: glazing (1), PV module (2), absorber plate (3), pipes (4), back insulation (5), casing (6)

casing (6). In this case, the air gap between the glazing and the PV module acts as a thermal insulator. The absorber plate is attached to the back of the PV module to remove the heat from it and to transfer this heat to the liquid flowing through the pipes bounded to the absorber plate. The other components have the same role as in the air-cooled PVT modules. This solution is more expensive than that using air as cooling fluid, but it is more effective due to the higher specific heat capacity of the liquids as compared to air. Water can be circulated directly through the PVT module when the outdoor temperature does not drop below 0 °C in the implementation location or an antifreeze solution can be used in the components exposed to the outdoor freezing conditions.

5.2 PVT Integration in the Built Environment

PVT systems represent a good solution when a high solar fraction is targeted and the available surface for installing the solar energy converters is scarce. This is the case of multifamily buildings with a large number of floors thus a reduced roof surface/inhabitant ratio also having a thermal energy demand larger than the electrical one [4]. When the building's roof is not available or not enough to install the PVT modules, the southern facade of the building can be further considered. The installation of the PVT modules on the building facade brings advantages by simplifying the piping system as the PVT modules can be directly connected to the storage tanks distributed in each apartment in the building thus reducing the length and size of the pipes, the associated thermal losses and the cost of the entire system. Researchers also investigated façade integrated air-cooled PVT modules as the review of Wang et al. [13] shows. It was observed that a higher airflow rate faster removed the thermal energy decreasing its temperature with a positive effect on the PVT efficiency.

An example of a PVT system installed on the rooftop of a laboratory building in the R&D Institute of the Transilvania University of Brasov, Romania, is presented in Fig. 5.5. This system consists of two identical PVT modules (1 and 2) and a PV module (3) used as a reference; the system is tested to evaluate the electrical

Fig. 5.5 PVT outdoor testing rig installed on the rooftop of a laboratory building consisting of two PVT modules (1 and 2) and a reference PV module (3)

and thermal efficiencies of the PVT modules in different operating conditions. All modules are fixed, installed at a tilt angle of 40°, facing South.

The functional scheme of this testing rig is detailed in Fig. 5.6. Both PVT modules (1 and 2) and the reference PV module (3) are serially connected, along with other 20 PV modules installed on the rooftop of the building, through individual power optimizers (4) to a 5 kW grid-connected inverter (5). The power optimizers are used both to increase the electrical power output and to monitor the functional parameters (e.g. power, voltage) of each PV/PVT module. The monitored data are stored in a database wherefrom these can be downloaded and analysed. The PVT modules (1 and 2) can be serially or parallel interconnected to provide the thermal energy to a storage tank (6) of 140 L capacity. An antifreeze solution is used as a thermal fluid, and the PVT system being also used during the wintertime when the outdoor air temperature drops below 0 °C. A hydraulic pump (7) is used to circulate the thermal fluid through the solar circuit. The expansion vessel (8) compensates the volume variation of the thermal fluid during the heating/cooling phases. Six temperature sensors (T_1–T_6) are used to measure the temperature of the thermal fluid at each inlet and outlet of the PVT modules and of the storage tank. Two flowmeters (F_1 and F_2) measure the thermal fluid flow rates through each PVT module. The temperature sensors and flowmeters are connected to a data logger where the recorded values are stored and can be accessed.

The specific parameters of the PV and PVT modules are included in Table 5.1. Similar electrical power and cell type were selected for the PV and PVT modules to comparatively assess their output.

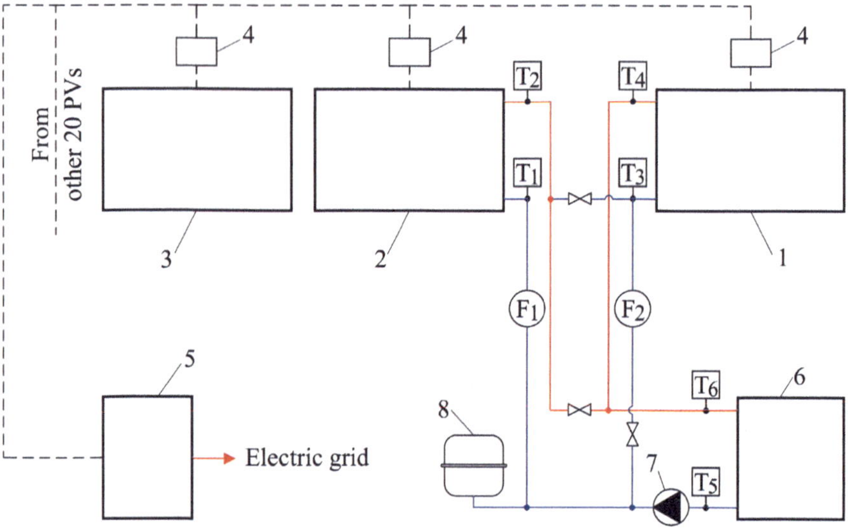

Fig. 5.6 Functional scheme of the PVT testing rig: PVT modules (1 and 2), PV module (3), power optimizers (4), inverter (5), storage tank (6), hydraulic pump (7), expansion vessel (8), temperature sensors (T_1–T_6), flowmeters (F_1 and F_2)

Table 5.1 Technical parameters of the PVT and PV modules

Parameter	PVT	PV
Producer	Solimpeks	Trina
Model	Volther PowerVolt	TSM-DC01A.05
PV cell type	Monocrystalline Si	Monocrystalline Si
Electrical power (W)	190	190
Nominal electrical efficiency (–)	0.143	0.149
Temperature coefficient of P_{MPP} (%/°C)		−0.45
Temperature coefficient of V_{OC} (%/°C)		−0.35
Temperature coefficient of I_{SC} (%/°C)		0.05
Thermal power (W)	460	
Nominal thermal efficiency (–)	0.478	
First-order thermal losses (W/m^2 K)	8.43	
Second-order thermal losses (W/m^2 K^2)	0.59	
Length/width/depth of the converter (m)	1.601/0.828/0.090	1.581/0.809/0.040
Surface (m^2)	1.33	1.33
Weight (kg)	24.4	15.6

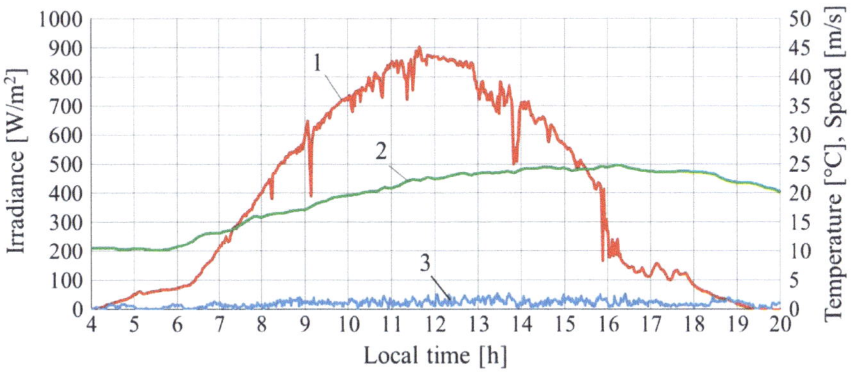

Fig. 5.7 Meteorological parameters on-site measured in Brasov, Romania on 1 June 2016: the received global solar irradiance (1), the outdoor air temperature (2), the wind speed (3)

The output of the reference PV and of the PVT modules is further outlined during a sunny summer day (1 June 2016). The variation of the meteorological parameters locally measured with a SOLYS2 Sun Tracker and a Delta-T weather station during the selected day is represented in Fig. 5.7.

The PVT modules were parallelly connected to the storage tank, and therefore, similar electrical and thermal energy outputs were recorded for both modules. This is why the results are presented for the reference PV and for one of the PVT modules. The specific power output of the PV and PVT modules, obtained by dividing the power output to the active surface of each module, is comparatively presented in Fig. 5.8. Both PV and PVT specific power follow the solar radiation profile, the PV module producing slightly more specific power than the PVT one(s) all over the day.

Based on the specific electrical power measured for each module, the cumulative specific electrical energy is calculated, and its variation is plotted in Fig. 5.9, for the PV module and for one of the PVT modules. At the end of the day, the reference PV

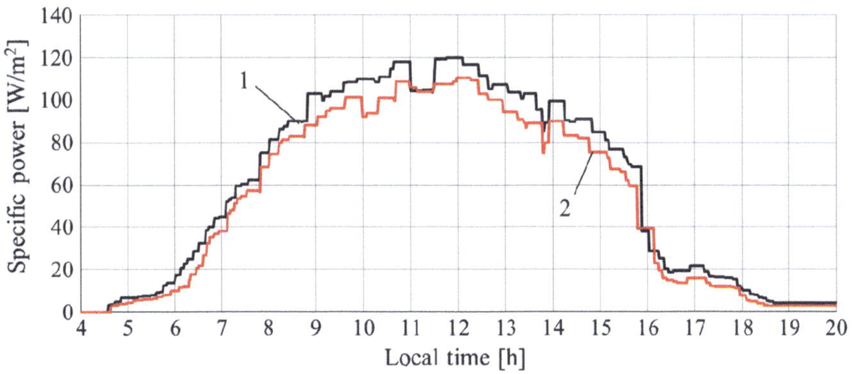

Fig. 5.8 Specific electric power output of the reference PV module (1) and of the PVT module (2)

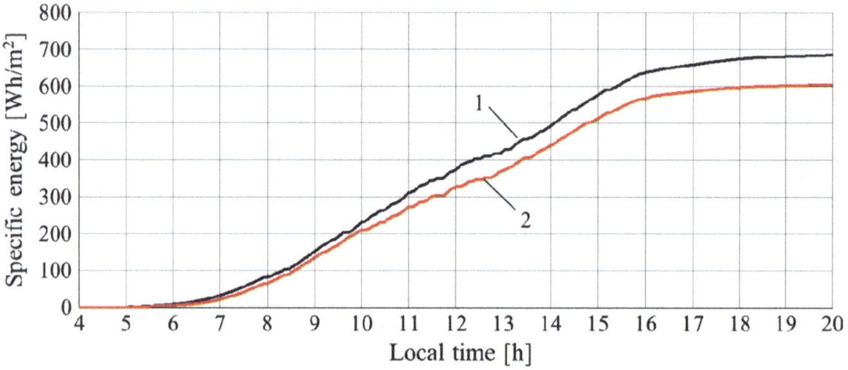

Fig. 5.9 Cumulative specific electrical energy output of the PV (1) and PVT (2) modules

module shows a specific electrical energy output with 12% higher than that of the PVT module.

The efficiency drop of the PVT module is the result of its temperature that is higher than that of the PV module all over the day (Fig. 5.10) negatively affecting its electrical power output. This temperature increase is caused by the temperature of the thermal fluid entering the PVT module. Moreover, the PVT module back insulation does not allow the dissipation of the thermal energy to the outdoor environment as in the case of PV module where the PV cells are up to 10 °C cooler.

The specific thermal power output of the PVT module was calculated based on the measured temperatures at the PVT module inlet/outlet, and its variation is plotted in Fig. 5.11 along with the electrical and the total (electrical and thermal) specific output. The thermal power output was almost two times higher than the electrical one at noon with total values up to 300 W/m^2 recorded during this period of the day.

Based on the specific electrical, thermal and total power output measured for each module, the cumulative specific electrical, thermal and total energies are calculated,

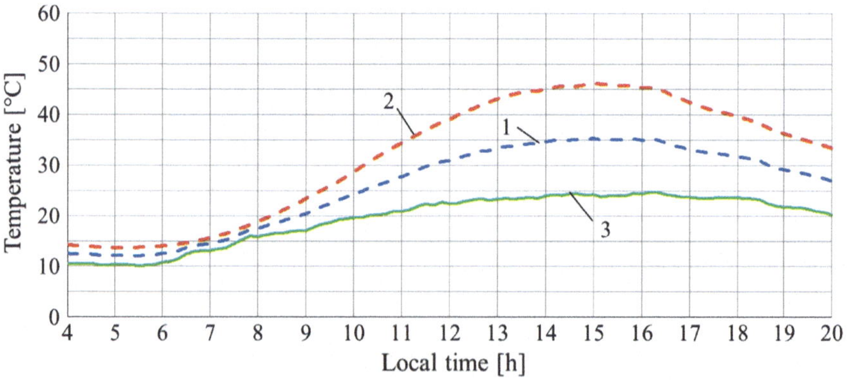

Fig. 5.10 Temperature variation of the PV (1) and PVT (2) modules and of the outdoor air (3)

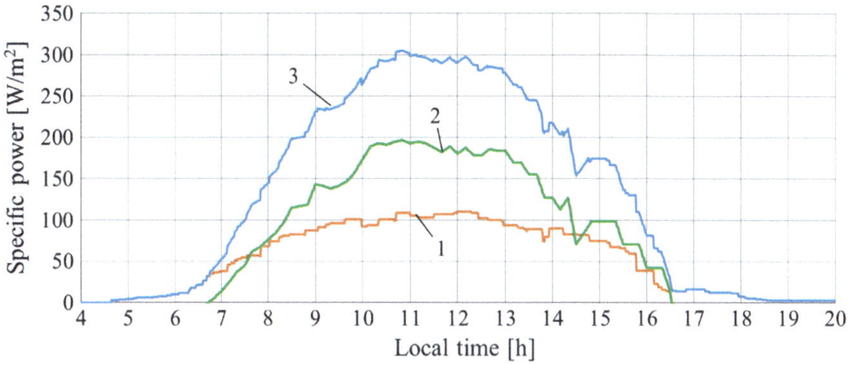

Fig. 5.11 Specific output power of the PVT module: electrical (1), thermal (2) and total (3)

and their variations are plotted in Fig. 5.12. At the end of the day, the PVT module has a specific total (electrical and thermal) energy output that is 2.67 times higher than that of the reference PV module.

Based on these results, it can be observed that during a sunny summer day, temperatures ranging from 25 to 45 °C on the PVT module decrease the daily specific electrical output with about 12%. This can be avoided by decreasing the storage temperature of the thermal fluid, and this is possible for the heating systems of, e.g. a swimming pool (during summer) or in the space heating systems with increased storage capacities (during winter). Another option for increasing the photovoltaic conversion efficiency consists of further decreasing the inlet temperature in the PVT module (e.g. by using a ground storage system or a heat pump having the PVT module included in its primary circuit).

By comparing the overall specific energy output of the reference PV and PVT modules, it can be stated that the electrical efficiency drop of the PVT module is fully compensated by its thermal energy output. Thus, the occupied implementation

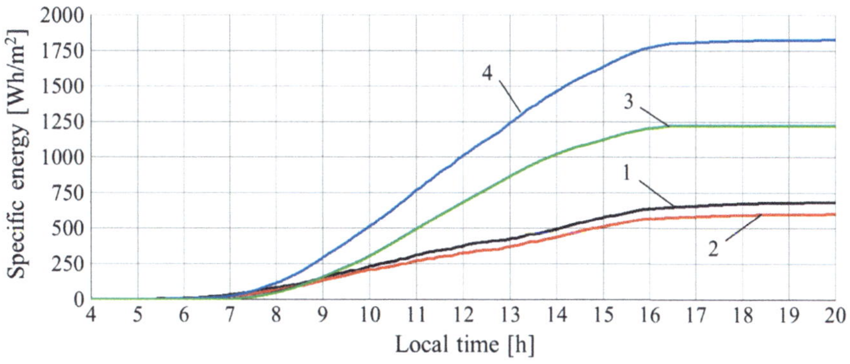

Fig. 5.12 Cumulative specific electrical energy output of the PV module (1) and cumulative specific energy output of the PVT module: electrical (2), thermal (3) and total (4)

area on or nearby a building would be fully used, especially for buildings where the entire roof or other available surfaces have to be fully covered to meet the electrical energy demand.

To optimally design a PVT system, these aspects must be further assessed over an entire year, considering the variations in the solar irradiance values and in the outdoor air temperature in the implementation location, and in the inlet temperature of the thermal fluid in the PVT modules.

5.3 Economic and Financial Aspects of Building Integrated PVT Systems

According to Weiss and Spörk-Dür [14], there are over one million square metres of PVT systems worldwide installed. The total capacity of these PVT systems reaches 525 MW on thermal power and 178 MW on electrical power with more than half of the systems installed in France (that is also the global market leader), but also in other different locations as in South Korea, China or in Germany.

Since the cost of the electrical energy is roughly four times higher than the cost of the thermal energy supplied using conventional gas combustion systems, the electricity supplied from renewable energy sources should not be used for heating, and the thermal energy should be supplied using different renewable energy systems. Thus, PVT systems are highly useful in applications when both electrical and thermal energy are required.

The cost for installing the PVT systems is about 1.5–2 times higher than that of similarly rated individual PV modules or solar thermal collectors, but slightly lower than the cost for separately installing both solar energy converters. These cost differences along with the high overall conversion efficiency of the PVT systems have as a direct result a lower cost per kWh produced [10]. This cost could be even lower if various photovoltaic and solar thermal subsidies or incentive schemes would be also applied to PVT systems. Unfortunately, in many countries, PVT system owners can qualify for only one subsidy (usually the *feed-in tariff*, but in rare cases also for a *Renewable Heat Incentive*) or for neither of them [1, 10].

The payback time of a PVT system is strongly influenced by the type of implementation. Lamnatou and Chemisana [9] reviewed the literature on PVT systems and concluded that building added PVT systems show payback times ranging between 1 and 4 years over their lifespan of 20 years while for building integrated PVT systems this indicator increases up to 14 years for a lifespan up to 30 years. These values also depend on the materials used for the PV cells, the absorber plate and the pipes, the casing, the heat transfer fluid, the use of a sunlight concentration system, etc. New types of PVT modules using non-conventional materials, as polymers [11] instead of copper (the most widespread material used as absorber plate in PVT modules) and different design features are investigated aiming at decreasing the payback time with up to 37.7% [5].

References

1. Barrett A (2017) Solar PV-T systems—what are the pros and cons? http://www.yougen.co.uk/. Accessed 1 May 2019
2. DualSun (2019) DualSun Spring—a revolutionary hybrid solar panel that produces simultaneously electricity and hot water. https://academy.dualsun.com/wp-content/uploads/DualSun-EN-Datasheet-Spring.pdf. Accessed 1 May 2019
3. Epp B (2018) Increasingly popular: heat and power from the same roof (PVT technology). http://solarheateurope.eu/2018/04/30/increasingly-popular-heat-and-power-from-the-same-roof/. Accessed 1 May 2019
4. Good C, Chen J, Dai Y, Grete A (2015) Hybrid photovoltaic-thermal systems in buildings—a review. Energ Proc 70:683–690
5. Herrando M, Ramos A, Zabalza I, Markides CN (2019) A comprehensive assessment of alternative absorber-exchanger designs for hybrid PVT-water collectors. Appl Energ 235:1583–1602
6. Kalogirou S (2001) Use of TRNSYS for modelling and simulation of a hybrid PV-thermal solar system for Cyprus. Renew Energ 23(2):247–260
7. Kalogirou S, Tripanagnostopoulos Y (2007) Industrial application of PV/T solar energy systems. Appl Therm Eng 27(8–9):1259–1270
8. Kalogirou SA (2009) Solar energy engineering. Processes and Systems, 2nd edn. Academic Press, London
9. Lamnatou C, Chemisana D (2017) Photovoltaic/thermal (PVT) systems: a review with emphasis on environmental issues. Renew Energ 105:270–287
10. Pullen T, Solar PVT Panels Guide (2017) Home building. https://www.homebuilding.co.uk/solar-pvt-guide/. Accessed 1 May 2019
11. Sandeep SJ, Dhoble AS (2018) Photovoltaic-thermal systems (PVT): Technology review and future trends. Renew Sust Energ Rev 92:848–882
12. Solimpeks (2019) Solimpeks catalogue. http://www.solimpeks.com/Files/Upload/Solimpeks%20Catalog-19.pdf. Accessed 1 May 2019
13. Wang Y, Shukla A, Liu S (2017) A state of art review on methodologies for heat transfer and energy flow characteristics of the active building envelopes. Renew Sust Energ Rev 78:1102–1116
14. Weiss W, Spörk-Dür M (2019) Solar heat worldwide. https://www.iea-shc.org/Data/Sites/1/publications/Solar-Heat-Worldwide-2019.pdf. Accessed 1 May 2019

Chapter 6
Sustainable Communities

The term *community* usually covers two main distinct groups: *interest communities* and *geographic communities* [73].

An *interest community* is defined as a group of people having a common interest, e.g. academic community, ethnic community. These communities are not usually related to a well-defined space.

The *geographic community* is defined by the geographic area and has specific limits (geographic boundaries). Greenius et al. [41] gave a definition as "a social group of any size with geographical proximity amongst its members, such as a neighbourhood, town, district or city where people can interact face to face". People in this community type have a common geographical location, a common built environment, common ties, commitment and involvement, and interact socially. Thus, the size of a geographic community ranges from an apartment or a house to a group of houses in a district, from a village to a city, to a region or a country, etc. In this chapter, the term *community* refers to a *geographic community*.

To become sustainable, a community has to implement the concept of *sustainable development* (SD), defined by the Brundtland Report *Our Common Future* [8] and formulated as "the development that meets the needs of the present without compromising the ability of the future generations to meet their own needs". A SD model that integrates the *economic*, *environmental* and *social pillars* was defined. Thus, to reach sustainability as a main goal of the SD process, all these pillars must be strong enough. If any of the pillars is weak, then the system as a whole is unsustainable. The traditional conceptual framework of SD based on three pillars was expanded to four pillars: *economy*, *environment*, *society* and *culture* [93]. The new pillar aims at developing the capacity of the organizations and institutions to promote sustainability by improving the knowledge, skills and creativity of the human capital. Further on, Bervar and Bertoncelj [7] proposed a new concept, based on five pillars, also including *security*. Further on, education has a significant role in the holistic understanding of the concept and its implementation in practice [87].

I. Visa et al., *Solar Energy Conversion Systems in the Built Environment*,
Green Energy and Technology, https://doi.org/10.1007/978-3-030-34829-8_6

Concrete strategies were developed at world level, followed by regional, national and local levels as steps forward in implementing SD in communities. The *United Nations Conference on Environment and Development* held in Rio de Janeiro in 1992 proposed a comprehensive action plan, called the *Agenda 21,* as a strategy to implement sustainable development at global, national and local level by organizations, governments, regional and local authorities [94]. In December 1992, the *Commission for Sustainable Development* (CSD) was created to implement, monitor and report on the *Agenda 21* implementation at international, national and local levels. The results and the new measures to increase the efficiency in implementing SD were analysed at the *World Summit on Sustainable Development* (WSSD) in Johannesburg, South Africa [95].

Strategies for SD and adequate structures were defined at regional, national and local communities, after 1992, to implement *Agenda 21*. As an example, in Europe in 2001, the *Gothenburg European Council* approved the European Union Strategy for SD [15] and invited the *member states* to develop their own national SD strategies.

Five years later, in 2006, the *European Council* adopted a renewal of the SD strategy, and in 2010, the *European Commission* proposed *Europe 2020—A strategy for smart, sustainable and inclusive growth* [25], based on five measurable targets to be also implemented at national level: *employment*, *research and innovation*, *climate change and energy*, *education* and *reducing poverty*. For climate change and energy, the following targets are proposed: greenhouse gases (GHG) emissions with 20% lower than the 1990 level, 20% from the energy used should be obtained using renewables and a 20% increase in energy efficiency.

In September 2015, the *United Nations Sustainable Development Summit* adopted a new agenda *The 2030 Agenda for Sustainable Development* [96]. The document contains 17 sustainable development goals (SDGs) with 169 targets related to *economic growth*, *environmental protection* and *social inclusion* with the aim to end extreme poverty, to fight against inequality and injustice and to fix (and mitigate) climate change. This document extends the Brundtland report with two components: *partnership* and *peace*.

In 2016, the European Commission ratified the *2030 Agenda for Sustainable Development.* In this official document, one goal refers to *Sustainable Cities and Communities*, aiming at increasing the access to basic services, to energy, housing and transportation.

A policy framework for climate change and energy was approved [27] for the period 2020–2030. This framework formulates a new set of goals: 40% cut in the GHG emissions compared to the 1990 level; at least 27% share of renewable energy in the entire energy consumption; energy efficiency improvement at EU level with at least 27% (to be renewed by 2030 for 30%).

A long-term vision is the *Energy Roadmap 2050* that states the targets for 2050 as follows: reducing the GHG emissions by 80–95% compared to the 1990 level, a decarbonized energy system with a high share (75%) of renewables in the gross final energy consumption, reaching a renewables share of 97% in the electricity consumption [26].

To evaluate the progress towards SD, starting with 2005 Eurostat produces every two years *Monitoring Reports*, by using more than 100 indicators grouped in ten themes [30] related to socio-economic development, climate change and energy, sustainable transport, sustainable consumption and production, natural resources, public health, social inclusion, demographic changes, global partnership and good governance. After promoting the 2030 Agenda, the *reports* are based on 17 SDGs [31]. Some of these goals directly refer to communities as SDG 11—*Sustainable cities and communities* (living comfort, noise reduction, pollution reduction, increasing the recycling rate of local waste, etc.) or to the communities primary needs as SDG 6—*Clean water and sanitation* (water quality, water treatment, water depollution), SDG 7—*Affordable and clean energy* (increasing the share of renewable energies, reducing the energy consumption at home, reaching the 2020 energy efficiency target, etc.) and SDG 13—*Climate action* (climate change mitigation, GHG reduction, etc.).

Each SDG has specific indicators. The SDG 7 comprises three main indicators: *energy consumption* in household per capita, energy efficiency and GHG emissions intensity; *energy supply* focusing on the renewable energy share in the gross final energy consumption that was 17% in 2016 in EU and on the energy dependence; *access to affordable energy*, strongly associated with a low income level.

To become sustainable, a community has to implement the *SD goals* as a positive answer to the humanity global problems: resources depletion, climate change, global warming and GHGs. At local community level, this has to be done through specific actions following the concept of "local solutions to global problems". In this concept, Fien and White [35] defined 13 indicators for a sustainable community and ten principles for one planet community as a whole, related to energy, wastes, water, transport, food, social and cultural aspects.

Communities are divers considering the area, the shape, the geographical location, people density, the infrastructure, the built environment, the resources and the governance. However, the common interest refers to key issues that are common to all communities, as healthy environment, clean economy and well-being for the community inhabitants. Therefore, the path and the concrete steps to reach sustainability are different and require specific planning, design, implementation and periodical assessment coming from the key players and commitment from the entire community. Burton [11] detailed the potential for eco-neighbourhoods while Condon [16] defined seven rules related to the cities' design, strategies and techniques to reach sustainability based on integrating the specific prerequisites for the local environment, infrastructure, building types, mobility, commercial services, etc. Desai [20] analysed how to turn sustainability from an abstract concept into a framework based on the practical design, development, long-term management and governance decisions, to create a coherent approach on sustainable development. Friedman [39] outlined the essential need for sustainable strategies at neighbourhood level and for innovation in planning.

The *energy* issue is crucial for implementing the process and for reaching almost all sustainable development goals. Energy is also stated as the main issue in the *top ten* problems humanity has to solve [88]. Increasing the world population has as result an increased energy demand for more products, food, buildings, transport and social activities. However, fossil fuels-based energy, that accounts for over 80% of the energy used all over the world, is mainly responsible for global warming and climate change through the contribution to over 60% of the total GHGs (mainly CO_2) emissions [106]. Industry, agriculture, transportation, the residential (homes and apartments) and the commercial sectors represent the major energy consumers, and most of this energy is currently produced using fossil fuels. Therefore, promoting sustainability requires to produce clean energy, increasing the efficiency and reducing the energy losses, in fact promoting and implementing at community level the *sustainable energy* concept, based on the three key issues: *energy efficiency*, *energy saving* and *renewable energy systems*.

In a community, the built environment (residential, public and commercial buildings) consumes a high share of the energy, usually estimated at about 40–44%. Thus, energy-efficient buildings and local energy production using renewables will have a direct impact on the amount of energy consumed in the buildings and on the clean environment in the community. Reducing the GHGs emissions requires global, regional and local strategies to replace fossil fuels in the energy production. Applied at the community level, this means to implement the concept of *sustainable energy community* (further called *energy community*) according to which the current, centralized fossil fuel-based energy systems having large companies as main beneficiaries have to be partially or totally replaced by decentralized energy systems, involving the community in the energy production and benefits, towards energy autonomy [77]. Klein and Coffey [55] developed a review on these topics for over 70 communities around the world and on the top-down strategy employed by the USA towards a Community Energy Database. Moreover, Peters et al. [70] detailed the concepts, definitions and applications on solar energy communities in the USA (Minnesota) and Europe (Denmark, Ireland, Germany) and outlined the individual benefits of shared solar energy conversion systems.

The strategy for implementing renewable energy systems in a community leads to a *Community on Renewable Energy* (*CRE*). According to Greenius et al. [41], a *CRE* can be described considering three main features: energy production in the community by using renewable energy sources, community involvement through local stakeholders and people and community benefits from the energy produced by using renewables. The community benefits are detailed as economic development, new jobs in the community, cheaper energy and more secure energy. A *CRE* involves the energy production using renewable sources and technologies that are partly or totally owned by the community. The specific path to be followed depends on the stakeholders involved, on the resources and on the community involvement. Started in 1978 in Denmark by installing a wind turbine in a community, the *CRE* concept was extensively implemented in countries as Denmark, Germany and the UK. In 2012 in Germany, the community groups owned 34% of the renewable energy capacity.

Energy concepts for self-supplying communities based on local and renewable energy sources considered at community level have certain advantages when comparing to the case focusing on buildings and, in this context, municipalities play a key role. Petersen [71] describes a practical example for a community of 50,000 inhabitants from northern Germany (inner city of Elmshorn), while opportunities and barriers in implementing renewables were detailed by San and Ganguly [84].

Different legal forms are possible for an *energy community*, as cooperation, partnership, trust and foundations, public utility company and public–private partnership [46]. Cooperative energy is one of the most common forms of local energy communities [29]. Sahovic and Pereira da Silva [83] developed a literature review on cooperative energy schemes, focusing on topology distinction, governance models, financial characteristics and membership structure. The development of cooperative renewable energy communities aiming at production, consumption, distribution and trading using different renewables (biomass, wind, solar, geothermal, biogas, hydro) and different business models (local group of citizens, regional, national cooperatives, network of cooperatives) are analysed by Heras-Saizarbitoria et al. [43], with a focus on Spain. A *Community Company for Sustainability* and an innovative tool to plan, implement and operate the renewable energy systems at community level, involving also prosumers and decision makers were presented by Visa and Duta [102].

6.1 Nearly Zero Energy Community

The *Nearly Zero Energy Community* (nZEC) concept is defined and analysed in direct relation with the energy produced using renewables and its consumption in the built environment, consisting of residential, public and commercial buildings and the associated infrastructure between the nearby buildings.

In a community identified as a place to promote sustainability, one or more energy communities can be implemented. An energy community is identified by its own, specific energy project. When more energy communities are defined, these can be independent or can cooperate as clusters of energy communities in the energy production and consumption to avoid losses in each energy community.

An energy community can be defined, as presented in Fig. 6.1, as a residential house or a single apartment in a collective house, an administrative building, an education building, a commercial building, etc., or as a group of buildings, or at community level as a whole, mainly for small communities. Large communities, as cities, can be divided in a number of districts where one or more energy communities can be defined.

An energy community can focus on electrical or on thermal energy production, or on both energy types and can partially or totally cover the energy demand of the buildings at community level.

The *coverage degree* (CD_i) by renewables can be calculated using

$$\mathrm{CD}_i = \frac{E_{\mathrm{pi}}}{E_{\mathrm{di}}} \tag{6.1}$$

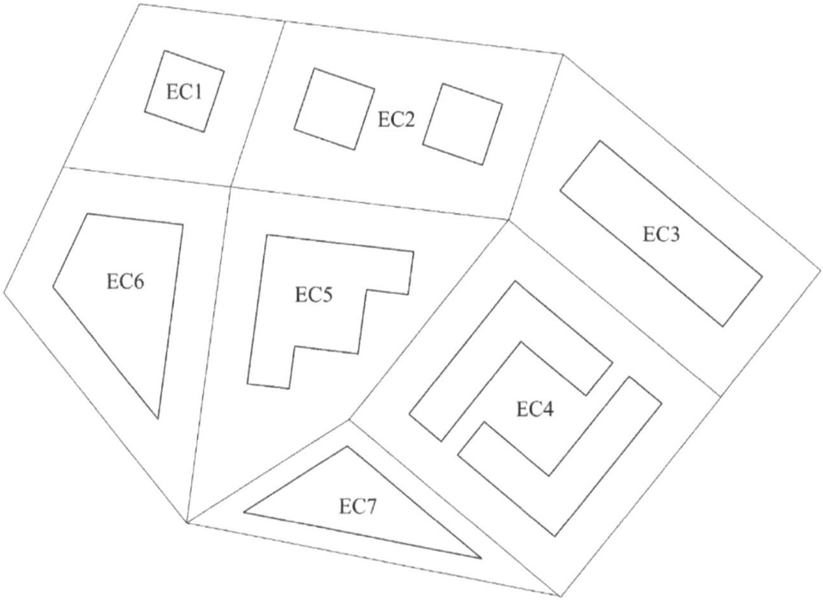

EC1 - residential house
EC2 - group of residential houses
EC3 - collective houses
EC4 - group of collective houses
EC5 - public building
EC6 - commercial buildings
EC7 - other consummers

Fig. 6.1 Energy communities within a community

where

- E_{pi} is the renewable energy produced in the *i*th energy community.
- E_{di} is the energy demand in the *i*th energy community.

Usually, $CD_i = 0$–1. However, the coverage degree can exceed the unitary value, $CD_i > 1$, if the energy community exports the energy produced in excess.

For more than one energy community within a community, the coverage degree at community level (considering all energy communities) can be calculated using

$$CD = \frac{E_p}{E_d} = \frac{\sum E_{pi}}{\sum E_{di}} \tag{6.2}$$

where

- E_p is the total energy produced in the energy communities using renewables.
- E_d is the total energy demand in the energy communities.

Obviously, there are three ways to increase CD: by decreasing E_d for a given E_p, by increasing E_p for a given E_d or by decreasing E_d and increasing E_p.

Decreasing the energy demand (E_d) requires, e.g. reducing the thermal energy for heating and cooling in the community buildings, by refurbishing the existing building stock and by developing new buildings as *low energy buildings* or *passive buildings*. Increasing the efficiency of the electrical equipment and implementing performant energy management systems represent possible paths for increasing the CD. Increasing the renewable energy produced at community level requires increasing the number of renewable energy systems implemented in that specific community based on an adequate strategy. Since 2008, there is a dynamic in implementing the *CRE* concept, especially in Europe, mainly in small communities. In larger urban areas, the potential for *CRE* implementation is still underdeveloped. According to REN21 [82], in Europe there were over 2800 energy cooperatives in operation in 2015, with a strong dynamic in Germany (from 67 in 2008 to 722 in 2014) and in the Netherlands (from 19 in 2008 to 500 in 2015).

Currently, all over the world, there are communities where the coverage degree is very low (CD = 0–0.5) and communities that already planned to fully cover their energy demand by using renewables (CD = 1). Moreover, there are communities that implemented renewables more than their energy demand (CD > 1).

The nZEB concept is well known as describing buildings with a very low energy demand that is covered in a significant extent by using renewable energy systems installed on or nearby the building [28]. Their main role in the transition process towards post-carbon cities was analysed by Baccio et al. [5] and reviewed by D'Agostino and Mazzarella [17]. Further on, an individual building is a *Zero Energy Building* (ZEB) if for a given period (usually one year) the produced energy using renewables is at least equal with the energy consumed [6]; moreover, a building can be a *Plus Energy Building*, if this energy balance is positive (the energy produced using renewables is higher than the energy consumed).

Based on the nZEB concept, the concept of nZEC—*Nearly Zero Energy Community* was formulated, considering $0.5 \leq \text{CD} \leq 1$ [104]. The same acronym was introduced for the cities, nZEC—*Nearly Zero Energy Cities* [100], as a transition concept towards sustainability.

To reach this level, appropriate strategies have to be formulated along with passive and active design solutions to get energy-efficient buildings and to implement renewable energy systems at community level [91].

In a community, the energy produced using renewable sources can be restricted at building's level, by installing small/medium-sized renewable energy systems on the buildings (on roofs, facades and/or terraces) or near the buildings, or it can be designed at community level, by large-scale renewable energy systems. These could be installed on individual buildings of any type (residential, collective, public and commercial) or on groups of buildings (at neighbourhood or at district level) or in the community outskirts using a new or the already existing infrastructure. At the community level, additional energy production can result from wastes collection and processing. As a result, the energy production in a community has to be approached in an integrate manner, implementing specific solutions [101].

In an energy community, the electrical energy production is mainly based on two renewable energy sources: solar radiation and wind. The coverage degree of the electrical energy in an energy community can be similarly defined as the total coverage degree for the energy community using

$$\mathrm{CDE}_i = \frac{\mathrm{EE_{pi}}}{\mathrm{EE_{di}}} \tag{6.3}$$

where

- $\mathrm{EE_{pi}}$ is the amount of electrical energy produced using renewable energy sources in the *i*th energy community.
- $\mathrm{EE_{di}}$ is the electrical energy demand in the *i*th energy community.

For all the energy communities, the coverage degree of the electrical energy (CDE) is calculated using

$$\mathrm{CDE} = \frac{\mathrm{EE_p}}{\mathrm{EE_d}} = \frac{\sum \mathrm{EE_{pi}}}{\sum \mathrm{EE_{di}}} \tag{6.4}$$

where

- $\mathrm{EE_p}$ is the electrical energy produced using renewables in all the energy communities.
- $\mathrm{EE_d}$ is the total electrical energy demand in all the energy communities.

Solar to electrical energy conversion uses in a community small- or medium-sized photovoltaic (PV) systems installed on or near the individual buildings and large-scale PV systems installed at community level, usually in the outskirts. PV systems are rather simple systems that can be adapted to the physical constraints or to the facilities in the communities. Currently, the solar electricity technology is based on PV modules, and there is a variety of grid-connected applications (on-grid) and stand-alone ones (off-grid). However, the solar energy variability has as direct consequence a variable electrical energy supply. Matching the demand with this supply is usually done by feeding the electrical energy into the national or local grid (mainly on islands) or by changing the energy demand profile using adequate control technologies, as the *Demand Side Management* (DSM), or using electric vehicles connected to the grid by smart charging systems [72].

In the stand-alone applications, a storage system is required (e.g. batteries, pumped hydrostorage, supercapacitors), according to the system power and to the type of application [77]. Usually, stand-alone systems have limited storage capacity, thus the energy exceedingly produced cannot be (entirely) stored [54]. On the other hand, from a financial and technical point of view, grid connection is beneficial as demonstrated by Rae and Bradlley [32]. To improve the load matching of individual buildings in a group of buildings, the concept of a *Cooperative Net Zero Energy Community* was introduced by Lopez et al. [58], having as result an increase in the electrical energy demand covered by on-site generation up to 21% and in the on-site used energy up to 15%.

To reduce the storage needs for large-scale photovoltaics implemented in the urban environment [38], there were used buildings surfaces with different tilt and orientation to improve the demand–supply match. By aggregating the demand profiles with the photovoltaics production and electricity storage in two districts in East Lisbon,

Portugal, containing collective houses and commercial buildings, it was outlined that the PV modules implemented only in high insolation locations lead to adverse impact on the grid. The diversity of tilt and surface orientation supports smoothing the electricity generation profile during the day. Self-consumption increased from 60–90% to close to 100% and self-sufficiency from 40% to 80–100%.

Awad and Gul [4] analysed, using simulation results, the PV systems based on micro-generation (at building level), when self-consumption does not exceed 25% in cold climate regions and the energy is exported during daytime to the grid, along with large PV systems integrated at community level, concluding that the second case is preferred compared to individual rooftop applications. This enhances the self-consumption and provides a uniform electrical energy distribution among all the community members. Based on the local energy production, storage and distribution systems, Rae and Bradley [77] introduced the concept of *degree of energy autonomy* in a community. McKenna et al. [62] used a technical–economical model to analyse the scale effect on the energy autonomy in residential buildings, considering the scale ranging from individual buildings to neighbourhoods and districts; they concluded that the electrical self-sufficiency is about 30% at the individual buildings' level and raises up to almost 100% in a district with 1000 single-family households.

In communities, the wind energy systems installed at building level usually consist of wind turbines integrated in the building design or near the buildings, in small capacity systems. The wind parameters have particular values in the built environment, as lower speed and higher turbulence due to the buildings' shapes. The turbulence model of the local wind parameters represents the key issue in assessing the energy yield and the wind turbines durability along with their performance and reliability [3].

Large wind turbines are used at community level installed far from the buildings, with large energy amounts feed into the grid to which they are connected. Small hydros and biomass gasifier plants can also represent solutions for the electrical energy obtained using renewables at community level, as synthetically presented in Fig. 6.2.

The feasibility of the transition towards a 100% renewable power in 2050 in Europe was discussed in the papers of Zappa et al. [108] and Child [13]. This is possible by increasing the capacity of the PV and wind energy systems, by developing an European smart grid and through balancing the variability of the electrical energy produced using PV and wind systems with additional renewable sources (hydropower, concentrated solar power, geothermal or biomass), or by installing energy storage capacities, to cover the yearly electrical energy demand.

The solar and wind energies will have a major contribution, estimated for 2050 to reach 69% for solar PV and at 18% for wind [79]. A comparative analysis of the electricity costs obtained using different renewables and other sources (fossil fuels, nuclear) showed that the *Levelized Cost of Electricity* (LCOE) analysed for the G20 countries that account for 85% of the renewable electrical energy consumption and storage will be far lower than that when using fossil and nuclear sources by 2030, even without considering external costs. The cheapest are the utility-scale solar PV followed by onshore wind, rooftop solar PV and offshore wind energy systems.

Sources	Systems installed at building level	Systems installed at community level
Solar	PV arrays PV platforms	PV parks
Wind	Small wind turbines	Large wind turbines Wind turbine farms
Hydro	-	Micro-hydro systems
Biomass	-	Gasifier plants

Fig. 6.2 Renewable energy sources and systems for electrical energy production in the energy communities

The thermal energy production in a community is mainly based on the solar, geothermal or biomass renewable energy sources (Fig. 6.3).

In the built environment, the thermal energy has to cover three main functionalities: domestic hot water (DHW) preparation, space heating and space cooling.

Sources	Systems installed at building level	Systems installed at community level
Solar	Flat plate solar collectors Evacuated tube solar collectors Small concentrating collectors	Large flat plate solar collectors arrays Large evacuated tube solar collectors arrays Large concentrating collectors arrays
Geo-thermal	Ground coupled heat pumps	Systems for direct use Large geothermal storage facilities
Biomass	Burning systems	Gasifier plants

Fig. 6.3 Renewable energy sources and systems for thermal energy production in the energy communities

A *coverage degree* for the thermal energy using renewables can be defined for an energy community as follows:

$$\mathrm{CDT}_i = \frac{\mathrm{TE_{pi}}}{\mathrm{TE_{di}}} \tag{6.5}$$

where

- $\mathrm{TE_{pi}}$ is the amount of thermal energy produced using renewable energy sources in the *i*th energy community.
- $\mathrm{TE_{di}}$ is the thermal energy demand in the *i*th energy community.

For all the energy communities in a given community, the *coverage degree* of the thermal energy (CDT) is calculated using

$$\mathrm{CDT} = \frac{\mathrm{TE_p}}{\mathrm{TE_d}} = \frac{\sum \mathrm{TE_{pi}}}{\sum \mathrm{TE_{di}}} \tag{6.6}$$

where

- $\mathrm{TE_p}$ is the amount of thermal energy produced using renewable energy sources in all the energy communities.
- $\mathrm{TE_d}$ is the amount of thermal energy required in all the energy communities.

While the thermal energy demand for DHW is almost constant over the year, the amount of thermal energy required for heating and cooling has a significant variation, depending on the geographical location of the community on the Earth, on the local climate, on the season and on the building's efficiency.

The solar thermal systems for DHW preparation in communities are based on small-scale, individual systems installed on or near the buildings or on large-scale systems installed on the ground, at district level. In terms of installation expenditures and energy cost over their entire lifetime, the solar thermal systems installed at district level proved to be cost effective. The installation of large-scale systems is more feasible than that of small units, on households, considering the energy savings and the initial cost per kWh. At world level, in China there are installed 70.5% of the solar thermal systems, in EU 12.3% and in Turkey 5% [78].

The solar thermal systems can feasibly cover the entire thermal energy demand for DHW production in regions with high solar energy and part of it in regions with medium and low solar energy levels. To mitigate the mismatch between the thermal energy production and its consumption, short-term thermal energy storage systems are required with a storage capacity that depends on the thermal energy that can be produced using solar thermal systems and on the climatic profile of the implementation location. Solar thermal systems for DHW preparation can operate alone or in hybrid heating/cooling/DHW systems based on solar and geothermal (heat pumps) systems, or on solar thermal and biomass plants [48].

Solar energy for heating and/or cooling is used in large-scale solar district heating plants, and Denmark is the front runner with more than 70% of the large solar district heating plants constructed [92]. For these types of systems, one of the main issues is the thermal energy storage to be used when the available solar radiation is low as during the wintertime, or is missing as during the night.

There are four different seasonal thermal energy storage technologies implemented in different plants, in operation in Germany as described by Rad and Fung [76]:

- Hot water thermal energy storage (the storage tank is constructed using reinforced concrete or steel);
- Aquifer thermal energy storage (storage in a porous media saturated with water);
- Gravel water thermal storage with the storage tank buried in the ground and using as storage media a mixture of gravel and water;
- Borehole thermal energy storage, with a number of vertical boreholes in the ground. The borehole thermal energy storage at community scale has the highest efficiency for the long-term energy storage. It is mainly used in cold climates due to the low level of winter solar radiation while in hot climates it can be skipped as a result of the high level of the solar radiation during the winter months.

The optimized design, the size of the systems and their layout as centralized or semi-decentralized (with a high-temperature tank, charged by an individual heat pump in each house) solar district heating systems installed at high latitudes in Finland were discussed by Rehman et al. [80] outlining that the decentralized systems are better performing than the centralized ones. Both can reach a close to 90% solar fraction but the solar thermal collector area can be reduced in a decentralized system while the losses in the centralized systems are with 12–40% higher compared with the decentralized systems.

The location and the community design (low-temperature heating systems, energy-efficient buildings and local ground properties) have an influence on the performance of the solar district heating system equipped with borehole seasonal storage [37]. These types of systems can reach a very high solar fraction and can cover almost the entire thermal energy demand in the community. As an example, the solar thermal system with borehole seasonal storage that provides thermal energy for 52 detached energy-efficient houses through a district heating network in the Drake Landing Solar Community, Canada, uses 34,000 m^3 (in the Earth) for storage and 144 boreholes with single U-tube heat exchangers. The performance analysis after five years of operation showed that the system provided 97% of the thermal energy required and a solar fraction over 90% was met [86].

Lund et al. [60] defined the concept of the *Fourth Generation District Heating* including the *smart thermal grid* as a network of pipes that connects the buildings in a neighbourhood using the energy produced in centralized plants and in decentralized systems, with low-temperature heat sources, in low energy buildings. The concept of *smart thermal grid*, along with the *smart electricity grid* concept, that integrates variable renewable energy systems, plays a major role in implementing renewables at community level.

In a Nearly Zero Energy Community, the *total coverage degree* of the energy (CD) is obtained by aggregating the electrical and thermal energy produced using renewables in the energy community (E_p) and dividing it with the total energy demand in the entire community (E_d) using

$$\mathrm{CD} = \frac{E_\mathrm{p}}{E_\mathrm{d}} = \frac{\mathrm{EE_p} + \mathrm{TE_p}}{E_\mathrm{d}} \tag{6.7}$$

If CD > 0.5, more than half of the energy demand is covered using renewables, and the community has the nZEC status [104]. Currently, all over the world, the main part of the communities did not reach this status, but there are communities that are producing all the energy they need using renewables (CD = 1) or export the excess energy produced using renewables (CD > 1).

6.2 Steps in Implementing Renewable Energy Systems in NZEB and in NZEC

The transition process to the energy production and consumption using renewables can be defined based on different models that mainly consider the community inhabitants and the key players' commitment, the technical, economic and financial aspects along with the specifics of the implementation and management steps. Several examples are further discussed.

The transition to a fully *renewable energy system* based on the *LUT model* [79] applied at world level (divided in nine major regions) contains the following major steps:

- Data preparation mainly focusing on: long-term and hourly energy demand across different sectors of power, heat, transport and desalination; renewable resources potential and a global database of power plants;
- Model set-up and simulation based on energy technologies, financial and technical assumptions, prosumers modelling by distributed self-generation and consumption in the buildings, best scenario for implementation;
- Model implementation.

Hicks et al. [44] proposed for a community-owned renewable energy system in Australia a guide in seven steps: initiation (defining the group, aims, start to develop); social feasibility (level of support in the community, technical and financial possibilities); technical feasibility (business plan); planning and approval; capital raising; construction; operation.

In the *Guide to Developing a Community Renewable Energy Project in North America* developed by the Commission for Environmental Cooperation [14], three major steps are defined: site analysis, design, permits and interconnection; site development, install and test the unit; operate the unit.

Doris [21], from the US Department of Energy (Office of Indian Energy), describes a very detailed process for the project development that contains five steps: project potential (data collection, opportunity, assessment); project options (strategy and details); project refinement (planning and development); project implementation (financing and construction); project operation and maintenance.

In the *Community Energy Planning Guidebook* [47], the process is described by seven steps:

- Identify an energy planner and a supportive council;
- Identify the community vision and goals;
- Describe the current state of the buildings along with the energy supply, consumption and cost;
- Identify the energy efficiency and the conservation (saving) opportunities in the community;
- Identify renewable opportunities as solar energy for buildings along with ground and wind potential, etc., and develop a working plan and a Gantt chart;
- Identify the economics and service requirements;
- Implementation.

Ten models of small-scale communities, solar feasible for the area of implementation, were analysed by Ziersch [109] considering the structure, differences, similarities and the common legal framework.

Visa and Duta [101] described a model based on three technical steps for implementing renewables in a community:

- Identify the electrical and the thermal energy demand for common uses in buildings and at community level (electrical energy for lighting and various appliances and thermal energy for DHW, heating and/or cooling), and for specific uses at community level (electrical energy for street lighting, or for auxiliary equipment as, e.g. for the fresh and wastewater treatment plants);
- Evaluate the on-site renewable energy potential of the variable resources (solar, wind, small water flows) and for site constant resources (geothermal energy for ground-coupled heat pumps, biomass). Additionally, the energy that can be obtained through wastes processing should also be considered;
- Design the optimum energy mix for the buildings and for the community, according to the available resources and to the local facilities, based on several scenarios considering the renewable energy systems installed at building level, at community level or at building and community level.

Visa et al. [103] formulated an extended flow for designing and implementing a Nearly Zero Energy Community, based on six steps:

- Defining the community and choosing the model for the energy production and distribution;
- Improving the built environment;
- Assessing the energy demand at community level;
- Assessing the renewable energy potential;
- Designing the sustainable energy mix;
- Developing the monitoring and control systems.

Following these results, a ten-steps flow can be the solution for implementing renewable energy systems at different levels, starting with the buildings up to large communities as presented in Fig. 6.4.

Step 1. Define the energy community limits

These limits have to be defined for variously sized communities (Fig. 6.5) by specifying the boundaries for a very small energy community (an apartment, or a house), or for a small energy community (a collective house, a group of houses or public/administrative/commercial buildings), or for an average energy community (neighbourhood), or for a large energy community (a district or a village) up to a very large community (e.g. a city). This can be done at the initiative of the energy community beneficiaries (bottom-up approach) or at the initiative of the community representatives (top-down) as part of the community commitment towards sustainability.

Step	Name of the step	Main outputs
Step 1	Define the energy community limits	- energy community boundaries - energy community built environment structure
Step 2	Building an energy community administrative and management structure to support renewables	- team structure - the tasks to plan, implement, operate
Step 3	Define the scenario for implementing the renewable energy system(s)	- scenario in implementing renewables
Step 4	Define the energy community in the built environment	- building characteristics and energy systems
Step 5	Assessing the energy demand in the energy community	- monthly and yearly energy demand
Step 6	Assessing the renewable energy potential in the energy community	- renewable energy potential by source
Step 7	Define and optimize the energy mix	- mix renewable sources - optimal mix
Step 8	Design and development of the renewable energy project	- detailed design
Step 9	Implementing the renewable energy project	- implementation of renewable energy systems
Step 10	System operation and maintenance	- operation procedure - monitoring and maintenance

Fig. 6.4 Steps in developing an energy community

Community size	Example
Very small	House
Small	Group of appartments
Average	Micro-community (neighborhood)
Large	District
Very large	City

Fig. 6.5 Energy community size and examples

Step 2. Building an energy community administrative and management structure to support renewables

The role of this structure is to promote and manage a renewable energy project in the community, using the legal frame and involving the community members in all the steps of development, implementing and operation of the energy systems, in the ownership, decisions, responsibilities, benefits and future developments. This structure has to cooperate with the local community authorities in the project development and approval, in using the community infrastructure, in promoting sustainability at community level.

According to the level of the energy community within the community, this structure may have one or more professionals as staff, up to a *Community Company for Sustainability* formed by members with technical expertise, communication and project management skills, education and training expertise, business and management competencies. The main tasks of this company at community level are related to planning the sustainable community based on the energy demand and resources (renewable energy mix) and to gain support and acceptance in implementing the *sustainable community plan* (including energy efficiency measures, renewable energy systems, water and waste management systems). After the plan is implemented, the company will be responsible for operating the systems (by monitoring and managing the demand—production balance), for giving use to any advantages for the community that raise from implementing the plan, thus gaining continuous acceptance, as presented in Fig. 6.6.

Task	Description
Task 1	Plan the Sustainable Community
Task 2	Implement the Sustainable Community
Task 3	Operate the implemented systems

Fig. 6.6 Main tasks of the *Community Company for Sustainability*

Step 3. Define the scenario for implementing the renewable energy system(s)

This scenario is formulated according to the energy community type and generally refers to the electrical and thermal energy production using renewables and their consumption at community level, as the path for promoting sustainability in the entire community by implementing the *Nearly Zero Energy* concept. For very small energy communities (e.g. a residential building) and for small energy communities (e.g. a group of residential houses, a collective house, a large building), the thermal energy demand for DHW preparation could be partially or totally covered using solar energy conversion systems with short-term thermal energy storage required to cover the mismatch between the production and the consumption. Alternatively, geothermal hot water sources (where available) represent an option. The thermal energy for heating and/or cooling can be covered by geothermal ground-coupled heat pumps, by using biomass or geothermal water. The electrical energy demand is covered using solar energy conversion systems (PV systems) and batteries storage (for stand-alone systems) or grid storage (for grid-connected systems). A backup source for the thermal energy produced by solar thermal systems and for the electrical energy produced using PV stand-alone systems is required, and it is usually relying on fossil fuels. In groups of buildings, the thermal energy exchange using the common infrastructure and the electrical energy balance among the buildings represent prerequisites for an optimum system. For an energy community defined at neighbourhood or district level, the thermal energy produced using the solar energy conversion requires adequate storage systems while the electrical energy requires grid storage. The infrastructure for the thermal energy and for the energy management systems has to be implemented targeting to optimize the energy consumption. As already outlined, at community level, except the solar energy conversion systems, the wind power and the hydropower systems or cogeneration systems can be used to produce thermal and electrical energy.

Considering the top-down approach, the *Community Company for Sustainability* has to develop several scenarios based on the use of renewables and, considering the technical–economical assessment, to select the optimum scenario. Visa et al. [103] outlined two different scenarios that can be considered: one based on renewable energy systems scattered implemented and with limited interconnections for the thermal energy storage and another scenario focused on the entire community, with specific solutions for each neighbourhood or district. The scenarios for thermal and electrical energy using renewables can differ according to the community size, involving centralized or decentralized systems.

Starting with the initial stages, the energy community administrative and management structure has to develop an *action plan* to reach the targeted coverage degree (CD) and, based on this, the future project is developed. A threshold for reaching the nZEC status is represented by minimum 50% of the total annual thermal and electrical energy demand to be covered using renewables implemented in the entire community. There can be defined more energy communities at community level with different renewable energy projects where the per cent of energy from renewables can be less or higher than the 50% threshold but the overall coverage degree in the nZEC has to be at least 50% (CD = 0.5). The model and the implementation duration are part of the community strategy for sustainability.

Step 4. Define the energy community in the built environment

The built environment is defined by the buildings' type (residential, public, administrative, commercial buildings, etc.) and the building's specific features (old or new buildings, geometry, construction materials, thermal and optical characteristics of the envelope, building's positioning towards the cardinal points, number of building users, etc.). Moreover, the energy efficiency of the buildings and the energy pattern in the community (centralized or decentralized thermal and electrical energy systems, renewables already implemented) are important.

Step 5. Assessing the energy demand in the energy community

Based on the initial data on the buildings and on the implementation location described at least by monthly average values of the meteorological parameters, the energy demand of each building and the energy required at community level have to be assessed and, based on, the energy efficiency in the buildings. The building codes can be used as data sources or local regulations based on standardized consumption, or predictions based on historical records, or actual consumption values based on the local energy consumption recorded data. These data should cover both, thermal and electrical energy.

Increasing the energy efficiency in an old building requires refurbishment up to the level of *Low Energy Building* through passive solutions (insulation, double or triple *low-E* windows, etc.) and through active solutions (low energy consumers, on-site thermal energy production, floor heating, thermal zoning, etc.). Affordable costs along with preserving or increasing the indoor comfort have also to be considered. The energy required for covering the losses in the district heating and cooling grid should be estimated along with the energy required for public utilities (fresh and wastewater treatment plants, street lighting systems, wastes disposal, etc.).

Step 6. Assessing the renewable energy potential in the energy community

The on-site renewable energies potential has to be evaluated for the available resources by using: outdoor meteorological data, infield measurements over at least one year, data generated by dedicated software or from available databases or mathematical models applied in the location. Predictions based on historical data in databases (e.g. Meteonorm) and prediction/simulation models for the renewable energies potential can also be used. Yearly and monthly values need to be analysed along with the peak values (highest, lowest) and their variability at community level (a significant variation is expected for the solar, wind and hydroenergy and lower variations are expected for the geothermal energy while the use of biomass is subject of planning). Under or overestimations of the renewable energies potential may have negative consequences on the final energy output and costs.

Step 7. Define and optimize the energy mix

Considering the mostly used renewable energy sources to produce thermal energy (solar, geothermal and biomass) and electrical energy (solar and wind), the energy mixes can be defined, as presented in Fig. 6.7, based on the conversion systems that can be installed on or near the buildings or at community level. The mostly used energy mixes to produce thermal energy are based on solar energy and geothermal energy or on solar energy and biomass while for electrical energy, solar energy and wind energy represent a possible energy mix in the built environment.

The renewable energy mix depends on the thermal or electrical energy demand and on the on-site renewable energy potential considering different resources. Moreover, the design of the energy mix has to consider the surfaces available for installing the renewable energy convertors.

At the building level, the thermal energy production is usually based on solar thermal collectors (flat plate or vacuum tube collectors) and on heat pumps or on biomass burning systems, while the electrical energy production is mainly based on PV arrays, platforms or strings and on small wind turbines.

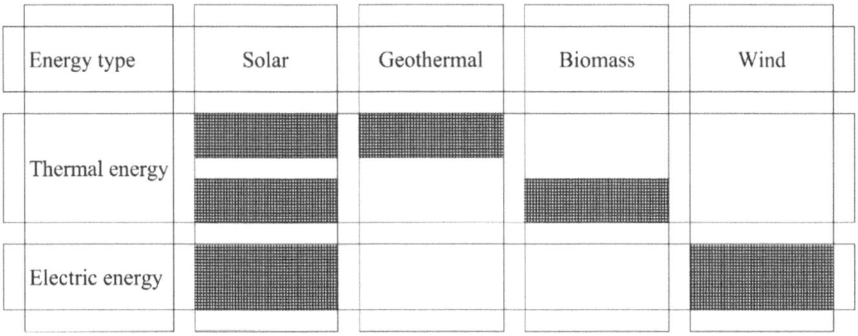

Fig. 6.7 Renewables-based energy mixes in the energy community

At the community level, the thermal energy can be produced using concentrating solar thermal collectors and/or heat pumps or biomass burning systems and the electrical energy using PV parks and/or wind farms. Moreover, biomass gasifier plants are recommended for thermal and electrical energy cogeneration. The structure of the energy mix and the size of its components are the results of a technical and economic (cost) analysis based on several scenarios for an optimum system. The contribution of the renewable energy system(s) on the community status as nZEC represents a key issue when selecting the energy mix.

Step 8. Design and development of the renewable energy project

The optimum system leads to the architectural and design details and, based on, files are prepared to get the necessary approvals for implementation. The specific design of the solar energy conversion systems will run according to the information in Chap. 3 (for the PV systems) and in Chap. 4 (for the solar thermal systems).

Step 9. Implementing the renewable energy project

Using the available financial resources, the renewable energy system project will be installed. Selecting the components of the system(s) requires professional counselling advising on the type of solar energy convertors best suited for the implementation location.

Step 10. System operation and maintenance

Monitoring the renewable energy system is required as it also is its periodical assessment; therefore, a *Building Energy Management System* (BEMS) is recommended. The results certify the system efficiency based on the amount of energy produced by the components in the energy mix and on the available renewable energy. These results can be used to improve the design of future similar systems.

6.3 Operation and Energy Management

Specific energy issues are currently missing in urban planning in many countries, as stated by the International Energy Agency [90], and this has to be changed by developing and implementing optimized energy strategies in cities, adapted to the local conditions. Once the vision and the targets are defined at community level, the project for implementing renewables can be formulated, according to the local needs and resources and, afterwards, its implementation starts. The renewable-based system(s) are monitored during their functioning to find out to what extent the implemented projects fit the initially stated goals. Moreover, beyond focusing on the system(s) efficiency, the monitoring phase should also consider how the project answers to the societal challenges initially identified.

The public acceptance of these projects is essential, as it also is the support of the stakeholders. Additionally, selected stakeholders should be part of a training process to fully understand the actions they are asked to support. Moreover, the leading person or organization should be clearly outlined as being responsible for the project's decisions and outcomes. Thus, the successful operation of the renewable-based projects has to consider the development of strong links with the municipality structures, either by directly involving these in the project implementation and operation or by developing a new structure that will be integrated in the existing municipality organization, e.g. the *Community Company for Sustainability*. The coordination process requires time and resources, and this should be considered from the very beginning. In this process, the energy suppliers and the network providers should be part of the stakeholders group, represented in the steering committees of the authorities. The best results were obtained when local initiatives were implemented, as in the Norwegian case presented by the Ministry of Foreign Affairs [63].

More options are available to develop energy management at community level when involving renewables. The variety comes from the different renewable energy sources that are considered and from the energy demand at community level that has to be met all over the year. It is to notice that the current business model (based on conventional grids fed up with non-renewable power) is mainly based on the interaction between the energy distribution and transmission, with little involvement of the end-users. Recent models that also involve the end-users were developed based on the smart grid concept. An *Advanced Metering Infrastructure* (AMI) that connects users and utilities, along with a *Home Energy Management System* (HEMS), is part of such smart grids. There are various methods to compensate the time periods when energy shortage occurs, including the energy produced using renewables.

Zafar et al. [107] detailed the concept of *energy prosumer* as an energy producer that uses the energy produced based on renewables and shares the excess energy with other consumers and/or with the grid. The prosumers can gather in an energy-sharing smart grid that links the consumers according to, e.g. the energy consumption behaviour. The energy-sharing network involves two key elements: optimization techniques and the information and communication technologies (as *Smart Metering* and the *Internet of Things*). The amount of energy-related data increases with the community size, thus data management centres should be defined as part of a wider community control centre or as stand-alone entities. These are further operated using the *Prosumers Energy Management and Sharing* (PEMS) systems. As communication inside the smart grid was outlined as the most important feature, the specific optimization technologies in PEMS are compared considering their efficiency and reliability.

Among the wireless communication technologies that can be used for energy management in the smart grids, there can be mentioned the *General Packed Energy Services* (GPRS), the *Worldwide Interoperability for Microwave Access* (WiMAX), Bluetooth or ZigBee.

The wired *Power Line Communication* (PLC) represents another promising option for PEMS with two alternatives: the *narrow band* (NB) and the *broad band* (BB).

The *energy management systems* support the prosumer to monitor the energy production and consumption. The study developed by Zafar et al. [107] shows that the individual prosumers may meet more challenges, as energy shortages, when compared to cooperative prosumer systems, thus the prosumers communities (the energy communities) represent a viable option, based on the prosumers energy-sharing behaviour. Further on, the energy produced but not consumed in the prosumers community (the excess energy) can be sold on the market, and many papers are discussing these new types of energy markets, as Cerquides et al. [12] showed for the energy in the smart grid.

In developing a prosumer community, essential is also the available space to implement the renewable energy systems, on the buildings façades or on the roofs and nearby the buildings. Moreover, energy storage systems need to be selected to reach the targeted energy independency and new protocols have to be developed to allow the smart grid management to support balancing the peak loads.

Combined cooling, heating and power (CCHP) technologies that are relying on renewables are based on multienergy systems that are currently developing as *smart multienergy systems* (SMES) increasing the efficiency of energy use and satisfying various energy demands met by using renewables. These types of systems are now part of various national strategies as the *Integrated Energy Systems Plan* in the USA [75], or the Energy System Catapult [24] in the UK, or the *E-Energy Plan* developed by the Federal Ministry for Economic Affairs and Energy [33] in Germany, or the *Internet + Smart Energy* for China [89]. According to Ma et al. [61], these SMES are designed aiming at giving use to all the available energy sources to meet the local energy demand at a minimal cost. The design uses as input data the profile of the energy demand, the production and the cost of different renewable energy systems along with the cost of the electricity and gas equipment, the climatic profile and the features of the energy production systems (capacity, efficiency or coefficient of performance, unit capital costs, maintenance costs and the maximum investment amount). Based on these data, the optimized project is defined in terms of type and capacity of the energy convertors and of the storage devices; further on, the optimal operation strategy is formulated and implemented. The main constraints that have to be considered are related to:

- The electricity, the cooling and the heating energy balance. This balance should consider the energy production systems and the usual consumption systems along with the loads corresponding to additional systems as the electricity storage devices, the electrical heaters or coolers, etc.;
- The capacity limits of the renewable energy convertors and sources, as these influence the maximum output that can be expected from a given system;
- Operation constraints of the energy storage devices, considering the capacity of the energy storage device and the maximum charging and discharging ratios of each device.

To develop a working model for the system and to keep the required information at a feasible level, four typical days are defined, one per each season (spring, summer, autumn and winter) and the electrical power, the heating and the cooling demands are calculated based on these. These energy demands are independently considered as loads per end-user. Moreover, the solar radiation and the outdoor temperatures variation can be modelled for these typical days, based on previous, infield records and, based on, the output energy produced by a solar energy conversion system can be predicted according to the type and size of the system.

Ma et al. [61] considered a practical example in Beijing, China, and demonstrated that the energy production–demand balance is reached by using renewables (PV systems) and gas combustion engines to cover the electricity demand. As the combustion engines also produce heat when generating electricity, the heating energy demand is mainly covered by these devices coupled with heat storage systems. The energy storage systems are involved in the SMES for directly storing the excess energy to be further used when needed but also to decrease the costs, thus energy production and storage during the low-cost periods and its use when the energy cost is high represents a strategic option. The results show that a SMES has a better performance and financial response when compared to the centralized systems or to the CCHP ones, showing a 35.21% reduction in the total annual costs compared to the centralized solution and a 55.34% reduction in the greenhouse gases emissions.

The CCHP systems also including renewables represent possible solutions for the new energy systems in the built environment and making these affordable represents a key prerequisite. Thus, the direct trade of the energy excess or shortage among buildings (considered as micro-grids) represents a suitable option. However, this is not the most feasible one as the energy in excess is sold at a lower price, and the energy required during shortage periods is bought at a higher price than the production costs; therefore, barter trading represents another possible option, as outlined by Bui et al. [9]. This trading system is discussed also at community level as a *Community Energy Management System*, C-EMS. The system consists of two subsystems that are managed by the C-EMS and by the distributed network operator(s), especially by the electricity operator. These two management structures decide the costs, based on the energy demand of each building (for the internal market, opened only to the buildings part of the community) and on the energy available at the entire network level (for the external market that controls the energy trading among the external network, the utility grid and buildings micro-grid). As outlined, the C-EMS manages the process at the building level (part of the community) aiming at minimizing the energy trade among the buildings. The external systems or the utility grid compensates the final surplus/shortage amounts.

This scheme is obviously a transition scheme that involves the gradual development of various energy production systems at building level and their integration in a specific structure at city level.

Once more sustainable cities will be developed, the actual problems related to their operation will get specific answers, and the development of functional energy management systems will be formulated.

6.4 Case Studies

Sustainable communities already exist in the world, and several representative case studies are further presented.

6.4.1 Rural Sustainable Communities, Europe

Doukas et al. [22] assessed the energy sustainability of eight rural communities in Europe based on the *Energy Sustainability Index*. Rural communities were assessed as they have a vast unexploited renewable energy potential (solar, wind, biomass, hydro, etc.) that can support their growth, jobs and sustainability.

The first place was awarded to the Natural Regional Park in the Ardeche Mountains in France, where a large share in the energy demand is covered using renewable energy sources. As described, 17 wood energy plants are functioning in 15 municipalities with a total installed capacity of 2675 kW and about 100 regular and small hydropower plants are producing 660 GWh/year that covers (twice!) the electricity demand in the region. Moreover, a small district heating wood-based energy system, a small wind farm and small photovoltaic developments and eco-constructions are mentioned.

The second placed in the rural communities ranking is Murau, Austria, where a complete self-sufficiency goal was set, about 30% covered from renewable energy sources by small, private hydropower and combined heat and power plants, the highest wind farm in the Alps along with biomass- and waste-to-heat systems.

Another rural community, Rouge in Estonia, developed an energy park aiming at a fossil fuel-free region; the share of biomass covers 100% the heating energy demand while for electricity five hydroplants can produce energy able to cover 15% of the community demand; the municipality aims at further updating the old hydroelectric plants while reducing the buildings energy consumption.

The other five assessed communities did not reach significant *Energy Sustainability Indexes* to classify them as sustainable communities, the renewable energy potential being exploited only through solar thermal collectors installed on residential buildings and house heating ground source heat pumps.

In a review developed by van der Schoor and Scholtens [99], thirteen communities in the northern Netherlands were investigated. In nine of these communities, solar energy is used (mainly for photovoltaic systems), and in four cases biomass systems are involved in the energy mix. Energy efficiency is targeted in six communities, and the sustainable use of water is considered in one case. Energy mixes were implemented in only three cases as photovoltaic and biomass systems in Oldehove, photovoltaic and solar thermal system in Zuidhorn and photovoltaic, biomass and solar thermal systems in Makkinga.

6.4.2 Frederikshavn, Denmark

Lund and Østergaard [59] describe a sustainable community developed in a relatively small town of Denmark, Frederikshavn, located on the northern coast of the Jutland peninsula. The town has about 25,000 inhabitants and the local government along with the local companies and the Aalborg University were included in the *Energy Town Frederikshavn Project*, aiming at 100% energy independency by using renewable energy systems.

After organizing community meetings and campaigns to raise the awareness among residents and encourage their participation, the city identified the key components of the new energy mix. This energy mix relies on bioenergy (vegetal and animal biomass), wind energy, solar energy and geothermal energy. The main component of the renewable-based energy mix is represented by a new 15 MW combined heat and power (CHP) plant, based on highly efficient burners using biogas, waste and straw to produce 91 GWh/year of electricity and 226 GWh/year thermal energy. The electricity produced by the CHP plant is fed in the local grid along with 130 GWh/year produced by 40 MW wind turbines. This local electricity grid is connected to the national grid to yearly balance about 25 GWh of import/export (zero net exchange of power). The electricity is mainly used in the industrial sector, followed by the heating systems based on heat pumps and by charging systems for electric cars. The thermal energy produced by the CHP plant is used, along with the thermal energy provided by the solar thermal systems (4 GWh) and geothermal heat pumps (95 GWh), for the space heating in residential and industrial buildings through a district heating grid. Individual solar thermal systems (1 GWh) and geothermal heat pumps (7 GWh) heat the residential buildings that were not connected to the district heating system. Additional heat pumps will be installed to recover the exhausted heat from the CHP plant(s) and from other sources as the wastewater. The biogas plant uses yearly about 34 million tons of manure. The biogas is used in the CHP plant (128 GWh) and to produce biomethanol (61 GWh) that is used along with biogas (10 GWh) in the transportation sector.

Thus, for transportation, biogas and biomethanol vehicles and electric cars are used, and refuelling stations are installed in the city for biofuels as well as recharging stations. Hybrid cars will combine the use of batteries with fuel cells based either on methanol or on hydrogen.

6.4.3 University of California Davis, West Village, California, USA

This community was launched in 2011 by the University of California Davis and includes 662 apartments and 343 single-family homes for students and campus employees [18]. It was designed as a net zero energy community, thus it generates as much energy as it yearly consumes.

The buildings are built based on "aggressive energy-efficient measures" including improved thermal insulation, radiant barrier roof sheathings, solar reflecting roofing, energy-efficient lighting with presence and crepuscular sensors, water-saving toilets. Recycled building materials and paints with low emissions of volatile organic compounds were used.

The energy is provided by a 4 MW solar photovoltaic system installed on the buildings' rooftop and on the parking canopies. A biodigester is installed to transform all the campus table scraps, the animal and the vegetal waste into energy.

The transport is based on an extensive bike network, on all-electric automated shuttle around a defined loop in the solar-panel-lined West Village community and on public transport powered by biogas.

6.4.4 Sino-Singapore Tianjin Eco-City, China

This sustainable community was developed in a Singapore–China cooperation aiming at building a sustainable eco-city as a pilot project for future cities [32]. It is located 150 km South from Beijing, in the Tianjin area, covering 30 km^2 for a projected population of 350,000 inhabitants. Its completion date is expected to be 2020. In the design process, two concepts were followed: *Three Harmonies* and *Three Abilities*. According to the first concept, people should live in harmony with other people, developing activities while protecting the environment, thus including the main three pillars of sustainable development (society, economy and environment). The second concept refers to the eco-city being practicable, replicable and scalable.

Based on these concepts, the newest green technologies and the best ideas are incorporated starting from the design stage to reach a green community. The carbon footprint is reduced by integrating the nature into the daily living, promoting walkability, hybrid cars and buses for general transportation. A vast *Eco-Valley*, obtained by restoring natural habitats and cleaning the water bodies and wetlands, serves as an ecological spine. The residential areas are integrated with commercial zones so that the residents live close to the working places, thus saving time and fuel consumption. Drinkable tap water will be available for the residents based on a sound water management. A collective waste management system is integrated with an incineration process producing energy and thus minimizing the landfills. The buildings are designed respecting the *Green Building Evaluation Standards* developed by experts from Singapore and China integrating cost-effective measures to enhance sustainability while maintaining their affordability.

A renewables-based energy mix is designed to provide clean energy, converted from solar and geothermal resources.

6.4.5 Saerbeck, Germany

Saerbeck is a village with 7200 inhabitants and an active life at community level. Everything started when the municipality won a 1 million EUR prize in a competition of all communities in the Federal State of Nordrhein-Westfalen with a project aiming at supplying, by 2030, the whole energy in Saerbeck by locally produced renewable energy. The first step consisted of encouraging the Saerbeck citizens to participate in the project by installing photovoltaic modules on their roofs. To this end, the municipality started to install photovoltaic modules on schools (Fig. 6.8) and disseminating the results through the pupils learning in those schools. This action led to nearly 400 photovoltaic modules newly installed on the roofs of the schools, of the residential buildings and of the farmhouses (8.8 MW).

The second part of the project addressed the replacement of the former gas boilers with a new central heating system, fed with wooden pellets for two schools, a kindergarten and several sport facilities.

To further extend the share of renewable energy covering the energy demand in the community, the third part of the project transformed a surface of 90 ha, located 1.5 km away from the village in a *Bioenergy Park* by installing a total of 29 MW renewable energy systems as follows: a 1 MW biomass power plant using vegetal waste obtained from 300 ha of 17 local farmers, a solar power park with 5.7 MW photovoltaic platforms and seven wind turbines of 3 MW each (Fig. 6.9). Due to the mismatch between the energy production (using wind and solar energy) and the energy consumption, several energy storage solutions were implemented from the beginning following a parallel strategy: different types of batteries (lithium-ion, sodium-sulphur), redox-flow and power to gas (hydrogen, methane) technologies.

The most important aspect is that the community worked out the entire planning process, the renewable energy systems are owned by the community, and the revenues belong to the community being reinvested in local projects. This may be a good business model for communities.

Fig. 6.8 Photovoltaic system installed on the roof of the school in Saerbeck, Germany

Fig. 6.9 Photovoltaic platforms and one of the seven wind turbines installed in the bioenergy park in Saerbeck, Germany

6.4.6 *The Genius Campus in the Transilvania University of Brasov, Romania*

In 2008, *Transilvania* University of Brasov started the design process of a novel structure for advanced research on sustainable energy: the *Green, ENergy Independent University CampuS* (GENIUS). The construction started in 2009, and the first phase was completed in 2013 by finalizing the R&D Institute [104] as a result of a structural funds project. Twelve low energy buildings were built incorporating state of the art materials and solutions to mitigate the thermal and electrical energy demand. The energy consumption is monitored for each building aiming at validating their low energy building status. Thus, each building is actually a 1:1 testing rig. Fully monitored renewable energy systems are implemented on the buildings' rooftop and nearby as presented in Fig. 6.10.

This complex research infrastructure consists of renewable-based thermal energy systems of ground-coupled heat pumps that extract the geothermal energy through vertical—(VGHX) and horizontal—(HGHX) geothermal heat exchangers and solar thermal systems (STS) with flat plate (FP) and evacuated tube (ET) collectors that deliver the thermal energy for DHW water preparation. Moreover, grid-connected *solar-tracked photovoltaic systems* (ST-PVS) are installed to yearly balance the electrical energy demand of the heat pumps along with the indoor lighting while stand-alone *wind turbine systems* (WTS) are delivering electrical energy for the outdoor lighting. An *outdoor testing rig* was developed to investigate the façade integration

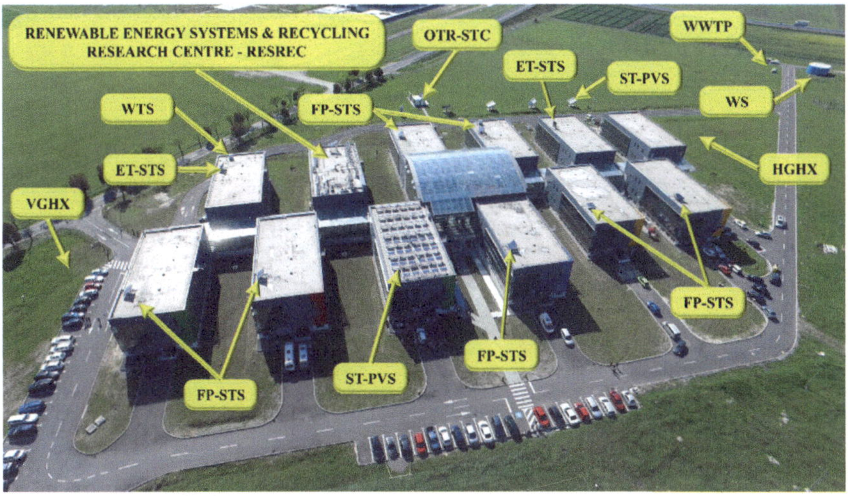

Fig. 6.10 Renewable energy systems installed in the R&D Institute of the Transilvania University of Brasov, Romania

of *novel flat plate solar thermal collectors* (OTR-STC) with trapeze and triangle shapes and various colours of the absorber plate.

The water is supplied from a local, 120-m-deep well (WS), and the wastewater is cleaned in a local *wastewater treatment plant* (WWTP) before discharge.

Ground-coupled heat pumps, of 22 kW, are installed in the basement of two buildings (Fig. 6.11), the first one being connected to a vertical geothermal heat exchanger with four boreholes at a depth of 90 m each and the second one to a horizontal geothermal heat exchanger of 1000 m^2 mounted at 2.5 m depth, below the ground surface.

On the rooftops of nine buildings, solar thermal collectors are installed to provide DHW, stored in 500 L tanks in the basement; as backup source, condensing gas boilers are used (Fig. 6.12).

Two solar thermal trough concentrators used in a solar cooling system and two photovoltaic thermal (PVT) modules, part of a 5 kW photovoltaic platform are installed on the rooftop of a laboratory building (Fig. 6.13).

A 12 kW photovoltaic solar-tracked system with a single-axis tracking mechanism is installed on the rooftop of another laboratory building, feeding the electricity into the local grid. A total of 48 photovoltaic modules (nominal power of 250 W/module) are serially connected to a central, three phases inverter, Fig. 6.14.

Other five photovoltaic platforms of about 2 kW each (Fig. 6.15) are installed on the ground, nearby the buildings. Four of them have different types of solar tracking mechanisms, and the fifth one is fixed, optimally tilted and is used as reference. Moreover, there are different types of PV modules installed on the platforms allowing to simultaneously monitor their photovoltaic response in the built environment conditions.

Fig. 6.11 Ground-connected heat pump system

(a) (b)

Fig. 6.12 a Solar thermal collectors installed on the rooftop; b storage tank and backup source installed in the basement

Fig. 6.13 Two solar thermal trough concentrators and two PVT modules installed on the rooftop of a laboratory building

Fig. 6.14 Tracked photovoltaic system (12 kW) installed on the rooftop of a laboratory building

Fig. 6.15 Tracked photovoltaic platforms installed nearby the buildings

Two groups, each of three small wind turbines are installed on the rooftops of two buildings, one group with a total capacity of 900 W (Fig. 6.16) and the second group with 1800 W, each of them feeding the electricity into rechargeable batteries, to be further used (e.g. during the night) to power the outdoor lighting system.

The energy consumption in this community is of about 1328.5 MWh/year [104] as presented in Table 6.1, considering the common uses.

By extending the already installed renewable energy systems to each building, the renewable energy that could be produced is estimated at 1317.2 MWh/year, as detailed in Table 6.2, corresponding to a renewable energy share of 99.15%; thus, the R&D Institute can be considered as a *Nearly Zero Energy Community*.

Fig. 6.16 Small wind turbines installed on the rooftop of one laboratory building

Table 6.1 Energy demand in the R&D Institute

Energy demand category	(kW h/year)
Thermal energy demand for space heating	556,956
Thermal energy demand for space cooling	47,832
Thermal energy demand for DHW preparation	50,928
Electrical energy demand for the lighting and IT systems	449,688
Electrical energy demand for the heat pumps	134,088
Electrical energy demand for fresh and wastewater	4455
Electrical energy demand for the backup sources	84,528
Total energy demand	1,328,475

Table 6.2 Energy planned to be supplied by renewable energy systems

Energy supply category	(kW h/year)
Geothermal heat pumps (264 kW, including space heating and cooling)	492,732
Solar thermal systems (43 kW)	32,316
Photovoltaic systems on the rooftops (44 kW)	549,164
Photovoltaic systems on ground (eight platforms × 2 kW)	228,096
Wind turbines	14,940
Total renewable energy supply	1,317,248

6.5 Emergent Trends in Using Solar Energy at Community Level

Sustainable cities and communities represent the focus of one of the seventeen sustainable development goals formulated by the United Nations during the New York Conference in 2015 as part of the *2030 Agenda for Sustainable Development* [96]. This goal addresses the issues raised by the urban growth as 54% of the world's population (about 4 billion) was living in cities in 2015, and this number is expected to increase up to 5 billion by 2030 [97]. Thus, cities and urban communities are facing enormous challenges, including an increased energy demand mainly at the buildings, industry and transportation levels with direct and negative consequences on the air quality.

To get inclusive, safe and sustainable human settlements, many countries (149) have developed, by May 2017, national strategies focusing on clean and affordable energy for all, addressing the following main challenges:

- New buildings have to be designed, from the very beginning, as low energy buildings while the existing building stock must be retrofitted aiming at decreasing the energy demand by applying new solutions for energy savings and energy efficiency in the built environment;
- Renewable energy mixes have to be designed to cover in a very large extent this reduced energy demand while the cost of renewable energy must be further decreased through technology advancements, increasing the conversion efficiency;
- The renewable energy mixes installed on each building in a community will be interconnected in micro-grids that will operate autonomously or connected to a larger grid. *Artificial Intelligence* will allow the continuous optimal operation of these micro-grids, giving priority to self-consuming the energy produced using renewable sources;
- The solar fraction must be increased at community level, solar energy being easily accessible and cost-effective;
- Energy storage represents a key issue to further balance the energy supply and the energy demand;
- An open energy market is required to empower the prosumers to buy and sell the energy, thus minimizing the payback time of the renewable-based energy mixes and maximizing the benefits;
- The communities' involvement in the efficient use of energy and in limiting the greenhouse gasses emissions represents an important aspect;
- Finally, yet importantly, communities in developing regions/countries must be supported to gain access to affordable, clean, resilient and sustainable energy.

New solutions that address these challenges are further described, focusing on those that support the increase in the solar energy share in the renewable energy mixes.

The energy demand at building level is commonly decreased by using building materials with low thermal conductivity. Common insulation materials are the polystyrene and polyurethane foams with thermal conductivities of about 0.045 W/m K and 0.025 W/m K, respectively. These materials can be easily used in the opaque parts of the building envelope where the thickness of the thermal insulation can be increased up to the required level. This is possible for the new and for the existing buildings, except heritage-protected buildings where specific conditions are imposed for the facades' structure and aspect.

New and advanced nanomaterials, developed for other sectors (e.g. in aerospace or military applications), can be translated to the built environment as the silica aerogel that, mixed with natural plaster, has a very low thermal conductivity, ranging between 0.014 and 0.016 W/m K [10]. Another option is to integrate the solar thermal collectors into the building's facade, the thermal insulation of the collector thus increasing the thermal resistance of the building while the facade is used to provide thermal energy. Precast concrete facade elements can be converted into solar collectors, thermally insulated on the backside, by embedding pipes in the concrete layer exposed to solar radiation [67]. After increasing the thermal resistance of the opaque elements in the building's envelope, another important task is to enhance the thermal resistance of the transparent parts (windows and glazed doors). Usual solutions consist of multiple glazing layers with advanced features such as low-E coatings, interior air replaced with inert gases, multichambered frames with advanced spacers. Research aims at further increasing the thermal resistance of these transparent elements, while allowing the solar energy to passively heat the building. Thin films of transparent or translucent materials such as polycarbonate, polypropylene, polymethylmethacrylate, cellulose triacetate and silica aerogel could decrease the heat transmittance down to 0.25 W/m^2 K [68].

The (reduced) energy demand of the buildings and of the community must be *further covered mainly using renewable energies*, to reduce the fossil fuels consumption and the related greenhouse gases emissions. The solar and wind energy conversion systems already proved to be comparable with the systems based on fossil fuels considering the costs of the produced energy, expecting that these costs will further drop in most regions below 0.1 USD/kWh by 2050 [19]. Due to the variability of the amount of renewable energy sources accessible in the built environment (as solar and wind energy), very high renewable energy shares covering the energy demand cannot be feasibly obtained relying only on one resource (e.g. the solar energy). This imposes the use of renewable energy mixes that include several types of renewable energy sources with at least one that does not fluctuate as much as the solar and wind energy are. Such an energy source can be the geothermal energy or biomass.

The main barrier in implementing a renewable energy mix is the investment cost for each component of the system. The challenge is to identify the optimal capacity of the components that will *minimize the cost of the provided energy using renewable(s)*. Once installed, these mixes will work as consumer-driven systems, with self-optimization algorithms to increase their efficiency while prioritizing the self-consumption. Combining the economic benefits (further expected to increase through evolving technologies) with the reduced environmental impact, renewable energy mixes are presumed to move from an acceptable energy source to a preferred one.

Further on, the renewable energy mixes implemented at building level will be interconnected in *micro-grids of distributed energy generation systems*, to strengthen the resilience of the traditionally large and centrally distributed energy production systems. Dedicated software tools were developed especially for the micro-grids design. The Sandia National Laboratories developed a *Micro-grid Design Toolkit* for the US Department of Energy [23]. This toolkit consists of different procedures for the optimal trade-space exploration, for preliminary sizing the distributed energy resources in a micro-grid and for simulating the micro-grid operation allowing optimizing its performance.

Micro-grids were developed all over the world. Recently, these micro-grids included also renewable energy systems and even renewable energy mixes. In the USA, the micro-grid concept was applied in many sites, being considered as a default power system for military facilities [65]. Feng et al. [34] reviewed seven micro-grids developed in the USA, five of them including renewable energy systems.

The micro-grid deployment trend is increasing, and an equilibrium is expected to be reached by 2035 between the distributed energy supplied by micro-grids and that provided by centralized power plants [74].

However, the micro-grids raise several challenges. Few studies are dealing with the management of the multimicro-grids, increasingly integrating renewable energy sources, to enhance the overall efficiency. Artificial intelligence and predictive control are required to control the different technologies involved in the renewable energy mix. New features of the cloud-based and *Internet of Things* platforms are expected to improve the integration and interoperability of the connected systems. Thus, the flexibility of the renewable energy mix is increased, allowing optimizing in near-real time the entire system [45].

Solar energy technologies are expected to gain a more consistent share in the renewable energy mixes. Settino et al. [85] reviewed the solar technologies for electricity, heating and cooling production and suggested that, for heating purposes, the solar thermal collectors still represent the best option while for cooling the combination of photovoltaic modules and heat pumps is highly promising considering the overlap of the moments when the energy demand and the energy supply occur. Stand-alone photovoltaic systems storing the electricity in batteries have a low environmental impact, but their use is limited by the high initial cost. Photovoltaic thermal systems could be a solution when there is a limited available implementation area; however, their market share is limited, and their cost is still high. Solar cooling systems using the thermal energy provided by concentrating solar thermal (CST) collectors and absorption or adsorption chillers are still expected to increase their conversion efficiency and reduce the costs [40, 64]. Lazzarin and Noro [56] outlined that the price of the solar thermal collectors must consistently decrease, and the efficiency of the thermally driven chillers must increase to obtain *energy efficiency rating* (EER) values higher than 1.4–1.5, to catch up with the photovoltaic heat pump solar cooling systems.

In large communities, a solution to increase the solar fraction in the renewable energy mix consists of using concentrating solar energy convertors providing both thermal energy and power. The most popular technologies are based on parabolic trough collectors and solar power towers already installed in many industrialized countries [53]. Research on these technologies increased during the last decade focusing on new, emerging topics as hybridizing with other technologies as the organic Rankine cycle, the supercritical CO_2 power cycle in concentrating solar power plants, etc.

Energy storage is mandatory when renewable energy systems are used, due to their variability, especially when considering the solar and wind energy sources. Several solutions to store the energy are currently investigated. Technology maturity is already reached for short-term energy storage, such as the storage tanks for domestic hot water or the rechargeable batteries for electrical energy, and their use is increasing at residential and communities scale. Effort is invested for increasing the technology readiness level of the new energy storage solutions while keeping the cost at a competitive level.

Hachem-Vermette et al. [42] explored the potential to reach a net zero energy status for a hypothetical large community in Calgary (Canada) having a mid-high latitude (52° N) and a rather cold climate. Residential buildings (detached and attached houses) and non-residential buildings (public schools, offices, retail and supermarket buildings) were considered in this mixed-use community. The renewable energy source(s) rely only on solar energy. The investigations showed that by applying energy efficiency measures to decrease the community energy demand and by implementing photovoltaic systems on each building rooftop, about 70% of the community total energy consumption can be covered. Moreover, installing solar thermal collectors on additional structures within the public areas (e.g. parking lots) with short-term storage tanks, all connected through a distribution system both to the buildings and to a borehole energy storage lead to reaching the net zero energy status.

Another solution to store the thermal energy into the ground, without investing in expensive boreholes and ground heat exchangers, consists of aquifer thermal energy storage. Various such systems were reported as being in operation for heating and cooling by Fleuchaus et al. [36]. In a recent study, Pellegrini et al. [69] showed that this type of storage systems has a good potential in Europe, leading to relevant economic and environmental benefits. Based on six pilot sites in five European countries (Spain, Italy, Belgium, Netherlands and Denmark), several solutions were investigated to overcome the technological and the non-technological barriers. Three types of barriers were identified: *general barriers*, *mature market barriers* and *novel market barriers*. In the first category, legal restrictions and a gap between the knowledge and the specific skills for implementing and exploiting the systems were identified in all the five countries. In the second category, the most important was the interference between the aquifer thermal energy storage systems and the polluted ground water. The novel market barriers refer mainly to the lack of public awareness and the limited knowledge on the characteristics of the underground environment. These first results showed that the aquifer thermal energy storage system can be applied in various climatic conditions through innovative technologies and, combined with renewable energy systems (e.g. solar thermal systems), could support economic and environmental benefits. These types of studies will continue with long-term monitoring and evaluation.

Building integrated thermal energy storage systems were also investigated at a smaller scale [66]. Two main directions were followed: the use of the walls, ceilings and floors of the building as thermal energy storage elements, and the integration of the heat storage water tanks in the building. Investigations on the first direction showed promising results, especially for ceilings and floors, as these are usually not covered as walls are. To increase the storage capacity, phase-change materials can be used as a storage medium. An advantage of this method relays in the possibility to prefabricate building components, thus decreasing the implementation cost of the thermal energy storage system. The main drawback of this solution is that it cannot be easily applied to already existent buildings. Suspended thermal energy storage panels were experimentally tested, and promising results were obtained indicating them as a potential solution for the refurbishment of the existing buildings. The second direction, aiming at integrating large, seasonal storage warm water tanks in the buildings, is facing the main issue of architectural integration. Due to the large water volume that requires seasonal storage, the storage tanks need plenty of space, limiting the potential of this solution. This type of seasonal storage systems is already installed in Central Europe, e.g. by integration in the building stairwells, proving that there are not technical issues, and only the size and the required installation space are raising problems.

Besides the thermal energy storage, another challenge is represented by the electrical energy provided by intermittent renewable sources as the solar and wind energy. Current and emerging electrical energy storage technologies include solid-state batteries, supercapacitors, hydrogen storage, compressed air and pumped hydro, among others. Solid-state batteries gained popularity in small-scale applications ranging from portable electronic devices to electric vehicles, but not in medium- and large-scale systems because of their poor scalability, high cost, short lifetime and safety concerns [13].

An alternative to energy storage is *the transaction of the exceeding energy on the energy market*. This will be possible through the interconnection of the renewable energy mixes in a network of energy hubs that integrate multiple forms of energy production, consumption and storage at community level. This concept is already investigated in modern power distribution systems where the transacted energy introduces a new type of autonomous electricity retail market along with new business and operation models of power generation, consumption and delivery (Cerquides 2015).

More and more devices used on a daily basis are now able to receive and convey information thanks to new platforms such as the *Internet of Things*. Several applications are developed to start these devices when a clean, renewable energy is available, thus maximizing the self-consumption. The micro-grids owners can also buy and sell energy at optimal moments. Due to the cyber-physical nature of the networked micro-grids, measures to reduce the cyber-security vulnerabilities have to be designed and applied.

A solution could rely on *the block chain technology*, originally developed for cryptocurrency transactions [57]. The block chain technology could provide a solution on the energy markets of renewable energy-based micro-grids, allowing their automatic and auditable interaction while maintaining a high degree of efficiency and reliability. This technology manages and records transactions through a peer-to-peer network, eliminating the centralization, energy inequality and inefficiency and empowers consumers to directly buy and sell energy from other consumers. This decentralized energy management will optimize both the energy and the financial flows, contributing to the advancement of the active distribution networks. The customers will have a more proactive and environmental friendly contribution in optimizing the settings of the renewable energy mixes in real time by monitoring and analysing the performance of the energy-consuming devices. Thus, the local communities will benefit on the locally produced renewable energy, on the improved environmental quality along with various financial advantages.

The communities' involvement will be very important in this transition phase towards sustainability. Over 40 cities worldwide are currently operating on 100% renewable electricity, and over 100 cities report that renewable energy is covering at least 70% of their energy demand. Moreover, many cities expressed their willingness to work towards a renewable-based net zero energy status. Recognizing their impact on climate change, major corporations have also introduced sustainability policies in their development strategies, as an important part of their commitment to 100% renewables. Many municipalities and companies started to formulate these commitments after the 2015 Paris Agreement, and based on the information provided into the report of the Intergovernmental Panel on Climate Change [49], an increased commitment to switch from fossil fuels to renewable energy sources is expected.

All these trends were mainly observed in developed countries with a mature industry, that significantly invest in research, development and implementation of renewable energy mixes to tackle the fossil fuels-related problems. Unfortunately, a significant percentage of the world's population has no access to energy at all, as one billion people are living without electricity and hundreds of millions are using unreliable or expensive energy sources. The communities in these developing regions/countries have to be supported to gain access to affordable, clean, resilient and sustainable energy. Renewable energy mixes interconnected in micro-grids represent a cost-effective way for providing affordable and reliable electrical energy to people currently living without it. Modular renewable energy systems might be an ideal choice for many communities in areas where conventional centralized forms of energy generation and delivery were not developed yet.

International organizations as the World Bank [105], the United Nations Environment Programme [98], the Renewable Energy Policy Network for the 21st Century [81], the International Renewable Energy Agency [51], the International Energy Agency [50], the International Solar Energy Society [52] and the Alliance for Rural Electrification [1, 2] are involved in promoting and developing renewable-based energy solutions for communities' electrification in developing countries.

References

1. Alliance for Rural Electrification (2017) Annual report 2016. https://www.ruralelec.org/sites/default/files/2017-10-26_are_annual_report_2016_final.pdf. Accessed 01 May 2019
2. Alliance for Rural Electrification (2018) Energy access from the bottom up: start-up and SME showcase. https://www.ruralelec.org/sites/default/files/ARE%20Energy%20Access%20from%20the%20Bottom%20Up%20Final.pdf. Accessed 01 May 2019
3. Anug KC, Whale J, Urmee T (2019) Urban wind conditions and small wind turbines in the built environment. A Rev Renew Energ 131:268–283
4. Awad H, Gul M (2018) Optimization of community shared solar application in energy efficient communities. Sustain Cities Soc 43:221–237
5. Baccio C, Corgnati SP, Delmastro C, Fabi V, Lombardi P (2016) The role of nearly—zero energy buildings in the transition towards Post-Carbon Cities. Sustain Cities Soc 27:324–337
6. Berardi U (2018) ZEB and nZEB (definitions, design, methodologies, good practices, and case studies). In: Desideri U, Asdrubali F (eds) Handbook of energy efficiency in buildings. Butterworth-Heinemann, Oxford, UK
7. Bervar M, Bertoncelj A (2016) The five pillars of sustainability: economic, social, environmental, cultural and security aspects. In: Proceedings of management international conference (MIC). Pula, Croatia, pp 1–4. June 2016
8. Brundtland GH (1987) Our common future. United nations world commission on environment and development. Oxford University Press, Oxford, UK
9. Bui V-H, Hussain A, Im Y-H, Kim K-M (2019) An internal trading strategy for optimal energy management of combined cooling, heat and power in building microgrids. Appl Energ 239:536–548
10. Buratti C, Moretti E, Belloni E, Agosti F (2014) Development of innovative aerogel based plasters: preliminary thermal and acoustic performance evaluation. Sustain-Basel 6(9):5839–5852
11. Burton H (2000) Sustainable communities. The potential for eco-neighbourhoods. Earthscan Publications Ltd, London
12. Cerquides J, Picard G, Rodríguez-Aguilar JA (2015) Designing a marketplace for the trading and distribution of energy in the smart grid. In: Bordini E, Weiss Y (eds) Proceedings of the 14th international conference on autonomous agents and multiagent systems (AAMAS 2015). Istanbul, Turkey, pp 1285–1293
13. Child M (2019) Flexible electricity generation, grid exchange and storage for the transition to a 100% renewable energy system in Europe. Renew Energ 139:80–101
14. Commission for Environment Cooperation (2010) Guide to developing a community renewable energy project in North America. http://www.communityplanning.net/pub-film/pdf/GuideToDevelopingACREProject.pdf. Accessed 01 July 2019
15. Commission of the European Communities (2001) A sustainable Europe for a better world: a European union strategy for sustainable development, COM(2001)264 final. https://eur-lex.europa.eu/legal-content/EN/TXT/PDF/?uri=CELEX:52001DC0264&from=EN. Accessed 05 July 2019
16. Condon P (2010) Seven rules for sustainable communities. Design strategies for the post carbon world. Island Press, Washington DC, US
17. D'Agostino D, Mazzarella L (2019) What is nearly zero energy building? Overview, implementation and comparison of definitions. J Build Eng 21:200–212
18. Davis (2013) http://www.ucdaviswestvillage.com. Accessed 23 June 2019
19. de Vries BJM, van Vuuren DP, Hoogwijk MM (2007) Renewable energy sources: their global potential for the first-half of the 21st century at a global level: an integrated approach. Energ Policy 35:2590–2610
20. Desai P (2009) One planet communities. A real life guide to sustainable living. John Wiley and Sons, UK

21. Doris L (2012) The five-steps process framework for project development. US Department of Energy, NREL, Office of Indian Energy. https://www.energy.gov/sites/prod/files/2016/08/f33/Doris-five-step-overview.pdf, www.energ.gov. Accessed 01 July 2019
22. Doukas H, Papadopoulou A, Savvakis N, Tsoutsos T, Psarras J (2012) Assessing energy sustainability of rural communities using principal component analysis. Renew Sust Energ Rev 16:1949–1957
23. Eddy J, Miner NE, Stamp J (2017) Sandia's Microgrid Design Toolkit. Electr J 30:62–67
24. Energy Systems Catapult (2019) https://es.catapult.org.uk/. Accessed 07 July 2019
25. European Commission (2010) A strategy for smart, sustainable and inclusive growth. COM (2010) 2020 final. https://eur-lex.europa.eu/LexUriServ/LexUriServ.do?uri=COM:2010:2020:FIN:EN:PDF. Accessed 05 July 2019
26. European Commission (2012) Energy roadmap 2050. COM (2011) 885 final. https://ec.europa.eu/energy/sites/ener/files/documents/2012_energy_roadmap_2050_en_0.pdf. Accessed 05 July 2019
27. European Commission (2014) A policy framework for climate and energy in the period from 2020 to 2030. COM (2014) 15 final. https://eur-lex.europa.eu/legal-content/EN/TXT/PDF/?uri=CELEX:52014DC0015&from=EN. Accessed 05 July 2019
28. European Parliament and Council (2010) Directive 2010/31/EU on the energy performances of buildings (recast). http://www.buildup.eu/en/practices/publications/directive-201031eu-energy-performance-buildings-recast-19-may-2010. Accessed 08 June 2019
29. European Union (2018) Models of local energy ownership and the role of local energy communities in energy transition in Europe. Commission for the Environment, Climate change and Energy. www.cor.europa.eu. Accessed 01 July 2019
30. Eurostat (2009) Sustainable development in the European Union. Monitoring report on the EU sustainable development strategy. https://ec.europa.eu/eurostat/documents/3217494/5703739/KS-78-09-865-EN.PDF/7ccc9240-03ae-40da-b2d8-2cc8a28df320. Accessed 05 July 2019
31. Eurostat (2018) Sustainable development in the European Union. Monitoring report on progress toward SDGs in an EU context—2018 edition. https://ec.europa.eu/eurostat/documents/3217494/9237449/KS-01-18-656-EN-N.pdf/2b2a096b-3bd6-4939-8ef3-11cfc14b9329. Accessed 05 July 2019
32. Fang Y, Zhang M (2012) A path to future urban planning and sustainable development: a case study of Sino-Singapore Tianjin Eco-city. Cross-disciplinary Scholars in Science and Technology Program, UCLA
33. Federal Ministry for Economic Affairs and Energy (2019) E-Energy plan. http://www.e-energy.de/.2018. Accessed 07 July 2019
34. Feng W, Jin M, Liu X, Bao Y, Marnay C, Yao C, Yu J (2018) A review of microgrid development in the United States—a decade of progress on policies, demonstrations, controls, and software tools. Appl Energ 228:656–1668
35. Fien J, White J (2010) Teaching and learning for a sustainable future. Module 17. Sustainable communities. http://www.unesco.org/education/tlsf/mods/theme_c/mod17.html?panel=2#top. Accessed 05 July 2019
36. Fleuchaus P, Godschalk B, Stober I, Blum P (2018) Worldwide application of aquifer thermal energy storage—a review. Renew Sust Energ Rev 94:861–876
37. Flynn C, Siren K (2015) Influence on local and design of the performance of a solar district heating system equipped with borehole seasonal storage. Renew Energ 81:377–388
38. Freitas S, Reinhart C, Brito MC (2018) Minimizing storage needs for large scale photovoltaics in the urban environment. Sol Energ 159:375–389
39. Friedman A (2017) Designing sustainable communities. Bloomsburg Visual Arts, London, UK
40. Ge TS, Wang RZ, Xu ZY, Pan QW, Du S, Chen X, Ma T, Wu XN, Sun XL, Chen JF (2018) Solar heating and cooling: present and future development. Renew Energ 126:1126–1140
41. Greenius L, Jagniecki E, Thompson K (2010) Moving towards sustainable community renewable energy: a strategic approach for communities. School of Engineering Blekinge, Institute of Technology, Karlskrona, Sweden

42. Hachem-Vermette C, Guarino F, La Rocca V, Cellura M (2019) Towards achieving net-zero energy communities: investigation of design strategies and seasonal solar collection and storage net-zero. Sol Energ. https://doi.org/10.1016/j.solener.2018.07.024. Accessed 3 July 2019
43. Heras-Saizarbitoria I, Saez L, Allur E, Morandeira J (2018) The emergence of renewable energy cooperatives in Spain: a review. Renew Sust Energ Rev 94:1036–1043
44. Hicks J, Ison N, Gilding J, Mey F (2014) Community–owned renewable energy. A how to guide. https://www.environment.nsw.gov.au/resources/communities/cpa-community-energy-how-to.pdf. Accessed 01 July 2019
45. Hossain MA, Pota HR, Hossain MJ, Blaabjerg F (2019) Evolution of microgrids with converter-interfaced generations: challenges and opportunities. Int J Electr Power 109:160–186
46. Hunkin S, Krell K (2018) Renewable energy communities. In: A policy brief from the policy learning platform on low-carbon economy. INTERREG Europe, Aug 2018. https://www.interregeurope.eu%2Ffileadmin%2Fuser_upload%2Fplp_uploads%2Fpolicy_briefs%2F2018-08-30_Policy_brief_Renewable_Energy_Communities_PB_TO4_final.pdf&usg=AOvVaw2lsmBMEVfBiOwIloOVJPil. Accessed 10 May 2019
47. Independent Electricity System Operator (2003) Community energy planning guidebook. http://www.kochiefs.ca/sites/default/files/16-10-03-DRAFTCommunity%20Energy%20Planning%20Guidebook.pdf. Accessed 01 July 2019
48. Ilie A, Visa I, Duta A (2017) Solar—thermal systems for domestic hot water production implemented in communities of collective households. J Energ Eng 143(6)
49. Intergovernmental Panel on Climate Change (2018) Global warming of 1.5 °C. https://report.ipcc.ch/sr15/pdf/sr15_spm_final.pdf. Accessed 05 July 2019
50. International Energy Agency (2008) Renewable energy services for developing countries. In: Support of the millennium development goals: recommended practice & key lessons. Report IEA-PVPS T9-09. https://www.iea-pvps.org%2Findex.php%3Fid%3D368%26eID%3Ddam_frontend_push%26docID%3D203&usg=AOvVaw1Dp9i0AUyBTvwsYKB1sRTT. Accessed 10 May 2019
51. International Renewable Energy Agency (2008) The role of IRENA in the context of other international organisations and initiatives. http://www.astrid-online.it/static/upload/protected/Role/Role_IRENA_dicembre_08.pdf. Accessed 03 July 2019
52. International Solar Energy Society (2005) Renewable energy future for the developing world—white paper. http://large.stanford.edu/courses/2016/ph240/thornton2/docs/ises-2005.pdf. Accessed 05 July 2019
53. Islam MT, Huda N, Abdullah AB, Saidur R (2018) A comprehensive review of state-of-the-art concentrating solar power (CSP) technologies: current status and research trends. Renew Sust Energ Rev 91:987–1018
54. Kaundinya DP, Balachandra P, Ravindranath NH (2009) Grid-connected versus stand-alone energy systems for decentralized power—review of literature. Renew Sust Energ Rev 13:1–8
55. Klein S, Coffey S (2016) Building a sustainable energy future, one community at a time. Renew Sust Energ Rev 60:867–880
56. Lazzarin RM, Noro M (2018) Past, present, future of solar cooling: technical and economical considerations. Sol Energ 172:2–13
57. Li Z, Bahramirad S, Paaso A, Yan M, Shahidehpour M (2019) Blockchain for decentralized transactive energy management system in networked microgrids. Electr J 32:58–72
58. Lopez RA, Martins J, Aelenei D, Lima CP (2016) A cooperative net zero energy community to improve load matching. Renew Energ 93:1–13
59. Lund H, Østergaard PA (2010) Climate change mitigation from a bottom up community approach: a case in Denmark. In: Clark WW II (ed) Sustainable communities design handbook. Elsevier Press, Amsterdam, NL
60. Lund H, Werner S, Wiltshire R, Svendsen S, Thorsen JE, Hvelplund F, Mathiesen BV (2014) 4th generation district heating (4 GDH). Integrating smart thermal grid into future sustainable energy systems. Energy 68:1–11

61. Ma T, Wu J, Hao L, Lee WJ, Yan H (2018) The optimal structure planning and energy management strategies of smart multi energy systems. Energy 160:122–141
62. McKenna R, Merkel E, Fichtner W (2017) Energy autonomy in residential buildings: a techno-economic model-based analysis of the scale effects. Appl Energ 189:800–815
63. Ministry of Foreign Affairs (2016) Norway's follow-up agenda 2030 and the sustainable development goals. https://www.regjeringen.no/en/dokumenter/follow-up-sdg2/id2507259/. Accessed 03 May 2019
64. Montagnino FM (2017) Solar cooling technologies. Design, application and performance of existing projects. Sol Energ 154:144–157
65. Naval Facilities Engineering Command (2015) Smart power infrastructure demonstration for energy reliability and security (SPIDERS), technology transition. In: Final Public report. https://www.energy.gov/eere/femp/downloads/smart-power-infrastructure-demonstration-energy-reliability-and-security-spiders. Accessed 10 May
66. Navarro L, de Garcia A, Colclough S, Browne M, McCormack SJ, Griffiths P, Cabeza LF (2016) Thermal energy storage in building integrated thermal systems: a review. Part 1. Active storage systems. Renew Energ 88:526–547
67. O'Hegarty R, Kinnane O, McCormack SJ (2017) Concrete solar collectors for façade integration: an experimental and numerical investigation. Appl Energ 206:1040–1061
68. Paneri A, Wong IL, Burek S (2019) Transparent insulation materials: an overview on past, present and future developments. Sol Energ 184:59–83
69. Pellegrini M, Bloemendal M, Hoekstra N, Spaak G, Andreu Gallego A, Rodriguez Comins J, Grotenhuis T, Picone S, Murrell AJ, Steeman HJ (2019) Low carbon heating and cooling by combining various technologies with aquifer thermal energy storage. Sci Total Environ 665:1–10
70. Peters M, Fudge S, High-Pippert A, Vincent C, Hoffman SM (2018) Community solar initiatives in the United States of America: comparisons with- and lessons for—the UK and other European countries. Energ Policy 121:355–364
71. Petersen JP (2016) Energy concept for self—supplying communities based on local and renewable energy sources: a case study from northern Germany. Sustain Cities Soc 26:1–8
72. Pfeifer A, Dobravec V, Pavlinek L, Krajacic G, Duic N (2018) Integration of renewable energy and demand response technologies, in interconnected energy systems. Energy 161:447–455
73. Poland M, Mare DC (2005) Defining geographic communities. Motu Economic and Public Policy Research Trust. Wellington, New Zeeland. http://motu-www.motu.org.nz/wpapers/05_09.pdf. Accessed 25 June 2019
74. Pullins S (2019) Why microgrids are becoming an important part of the energy infrastructure. Electr J 32(5):17–21
75. Quelhas A, Gil E, McCalley JD, Ryan SM (2007) A multiperiod generalized network flow model of the US integrated energy system: part I—model description. IEEE Trans Power Syst 22(2):829–836
76. Rad FM, Fung AS (2016) Solar community heating and cooling system with borehole thermal energy storage—review of systems. Renew Sust Energ Rev 60:1550–1561
77. Rae C, Bradley F (2012) Energy autonomy in sustainable communities—a review of key issues. Renew Sust Energ Rev 16:6497–6506
78. Raisul I, Sumathi K, Ullah Khan S (2013) Solar water heating systems and their market trends. Renew Sust Energ Rev 17:1–25
79. Ram M, Bogdanov D, Aghahosseini A, Gulagi A, Oyewo AS, Child M, Caldera U, Sadovskaia K, Farfan J, Barbosa LSNS, Fasihi M, Khalili S, Dalheimer B, Gruber G, Traber T, De Caluwe F, Fell H-J, Breyer C (2019) Global energy system based on 100% renewable energy—power, heat, transport and desalination sectors. Study by Lappeenranta University of Technology and Energy Watch Group, Lappeenranta, Berlin. http://energywatchgroup.org/new-study-global-energy-system-based-100-renewable-energy. Accessed 07 July 2019
80. Rehman HU, Hirnoven J, Siren K (2018) Performance comparison between optimized design of a centralized and semi-decentralized community size solar district heating system. Appl Energ 229:1072–1094

81. REN21 (2012) Renewables 2012 global status report. http://ren21.net/Portals/0/documents/Resources/GSR2012_low%20res_FINAL.pdf. Accessed 05 July 2019
82. REN21 (2016) Renewables 2016. In: Global status report. https://www.ren21.net/reports/global-status-report/. Accessed 05 July 2019
83. Sahovic N, Pereira da Silva P (2016) Community renewable energy—research perspectives. Energ Proc 106:46–58
84. San S, Ganguly S (2017) Opportunities, barriers and issues with renewable energy development—a discussion. Renew Sust Energ Rev 69:1170–1181
85. Settino J, Sant T, Micallef C, Farrugia M, Staines CS, Licari J, Micallef A (2018) Overview of solar technologies for electricity, heating and cooling production. Renew Sust Energ Rev 90:892–909
86. Sibitt B, McClenahan D, Djebbar R, Thornton J, Wong B, Carriere J, Kokko J (2012) The performance of a high solar fraction seasonal storage district heating system—five years of operation. Energ Proc 30:856–865
87. Sinakou E, Boeve-de Pauw J, Van Petegem P (2019) Exploring the concept of sustainable development within education for sustainable development: implications for ESD research and practice. Environ Dev Sustain 21(1):1–10
88. Smalley RE (2003) Top ten problems of humankind for next 50 years. In: Energy & nanotechnology conference. Rice University, May 2003
89. State Council (2018) Guidelines of the state council on actively pushing "Internet+" action. http://www.gov.cn/zhengce/content/2015-07/04/content_10002.htm. Accessed 07 July 2019
90. Strasser H, Kimman J, Koch A, am Tinkhof M, Muller D, Schiefelbein J, Schiefelbein C (2018) IEA EBC annex 63-implementation of energy strategies in communities. Energ Build 158:123–134
91. Suh SH, Kim DD (2018) Energy performance assessment toward nearly zero energy community buildings in South Korea. Sustain Cities Soc 44:488–498
92. Tian Z, Zhang S, Deng J, Fan J, Huang J, Kong W, Perers B, Furbo S (2019) Large scale solar district heating plants in Danish smart thermal grid: developments and recent trends. Energ Convers Manage 189:67–80
93. United Cities and Local Governments (2010) Culture: fourth pillar on sustainable development. http://www.agenda21culture.net/sites/default/files/files/documents/en/zz_culture4pillarsd_eng.pdf. Accessed 05 July 2016
94. United Nations (1992) Conference on Environment and Development (UNCED), Agenda 21. Rio de Janeiro, Brazil. https://sustainabledevelopment.un.org/content/documents/Agenda21.pdf. Accessed 05 July 2019
95. United Nations (2002) Report of the World Summit on Sustainable Development. Johannesburg, South Africa. www.digitallibrary.un.org. Accessed 03 July 2019
96. United Nations (2015) Transforming our world: the 2030 Agenda for Sustainable Development, New York. https://www.un.org/ga/search/view_doc.asp?sybol=A/RES/70/1&Lang=E. Accessed 05 July 2019
97. United Nations Economic and Social Council (2017) Progress towards the sustainable development goals. Report of the secretary-general, E/2017/66. https://unstats.un.org/sdgs/files/report/2017/secretary-general-sdg-report-2017–EN.pdf. Accessed 05 July 2019
98. United Nations Environment Programme (2017) Renewable energy and energy efficiency in developing countries: contributions to reducing global emissions. https://www.unenvironment.org/resources/report/renewable-energy-and-energy-efficiency-developing-countries-contributions-0. Accessed 01 July 2019
99. van der Schoor T, Scholtens B (2015) Power to the people: local community initiatives and the transition to sustainable energy. Renew Sust Energ Rev 43:666–675
100. Villa-Arrieta M, Sumper A (2019) Economic evaluation of nearly zero energy cities. Appl Energ 237:404–416
101. Visa I, Duta A (2014) The built environment in sustainable communities. In: Visa I (ed) Sustainable energy in the built environment—steps toward nZEB. Springer Proceedings in Energy, pp 3–30

102. Visa I, Duta A (2015) Community company for sustainability—an innovative tool for implementing renewables. In: Proceeding of the Xth ASTR conference, Galati, Romania. 9–10 Oct 2015, pp 56–61
103. Visa I, Duta A, Moldovan M (2017) Nearly zero energy community—a transition concept toward sustainability. J of Eng Sci Innov 2(3):69–79
104. Visa I, Duta A, Moldovan M, Burduhos B (2017b) Implementing renewable energy systems in nearly zero energy communities. In: Visa I, Duta A (eds) (2017) Nearly zero energy communities. Springer Proceedings in Energy, pp 3–24
105. World Bank (2015) Toward a sustainable energy future for all: directions for the World Bank Group's Energy Sector. http://documents.worldbank.org/curated/en/745601468160524040/pdf/795970SST0SecM00box377380B00PUBLIC0.pdf. Accessed 03 July 2019
106. World Resources Institute (2017) Climate analysis indicators tool. https://www.wri.org/our-work/project/cait-climate-data-explorer. Accessed 05 July 2019
107. Zafar R, Mahmood A, Razzaq S, Ali W, Naeem U, Shezad K (2018) Prosumer based energy management and sharing in smart grid. Renew Sust Energ Rev 82:1675–1684
108. Zappa W, Junginger M, Van den Broek M (2019) Is a 100% renewable European power system feasible by 2050? Appl Energ 233–234:1027–1050
109. Ziersch S (2017) C4CE—small scale community solar guide v2. Coalition for community energy. http://c4ce.net.au/wp-content/uploads/2017/09/C4CE-Small-Scale-Community-Solar-Guide-v2.pdf. Accessed 03 July 2019

The manufacturer's authorised representative in the EU is Springer
Nature Customer Service Centre GmbH, Europaplatz 3, 69115 Heidelberg,
Germany. If you have any concerns regarding our products, please
contact ProductSafety@springernature.com

Printed and bound by CPI Group (UK) Ltd, Croydon, CR0 4YY
30/09/2026
02225647-0012